Lightweight Materials for Electric Vehicles

Lightweight Materials for Electric Vehicles serves as a comprehensive guide to the seamless integration of sustainable materials, advanced manufacturing techniques, and innovative solutions to tackle industry challenges. It underscores the pivotal role of academia in equipping the next generation of engineers with expertise in electric vehicle (EV) technologies. The book provides an in-depth exploration of the latest advancements in lightweight materials for EVs, focusing on their properties, applications, processing methods, and the challenges they present.

Furthermore, it examines emerging trends such as artificial intelligence-driven design, additive manufacturing, and the adoption of circular economy principles in automotive materials engineering. By addressing the evolving complexities of material selection and manufacturing, the book offers valuable insights into the future of EV innovation, extending its relevance to sectors beyond automotive, including aeronautics and aerospace.

Through case studies and research findings, the book also provides practical implementation strategies and actionable recommendations, while reinforcing the critical role of academia in driving progress and innovation in this transformative field.

Features:

- Discusses contemporary issues and challenges in the selection criteria of materials for EVs.
- Explores various materials employed for EV batteries, such as lithium-ion, nickel-metal hydride, and solid-state batteries.
- Overviews the use of lightweight materials, such as aluminum, carbon fiber-reinforced polymers, and magnesium alloys, in EV bodies.
- Reviews different types of materials, their sustainability, feasibility of usage, and challenges.
- Presents case studies on the application of different materials in EVs.

This book is an essential resource for practitioners, research scholars, academicians, and students involved in the study and development of materials for EVs.

Emerging Materials and Technologies

Series Editor: Boris I. Kharissov

The *Emerging Materials and Technologies* series is devoted to highlighting publications centered on emerging advanced materials and novel technologies. Attention is paid to those newly discovered or applied materials with potential to solve pressing societal problems and improve quality of life, corresponding to environmental protection, medicine, communications, energy, transportation, advanced manufacturing, and related areas.

The series takes into account that, under present strong demands for energy, material, and cost savings, as well as heavy contamination problems and worldwide pandemic conditions, the area of emerging materials and related scalable technologies is a highly interdisciplinary field, with the need for researchers, professionals, and academics across the spectrum of engineering and technological disciplines. The main objective of this book series is to attract more attention to these materials and technologies and invite conversation among the international R&D community.

Materials for Dye Degradation
Edited by Tarun Parangi and Soyebkhan Pathan

Circularity of Cementitious Materials
A Practical Approach
Christian Paglia

Machine Learning for Semiconductor Materials
Edited by Neeraj Gupta, Rashmi Gupta, Rekha Yadav, Sandeep Dhariwal and Rajkumar Sarma

Multilayer Nanostructured Wear-Resistant Coatings for Metal-Cutting Tools
A.A. Vereschaka, S.N. Grigoriev and V.P. Tabakov

Multifunctional Inorganic Nanomaterials
Synthesis, Properties and Applications
Edited by H.C. Ananda Murthy, Velumani Subramaniam, Mika Sillanpää, R. Balachandran and Gurumurthy Hegde

Handbook of High Entropy Alloys
Fundamentals to Applications
Edited by Shashanka Rajendrachari

Emerging Materials for Biofuel Developments
Edited by Selvakumar Periyasamy, Eduardo Alberto López-Maldonado, and R. Balachandran

Nanostructured Materials for Energy Applications
Edited by Abdullah M. Al-Enizi, Mohd Ubaidullah, Mohd Shkir and Abhay Kumar Singh

Boron Nitride Nanostructures
Applications in Energy and Healthcare
Indranil Lahiri, Amrita De Adhikari, Souvik Ghosh, Rita Joshi, Satish Jaiswal, and Debrupa Lahiri

Smart Polymers
An Emerging Vista for Nanotheranostics
Dipankar Chattopadhyay, Adrija Ghosh, Jonathan Tersur Orasugh and Suprakas Sinha Ray

Lightweight Materials for Electric Vehicles
Sustainable Materials, Production Process, and Modeling Techniques
Edited by R. Suresh, C. Durga Prasad, Satish Kumar, K.N. Bharath, and Ajith G. Joshi

For more information about this series, please visit: www.routledge.com/Emerging-Materials-and-Technologies/book-series/CRCEMT

Lightweight Materials for Electric Vehicles

Sustainable Materials, Production Process, and Modeling Techniques

Edited by
R. Suresh, C. Durga Prasad, Satish Kumar,
K.N. Bharath, and Ajith G. Joshi

CRC Press
Taylor & Francis Group
Boca Raton London New York

CRC Press is an imprint of the
Taylor & Francis Group, an **informa** business

Designed cover image: © Shutterstock Images

First edition published 2026
by CRC Press
2385 NW Executive Center Drive, Suite 320, Boca Raton FL 33431

and by CRC Press
4 Park Square, Milton Park, Abingdon, Oxon, OX14 4RN

CRC Press is an imprint of Taylor & Francis Group, LLC

ISBN: 9781032975054 (hbk)
ISBN: 9781032979151 (pbk)
ISBN: 9781003596097 (ebk)

DOI: 10.1201/9781003596097

Typeset in Times
by Newgen Publishing UK

Contents

Preface

The automotive industry is experiencing a paradigm shift, as the transition toward electric mobility becomes not only a technological ambition but an environmental necessity. In this rapidly evolving landscape, the materials that shape our vehicles are as important as the energy that powers them. As electric vehicles (EVs) and hybrid EVs (HEVs) continue to rise in prominence, the demand for lightweight, high-performance, and sustainable materials has reached an all-time high.

This book, *Lightweight Materials and Processing Technologies for Electric Vehicles*, arrives at a critical moment. It provides a thorough and well-structured exploration of the materials and manufacturing innovations that are defining the future of mobility. From lightweight alloys to composite materials, from additive manufacturing to nanomaterials for energy storage, the authors have compiled a comprehensive and timely resource that addresses both the promise and the complexity of next-generation vehicle materials.

What sets this book apart is its balance between foundational knowledge and advanced research. The discussions on material preparation, mechanical behavior, wear resistance, and fabrication challenges are not only academically rigorous but also deeply relevant to real-world engineering problems. Whether you're a student entering the field, a researcher pushing the boundaries of materials science, or an industry professional solving practical design challenges, this book offers valuable insights.

The editors commend the authors for their thoughtful contributions to this dynamic and critical area of research. As we collectively advance toward cleaner, smarter, and more efficient transportation systems, resources like this will serve as essential guides for innovation and collaboration.

R. Suresh
C. Durga Prasad
Satish Kumar
K.N. Bharath
Ajith G. Joshi

About the Editors

R. Suresh is a full-time Professor in the Department of Mechanical Engineering at M.S. Ramaiah University of Applied Sciences (RUAS), Bengaluru, India. He earned his M.Tech and Ph.D. from Kuvempu University in 2002 and 2013, respectively. With 28 years of experience spanning academia, research, and industry, his expertise lies in hard and refractory materials, composites, traditional and non-traditional machining processes, surface engineering, solid-state joining of metals, and additive manufacturing. He has an extensive research record, with over 150 publications in international and national journals/conferences, along with book chapters and edited books and two conference proceedings. His contributions have earned him prestigious awards, including the "Outstanding Scientist Award" from the Centre for Advanced Research and Design, Chennai (2015) and the "Annual Exemplary Faculty-Research" award from RUAS (2020). He has received research funding from agencies including Department of Science & Technology (DST), All India Council for Technical Education (AICTE), and Karnataka Council of Technological Upgradation (KCTU)-Government of Karnataka and has collaborated on various industrial projects. He is also a reviewer for reputed journals such as Taylor & Francis and MDPI journals. He has supervised Ph.D. and post-graduate scholar research projects. Additionally, he is a sought-after speaker at leading universities and institutions, delivering talks on materials and manufacturing processes. He is an active member of professional societies such as Member of Indian Society of Technical Education (MISTE) and Society of Association Executives (SAE) India and serves as an official reviewer for research articles in materials and manufacturing fields.

C. Durga Prasad is an Assistant Professor in the Department of Mechanical Engineering at RV Institute of Technology and Management (RVITM), Bengaluru, since 2019. He earned his Ph.D. from National Institute of Technology (NIT) Karnataka and an M.Tech in Manufacturing Science from Siddaganga Institute of Technology. He has six years of teaching and eight years of research experience, specializing in thermal spray coatings, microwave processing of materials, wear, tribo-corrosion, bio-composites, and biofuels. He has an extensive research portfolio with over 70 publications in reputed international journals, 30 conference proceedings, five book chapters, and two textbooks. He has guided 10 UG batch projects and is currently supervising two Ph.D. scholars. His research has attracted five funded projects from Science and Engineering Research Board (SERB), DST, and Visvesvaraya Technological University (VTU). Dr. Prasad is a recipient of the Indian Academy of Science Summer Fellowship (2019–2020) and the Academic Achiever Award (2020–2024) from Rashtreeya Sikshana Samithi Trust (RSST). He actively collaborates with industries as a technical consultant in materials and biofuels. He has delivered several technical talks at national and international levels, and serves as a reviewer for prestigious journals.

Satish Kumar is an Associate Professor in the Department of Robotic Engineering Symbiosis Institute of Technology at Symbiosis International (Deemed University), Pune, India. He is captivated by technology-oriented innovative concepts and always looks ahead to execute interdisciplinary projects. He has expertise in the application of artificial intelligence in a variety of crucial manufacturing processes, including quality enhancement, process improvement, and optimization. He is now working on several Industry 4.0 initiatives, including digital twin smart manufacturing and remaining usable life estimation. He has authored more than 45 Scopus/Web of Science (WoS) indexed international/national journal and conferences publications. He has recently published a patent on a "Micro-oxidation Coating Device." He is also currently editing a Taylor & Francis CRC Press book titled *Industry 4.0 in Small and Medium-Sized Enterprises (SMEs): Opportunities, Challenges, and Solutions*. He has completed various research-funded proposals at Symbiosis International Deemed University and is also currently working on two research proposals

"Inferring quality and fault localization of 3D printer products using digital twin" and "Artificial Intelligence-based fault detection in bearing using vibration and sound sensor." One student was recently awarded a Ph.D. under his supervision and he is currently guiding five Ph.D. research scholars. Dr. Satish Kumar is also an expert reviewer on board for *IEEE Access*, *Silicon*, *PeerJ Computer Science Surface Topography: Metrology and Properties*, *Material Research Express*, and other reputed journals. He has also deputed for research initiatives at the Symbiosis Centre for Research and Innovation (SCRI). He has delivered many talks at leading universities on Industry 4.0 and Smart Manufacturing.

K.N. Bharath is a Professor and Dean (Research and Innovation) at GM University, Davangere. He earned his Bachelor's and Master's degrees in Mechanical Engineering from Kuvempu University and completed his Ph.D. from VTU at the age of 32. He further pursued a Post-Doctoral Fellowship at the Composite Materials & Engineering Center, Washington State University, USA. With 16 years of teaching and five years of R&D experience, his expertise includes composite materials, bioplastics, and machine design. He has secured ₹60.20 lakhs in research funding from agencies like Vision Group on Science and Technology (VGST), Karnataka State Council for Science and Technology (KSCST), and National Science Foundation (NSF), receiving the "Seed Money to Young Scientist" grant in 2015. He has collaborated with researchers from the USA, Malaysia, Indonesia, United Arab Emirates (UAE), Thailand, and Russia, contributing to projects funded by CB2 and United States Department of Agriculture (USDA). He has completed three funded research projects worth ₹54 lakhs. He has over 55 publications in renowned international journals with more than 1,250 citations. He is a recognized reviewer for Elsevier, a guest editor, and an advisory committee member for international journals and conferences. He has also served as a technical reviewer for book proposals and research articles. Additionally, he has delivered 13 invited talks at engineering colleges, R&D institutions, industries, and international platforms.

Ajith G. Joshi is an Assistant Professor in Mechanical Engineering at Kammavari Sangha (KS) Institute of Technology, India. He holds a B.E. from University B.D.T. College of Engineering and an M.Sc. (Engineering by Research) from VTU, where he is currently pursuing a Ph.D. With extensive teaching experience, his research focuses on composite materials, tribology, and machining of difficult-to-cut alloys. He previously taught at Madanapalle Institute of Technology and Science, Canara Engineering College, and Tontadarya College of Engineering, supervising several B.Tech. theses. He has published 25 research articles in reputed journals, nine book chapters, and presented at national and international conferences. He has secured research grants from national funding agencies and conducted faculty development programs as a speaker. He also serves as a reviewer for international journals and conferences.

Contributors

Ahash A.
Dept. of CSE and AI&ML,
MLR Institute of Technology,
Dundigal, Medchal District,
Hyderabad, Telangana, India

Rajesh Kannan A.
Department of Mechanical Engineering,
Incheon National University,
Incheon, South Korea

Adem Abdirkadir Aden
Department of Mechanical Engineering,
Bulehora University,
Ethiopia

Varun Kumar Arulvizhi
Department of Mechanical Engineering,
B. S. Abdur Rahman Crescent Institute of Science and Technology,
Vandalur, Chennai, India

Veerabhadraswamy B.N.
Department of Chemistry,
GM University,
Davanagere, Karnataka, India

Omkaresh B.R.
Department of Mechanical Engineering,
Siddaganga Institute of Technology,
Tumakuru, Karnataka, India

Madhu Basavaraj
PACCAR North Texas,
Lewisville, USA

Pritam Bhat
Department of Mechanical Engineering
M.S. Ramaiah University of Applied Sciences
Bangalore, India

Durga Prasad C.
Department of Mechanical Engineering,
RV Institute of Technology and Management,
Bengaluru, Karnataka, India

Prakash C. H.
Department of Mechanical Engineering
Akshaya Institute of Technology,
Tumkur, Karnataka

Rahul Murali Cadambi
Department of Mechanical Engineering
M.S. Ramaiah University of Applied Sciences
Bangalore, India

Shrinivasa Mayya D.
Srinivas Institute of Technology Karnataka,
Mangalore, Karnataka, India

Subhankur Dutta
Department of Mechanical Engineering,
National Institute of Technology Nagaland,
Dimapur, India

Chandrakumar Elumalai
Department of Electrical and Electronics Engineering,
K.S. Rangasamy College of Technology,
Namakkal, Tamilnadu, India

Veeresha G.
Department of Mechanical Engineering,
New Horizon College of Engineering,
Bengaluru, Karnataka, India

Kiran Kumar G.R.
Department of Electrical and Electronics Engineering,
PES Institute of Technology and Management,
Shivamogga, Karnataka, India

Pradeep Kumar G.S.
Department of Mechanical Engineering,
Christ University,
Bengaluru, Karnataka, India

Nimona Hailu
Department of Mechanical Engineering,
Bulehora University,
Ethiopia

Mohamad Zaki Hassan
Universiti Teknologi Malaysia
Johor Bahru,
Malaysia

Mohamad Hamid Hussain
Dept. of Mechanical Engineering,
University of Technology and Applied Sciences
Nizwa, Oman

Ananth S. Iyengar
Department of Mechanical Engineering
M.S. Ramaiah University of Applied Sciences

Sivakrishna K.
Dept. of CSE and AI&ML,
MLR Institute of Technology,
Dundigal,
Medchal District,
Hyderabad, Telangana, India

Vaishnavi K.
Dept. of CSE and AI&ML,
MLR Institute of Technology,
Dundigal,
Medchal District,
Hyderabad, Telangana, India

Kavitha K.J.
Department of Electronics and Communication,
GM University, Davanagere,
Karnataka, India

Rashmi K.R.
Department of Mathematics,
Nitte Meenakshi Institute of Technology
Benagaluru, Karnataka, India

Varun K.R.
Department of Mechanical Engineering,
The Oxford College of Engineering,
Bengaluru, Karnataka, India

Gangadharan K.V.
Dept. of Mechanical Engineering,
National Institute of Technology Karnataka,
Mangalore, Karnataka, India

Sohit Reddy Kalluru
MYGO Consulting Inc,
Plainfield,
Illinois, USA

Nagaraj Kamath
Dept. of Mechanical Engineering,
National Institute of Technology Karnataka,
Mangalore, India

Sathickbasha Katharbasha
Department of Mechanical Engineering,
B S Abdur Rahman Crescent Institute of Science and Technology,
Vandalur, Chennai, India

Jitendra Kumar
Department of Mechanical Engineering,
Institute of Engineering and Technology
Faizabad, Ayodhya U.P., India

Leela Sai Sri Kirshna L.
Dept. of CSE and AI&ML,
Institute of Technology,
Dundigal, Medchal District,
Hyderabad, Telangana, India

Sunith Babu L.
Department of Mechanical Engineering,
Ramaiah Institute of Technology,
Bengaluru, Karnataka, India

Shanmuka M. C.
Department of Mathematics,
PES Institute of Technology and Management,
Shivamogga, Karnataka, India

Niyaz Ahamed M.B.
Mechanical Engineering Department,
University of Technology and Applied Science,
Shinas, Sultanate of Oman

Shashi Kumar M.E.
Department of Mechanical Engineering,
Amrita School of Engineering,
Amrita Vishwa Vidyapeetham,
Bengaluru, Karnataka, India

Ananda M.K.
Department of Electronics and Communication,
GM University, Davanagere,
Karnataka, India
Bangalore, India

Paswan M. K.
Department of Mechanical Engineering,
National Institute of Technology
Jamshedpur,
Jharkhand, India

Rajesh M.N.
Department of Mechanical Engineering,
GITAM (deemed to be) University,
Bengaluru, Karnataka, India

Kiran M.S.
Department of Chemistry,
GM University,
Davanagere, Karnataka, India

Basir Maleki
Department of Chemical Engineering,
Esfarayen University of Technology,
Esfarayen, North Khorasan,
Iran

Raju Maliger
Chemical Engg Dept.,
University of Technology and Applied Sciences
(UTAS),
Muscat,
Oman

Anarghya Ananda Murthy
Department of Mechanical and Mechatronics
Engineering,
The University of Auckland, Auckland,
New Zealand

Abhilash N.
Department of Mechanical Engineering
M.S. Ramaiah University of Applied Sciences
Bangalore, Karnataka, India

Hareesha N.
Department of Chemistry, GM University,
Davanagere, Karnataka, India

Jeyaprakash N.
Mechanical Engineering Department,
Global College of Engineering and
Technology,
Muscat, Oman

Yuvarajagouda P.
Department of Chemistry,
GM University,
Davanagere,
Karnataka, India

Usha Raghunathagouda Patil
Department of Electrical and Electronics
Engineering,
PES Institute of Technology and
Management,
Shivamogga, Karnataka, India

Hariharasakthisudhan Ponnarengan
Department of Mechanical Engineering,
Dr. Mahalingam College of Engineering and
Technology,
Pollachi, Coimbatore, Tamil Nadu, India

Krupa R.
Department of Mechanical Engineering,
Nitte Meenakshi Institute of
Technology,
Bengaluru, Karnataka, India

Sekar R.
Department of Electrical and Electronics
Engineering,
PES Institute of Technology and
Management,
Shivamogga, Karnataka, India

Suresh R.
Department of Mechanical Engineering
M.S. Ramaiah University of Applied
Sciences
Bangalore, India

Gopalakrishnan Radhakrishnan
Department of Electrical and Electronics
Engineering,
K.S. Rangasamy College of Technology,
Namakkal, Tamilnadu, India

Satheesh Kumar Radhakrishnan
Department of Electrical and Electronics Engineering,
Sona College of Technology,
Salem, Tamilnadu, India

Dinesh Kumar Rao
Department of Mechanical Engineering,
Institute of Engineering and Technology Faizabad
Ayodhya, Uttar Pradesh, India

Pavan Kumar Reddy
Department of Mechanical Engineering
M.S. Ramaiah University of Applied Sciences
Bangalore, India

Prasanna Kumar Reddy Gurijala
MYGO Consulting Inc,
Plainfield, Illinois,
USA

Mohan Kumar S.
Department of Mechanical Engineering,
Amrita School of Engineering,
Amrita Vishwa Vidyapeetham,
Bengaluru, Karnataka, India

Prashanth S.
Department of Mechanical Engineering,
National Institute of Technology Nagaland,
Dimapur, India

Prashantha S.
Department of Mechanical Engineering,
Siddaganga Institute of Technology,
Tumakuru, Karnataka, India

Rudresha S. J.
Department of Electrical and Electronics Engineering,
PES Institute of Technology and Management,
Shivamogga, Karnataka, India

Praveen Shenoy Kaup
Dept. of Aeronautical Engineering,
Srinivas Institute of Technology,
Valachil, Mangalore, India

Amit Kumar Singh
Department of Mechanical Engineering,
National Institute of Technology Nagaland,
Dimapur, India

Laishram Dhanabir Singh
Department of Mechanical Engineering,
National Institute of Technology Nagaland,
Dimapur, India

Sandeep Kumar Singh
Department of Mechanical Engineering,
National Institute of Technology
Nagaland, Dimapur, India

Thingujam Jackson Singh
Department of Mechanical Engineering,
National Institute of Technology Nagaland,
Dimapur, India

Hargovind Soni
Department of Industrial Engineering,
College of Science, Engineering and Technology,
University of South Africa,
Pretoria, South Africa

Puneet Kumar Sonker
Department of Mechanical Engineering,
National Institute of Technology
Nagaland, Dimapur, India

Mohamed Azlan Suhot
Universiti Teknologi,
Malaysia

Indran Suyambulingam
Department of Mechanical Engineering
Alliance University,
Bengaluru, India

Nicholas Kuan Hoo Tien
Universiti Malaysia Sarawak,
Malaysia

Upendra U.
Dept. of CSE and AI&ML,
MLR Institute of Technology,
Dundigal, Medchal District,
Hyderabad, Telangana, India

Mahesh K. Varpe
Department of Aerospace and Automotive Engineering,
M.S. Ramaiah University of Applied Sciences,
Bangalore, India

Yogendra Kumar Verma
Department of Mechanical Engineering,
National Institute of Technology Nagaland,
Dimapur, India

Introduction

R. Suresh, C. Durga Prasad, Satish Kumar, K.N. Bharath, and Ajith G. Joshi

The growing demand for energy-efficient and high-performance electric vehicles (EVs) has led to an increased focus on lightweight materials that enhance vehicle efficiency, durability, and sustainability. Lightweight alloys and composite materials have become essential in EV manufacturing due to their superior mechanical properties, machinability, and resistance to wear and corrosion. However, the processing and fabrication of these materials present significant challenges, including their heterogeneity and the abrasive nature of composite reinforcements.

This book explores the latest advancements in lightweight materials and processing technologies, with a particular emphasis on their application in EVs and hybrid EVs (HEVs). It covers critical aspects such as material preparation techniques, mechanical and physical properties, and manufacturing challenges. Additionally, it highlights emerging trends in additive manufacturing, nanomaterials for battery applications, and computational modeling for EV components. Designed as a comprehensive resource, this book serves as a valuable reference for undergraduate and postgraduate students, researchers, academicians, and professionals in the fields of materials science, automotive engineering, and EV manufacturing. By addressing the evolving needs of the industry, this book contributes to the ongoing development of innovative materials and technologies that drive the future of sustainable transportation.

Seeking to promote research on the selection and implementation of lightweight materials and processing technologies for EVs, the book *Lightweight Materials for Electric Vehicles* is divided into 18 chapters. Chapter 1 introduces the development of advanced lightweight materials, composites reinforced with particles, natural and synthetic fibers, aimed at replacing heavier materials while maintaining strength and safety. It explores material properties, processing methods, artificial intelligence (AI)-driven design, and emerging innovations. Additionally, the chapter discusses key challenges such as machinability, corrosion resistance, and recyclability, emphasizing additive manufacturing (AM) and circular economy principles. These advancements have broader applications beyond EV technologies.

Chapter 2 explores advancements in epoxy-enhanced bamboo nanocomposites, emphasizing their synthesis, structure, and applications. Bamboo's high strength-to-weight ratio and sustainability make it an ideal reinforcement, while epoxy resins enhance mechanical and chemical properties. The chapter covers nanofiller integration, preparation techniques, and structural characterization methods like Fourier transform infrared (FTIR) spectroscopy, scanning electron microscopy (SEM), X-ray diffraction (XRD), and transmission electron microscopy (TEM). It discusses thermal stability, viscoelastic properties, and applications in automotive, aerospace, and construction. Challenges such as scalability, cost, and sustainable processing are also examined to improve commercial viability.

Chapter 3 presents a comprehensive overview of EV battery technologies, emphasizing innovations in energy storage and management systems essential for the future of sustainable transportation. The rapid growth of EVs represents a significant step toward an eco-friendly future, addressing the environmental challenges posed by traditional automobiles. Battery technology plays a crucial role in enhancing EV performance, with lithium-ion batteries (LIB) being the preferred choice due to their high energy density and long life cycle. However, emerging battery technologies, such as lithium-metal and solid-state batteries, present promising alternatives with significantly higher energy storage potential. A robust battery management system (BMS) is vital for ensuring battery efficiency, safety, and longevity, with advancements incorporating Internet of Things (IoT) and AI-driven solutions.

The rapid evolution of energy storage technology has made the development of advanced nanomaterials for next-generation batteries a necessity. Chapter 4 explores how nanomaterials, with their unique properties, such as high surface area, enhanced electron transport, and tunable electrochemical characteristics, can significantly improve battery performance. It examines their role in lithium-ion, sodium-ion, and solid-state batteries, addressing key challenges like energy density, cycle stability, and safety. Additionally, the chapter highlights innovations in nanomaterial-based batteries, including two-dimensional (2D) materials, quantum dots, and metal–organic frameworks (MOFs), and their potential to revolutionize large-scale energy storage, EVs, and portable devices. Ensuring the scalability and sustainability of nanomaterials for widespread battery applications remains a crucial focus for the future.

The growing adoption of EVs requires advanced, lightweight BMSs for improved performance and safety. Chapter 5 explores innovative materials like polymer composites, MOFs, and nanomaterials that enhance energy density and thermal management. It also highlights bio-based polymers and recyclable composites for sustainability while maintaining durability. Case studies and AI-driven modeling provide insights into next-generation BMS design for EVs.

The advancement of EVs and HEVs demands innovative manufacturing solutions for improved performance and sustainability. AM plays a crucial role by enabling lightweight, complex, and durable components. Chapter 6 explores the integration of thermal spray coatings (TSCs) with AM to enhance wear resistance, corrosion protection, and energy efficiency in EV/HEV systems. By combining AM and TSC, manufacturers can optimize key vehicle parts, contributing to sustainable and high-performance automotive designs.

Chapter 7 delves into the growing significance of polymeric heat exchangers (PHEs) in thermal management systems, particularly for LIBs in EVs. It highlights the innovative designs, efficiency, and versatility of PHEs, showcasing their advantages over traditional metallic counterparts. By exploring computational and experimental analyses, along with current research, the chapter provides a comprehensive overview of PHE applications and advancements, emphasizing their role in fostering sustainable and high-performance technologies.

Chapter 8 presents the evolution of EVs, which has been driven by advancements in lightweight materials, improving performance and efficiency. This study explores the use of biomaterials, eco-friendly composites, and fire-retardant polymers in EVs and HEVs. By analyzing their impact on vehicle dynamics, energy consumption, and safety, along with their life cycle and recyclability, this research highlights the role of sustainable materials in the future of automotive design. With the rise of EVs and HEVs, the demand for durable and sustainable tribological materials is increasing. Chapter 9 delves into advanced materials such as biodegradable polymers, recycled metals, and self-lubricating coatings that enhance wear resistance, efficiency, and environmental sustainability. By reducing friction, improving recyclability, and minimizing carbon footprints, these materials play a crucial role in the evolution of greener automotive technologies. Additionally, the chapter discusses how innovations in sustainable tribology contribute to lightweight, high-performance vehicle components, supporting the industry's transition toward cleaner energy solutions.

Chapter 10 explores the concept of remaining useful life (RUL) prediction, which is critical for optimizing performance and ensuring safety in LIBs used in EVs, energy storage, and consumer electronics. Traditional RUL prediction methods face challenges in interpretability and computational efficiency. To address this, we propose a hybrid deep learning framework that integrates one-dimensional convolution neural networks (1D CNNs) with attention mechanisms while also evaluating alternative models like support vector machine (SVM), long short-term memory (LSTM), and Transformers. Feature selection is enhanced using SelectKBest with mutual information regression (MIR), and model performance is assessed through mean squared error (MSE), mean absolute error (MAE), and root mean squared error (RMSE). This approach enhances generalization and efficiency, making it highly relevant for industries relying on battery-powered systems.

The automotive industry is under growing pressure to adopt sustainable materials, with natural fiber composites emerging as a lightweight and eco-friendly alternative, particularly for EVs. Chapter 11 describes their sustainability, property enhancements through chemical treatments, and economic viability. It also discusses challenges and compares them with glass fiber composites to assess their potential in automotive applications. Finally, key insights and recommendations are provided to guide future research and industry adoption.

Chapter 12 explores 2D carbon-based polymer nanocomposites, highlighting their enhanced mechanical and thermal properties through interface engineering. By incorporating nanomaterials like graphene, carbon nanotubes (CNTs), and carbon nanofibers (CNFs) with functional agents such as 3-aminopropyltriethoxysilane (APTES) and dopamine, these composites improve performance in automotive applications. This review examines recent advancements in nanoparticle–matrix interactions, load-transfer mechanisms, and reinforcement strategies.

Chapter 13 presents the role of EVs and HEVs in achieving carbon neutrality and addressing the climate crisis. It highlights the importance of advancements in battery technology and BMSs for the future of electric mobility. The chapter traces the evolution of battery materials from lead-acid to lithium-ion, metal-air, and flow batteries driving progress in energy storage. Additionally, it examines emerging materials like zinc manganese oxide, lithium tungsten, and lithium-sulfur batteries, which offer fast charging, affordability, and sustainability, shaping the future of electrification.

With the growing demand for sustainable energy solutions, perovskite materials have emerged as promising candidates for next-generation battery technologies. Their unique properties, including high ionic conductivity, tunable band gaps, and excellent electrochemical stability, enhance battery efficiency, and lifespan. Chapter 14 explores advancements in perovskite-based batteries while addressing key innovations, challenges, and future prospects for high-performance energy storage solutions.

AI and machine learning (ML) are transforming material processing by enhancing efficiency, precision, and sustainability. Chapter 15 explores how AI and ML optimize techniques like casting, welding, machining, and additive manufacturing through predictive modeling and real-time control. By integrating AI-driven analytics and the IoT, these advancements minimize costs, reduce waste, and improve manufacturing processes, offering valuable insights for researchers and industry professionals.

Chapter 16 explores the use of magnetorheological elastomers (MREs) in vibration isolation systems for EVs. Unlike magnetorheological fluids (MRFs), MREs offer long-term stability by preventing sedimentation and leakage. Using a blocked transfer stiffness approach, the study evaluates MREs' dynamic performance under torsional vibrations. Results show that increasing the magnetic field enhances stiffness and damping, improving ride comfort and noise reduction. MREs' adaptability makes them ideal for next-generation EV vibration isolation systems.

Chapter 17 explores the critical role of BMSs in ensuring the safety, efficiency, and longevity of EV energy storage. While advances in lithium-ion, solid-state, and sodium-ion batteries drive innovation, challenges persist in thermal management, charge cycles, and safety. Emerging materials such as solid electrolytes, nanomaterials, and phase-change materials (PCMs) enhance BMS performance. Integrating these advancements with sensor technology and AI can further improve efficiency and sustainability. This chapter examines how materials science innovations are shaping the future of EV batteries.

The rapid shift toward EVs is revolutionizing the automotive industry, driving advanced developments in materials for enhanced performance and sustainability. Among these innovations, advanced polymers and composites play a pivotal role in reducing weight, improving energy efficiency, and optimizing thermal management. Their integration not only extends driving range but also enhances safety and structural integrity. Chapter 18 explores the crucial contributions of advanced polymers and composites in shaping the future of EVs and HEVs, addressing both their potential and the challenges that lie ahead.

The editors acknowledge their gratitude to CRC Press/Taylor & Francis Group for this opportunity and their professional support. Finally, we would like to thank all chapter authors for their interest and availability to work on this project.

1 Lightweight Materials Used in Battery Vehicles

Revolutionizing Mobility

Ananth S. Iyengar, R. Suresh, Raju Maliger, Rajesh Kannan A., and Hargovind Soni

1.1 INTRODUCTION

The emergence of electric vehicles (EVs) in the last decade is a consequence of the improvement of the energy density of EV batteries. EVs are still 20–30% heavier than internal combustion engine (ICE) vehicles of the same class with comparable power necessitating future improvements [1]. This combination of factors makes the efficiency of automobiles directly impacted related to their weight. A heavier vehicle requires more energy to move, and much of the engine effort is lost leading to slow pickup and low pulling power. This effect is more pronounced in the case of EVs since the energy per kilogram for gasoline is 50–100 times higher than the power density of state-of-the-art batteries [2]. The effect is observable from the graph shown in Figure 1.1. The prime mover to wheel efficiency, however, is much higher in the case of EVs, making them a good competitor to ICE vehicles. The weight of the vehicle also determines the travel range of EVs. The pursuit of lightweight solutions is aimed at improving energy efficiency, extending battery range, and maintaining structural integrity, while reducing environmental impact. Moreover, lightweight materials contribute to the agility and handling of EVs, enhancing driving dynamics and overall user experience. They also play a significant role in lowering production and operational costs by reducing wear and tear on tires, brakes, and other vehicle components.

The challenges faced by automotive designers include well-to-wheel emissions, and sustainability leads to the framing of conceptual design goals. Some of these design goals, such as long life time and adaptable, lightweight, comfortable, and aerodynamic, can be achieved with careful consideration. Other design challenges, such as high utilization, community transport model, low maintenance, and fuel costs are out of scope of this study [3]. Material flow analysis (MFA) is used at various levels of production to understand resource utilization and optimization [4]. It is reported that the adoption of EVs lowers the demand for cast alloys, while efficient scrap sorting significantly reduces both the surplus of aluminum scrap and the consumption of primary aluminum in vehicles. The use of aluminum in modern EVs for lightweighting also increases aluminum demand and aluminum end-of-service life scrap generation. The impact of technology choices related to automotive lightweighting, electrification, decarbonization, and scrap sorting on the use of metal alloys and composites in vehicles, and the overall materials demand by type, in the automotive industry, has not yet been thoroughly examined [4].

1.1.1 Mathematical Models for Lightweight Materials in EVs

EV performance is determined by a combination of external environmental factors, vehicular design elements, and drive cycle characteristics. External conditions such as ambient temperature, terrain, road quality, traffic patterns, and weather significantly influence EV operation. Similarly,

DOI: 10.1201/9781003596097-1

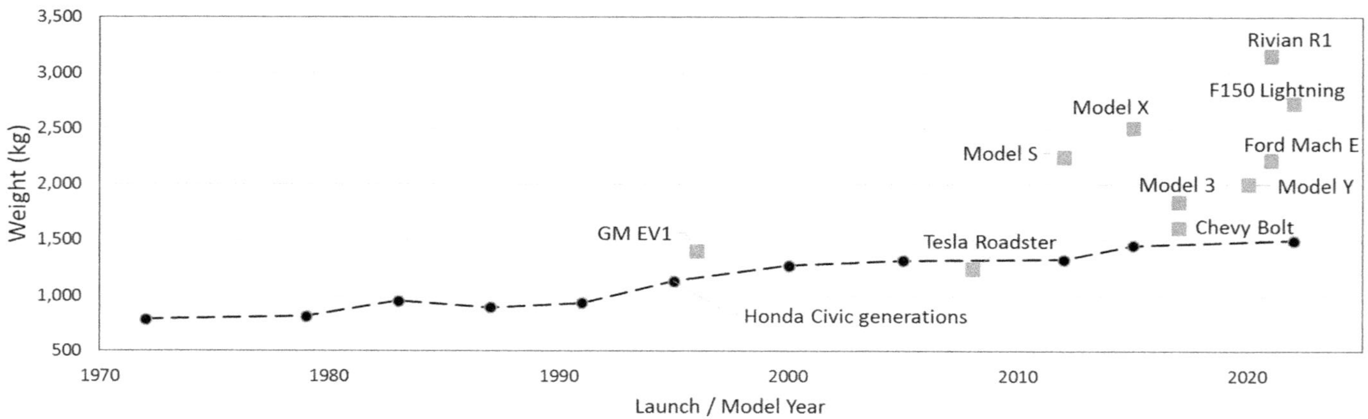

FIGURE 1.1 Comparison of weights for EVs with a representative ICE vehicle (dotted continuous line) over the decades.

the vehicle's design and engineering play a crucial role in its efficiency and performance. Key design parameters include aerodynamics, weight, tire specifications, powertrain efficiency, transmission design, energy recovery systems, and thermal management. Drive cycle characteristics, encompassing speed patterns, acceleration and deceleration behavior, trip duration, and passenger or cargo weight, also impact performance. Overall, EV performance metrics such as range, efficiency, and durability result from the interplay of these factors. However, the challenge lies in addressing the complexity introduced by the interdependencies among these variables, the impact of design modifications on manufacturing processes, and the need to balance multiple objectives simultaneously. Mathematical methods such as first principle modeling, statistical tools, and artificial intelligence (AI) tools looking at detailed weight calculations starting from components, subassembly, and assembly of part level manufacturing calculations have been published in the literature. The mathematical models may also include numerical simulations to understand the manufacturability and design criteria satisfaction. Mayyas et al. developed a first principles method to calculate the comparisons of ICE vehicles and EVs with similar lightweight materials [5]. The comparisons indicate that Body in White (BiW) builds using magnesium utilize more life cycle energy due to the emissions caused by the extraction of magnesium and alloying process, showing a higher carbon footprint compared to other consumption methods. Ballo et al. used multi-physics combined with a driving strategy optimization approach to model lightweight EVs [6]. They reported that the optimal gear ratio and driving profile allows energy savings of up to 9% [6]. Similarly, only drive strategy comparison provides 4% variations in energy consumption. A battery equivalent circuit and system of equations are used to represent the combination of electrical, thermal, and mechanical characteristics of goods vehicles [7]. The input parameters for such a model are electric current and ambient temperature. The heat generated during the operation was simulated. These kinds of models can be part of battery management systems in the future. Optimizing the driving strategy using a mathematical model of the EV has been shown to enhance its performance by 5.5% [8]. Further advancements in optimization are reported when incorporating regenerative braking, with the recommendation to utilize AI for achieving more precise optimization. A response surface model to optimize the racing frame is presented with a goal to minimize the mass of the frame [9]. This model uses the thickness of all frame members as design variables, aiming to improve frontal impact performance while minimizing mass. The model achieves a 14.38% weight reduction with only a slight increase in overall frame deflection, ensuring that frontal crash safety remains uncompromised.

An analytical calculation method has been developed to establish a relationship between energy consumption and vehicle mass, utilizing a parameter referred to as the energy reduction value [10]. This parameter is further linked to the impact reduction value, which accounts for the type of electricity source used for EV charging. The calculation is extended to analyze case studies from the literature, with the break-even point serving as the performance criterion. The model demonstrates a strong correlation between the break-even point and factors such as vehicle model and make, driving cycles, and high acceleration or deceleration scenarios, particularly in regions with fossil-fuel-intensive energy grids. The findings emphasize the importance of careful vehicle design, optimized material selection, and a shift toward cleaner energy grids to improve overall sustainability. Der et al. [11] delivered a well-rounded appraisal of lightweighting EV components, spanning from early design stages to end-of-life. The study highlights that lightweight materials often have a higher environmental impact during the early stages of their life cycle. Bottom-up modeling and process chain models are employed to analyze case studies. The assessment and comparison of power consumption for compression molding for door components made from glass fiber and carbon fiber sheet molding compounds with that of traditional steel counterparts shows a marked improvement. These studies emphasize a detailed view of the mass of components of EVs.

1.1.2 Weight Distribution in EVs

The lightweighting process is not straightforward material replacement, as many influencing factors have to be considered. Some of the influencing engineering factors include the cost of the new material, availability of the material, part complexity, part size, re-engineering effort, and manufacturing process. The other important properties required of the structural materials include high tensile strength, light weight, corrosion resistance, crash energy absorption, and machinability. Some of the market-driven factors include fluctuations in the cost of raw materials, annual volume of parts required, spares to be produced, capital investment, factory location, and labor. As most modern cars are getting optimized, the cost of lightweighting increases. The mass add-back due to additional safety features, performance enhancement, or noise vibration and harshness (NVH) improvement needs attention after initial test-track tests [12]. Mass add-back is a balancing act between maintaining the benefits of lightweighting, such as improved fuel efficiency or extended EV range, and ensuring the vehicle meets all functional, safety, and market requirements.

The drive to increase on-board technology devices, sensors, and control modules increases the weight of passenger cars. The major components in EVs include the battery pack, body, drivetrain, electrical components, interior, exterior, and other miscellaneous components. Figure 1.2 shows the component-wise weight of an EV. From Figure 1.2, it can be seen that the battery and electrical components constitute 31% and the drivetrain about 23%. The other components, such as the body, interior components, and exterior components, including closures and fenders are common to both EVs and ICE vehicles. The reduction of the gross or curb weight of an EV therefore should emphasize the following components in order, starting with (i) battery, (ii) exterior (closures and fenders), (iii) body, (iv) electricals, and (v) interiors and drivetrain [12].

1.1.3 Lightweighting Lithium Batteries

A variety of battery geometries are available in the market for EV cars. These geometries can be broadly divided into cylindrical, prismatic, and pouch-type batteries. In Figure 1.3, a cylindrical 18650 cell of an EV battery pack is shown. It contains a positive electrode called a cathode, a negative electrode called an anode, an electrolyte, and a separator. Typically, the anode consists of carbon

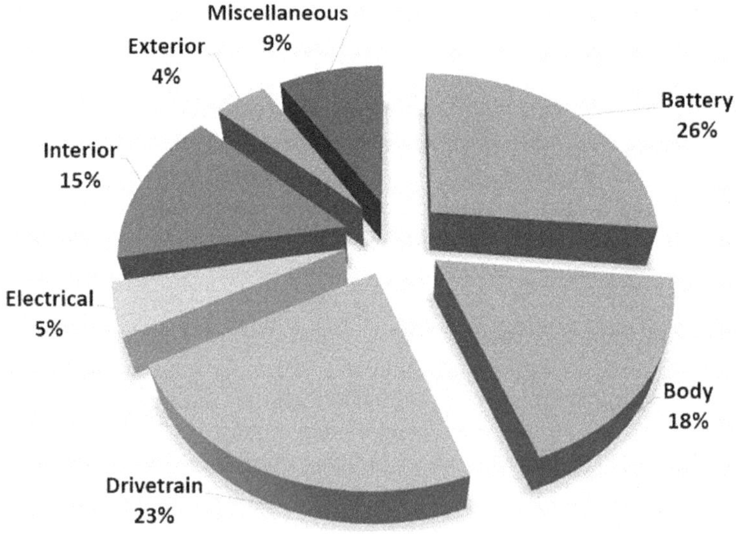

FIGURE 1.2 A chart illustrating the component-wise weight of EVs (Tesla Model S).

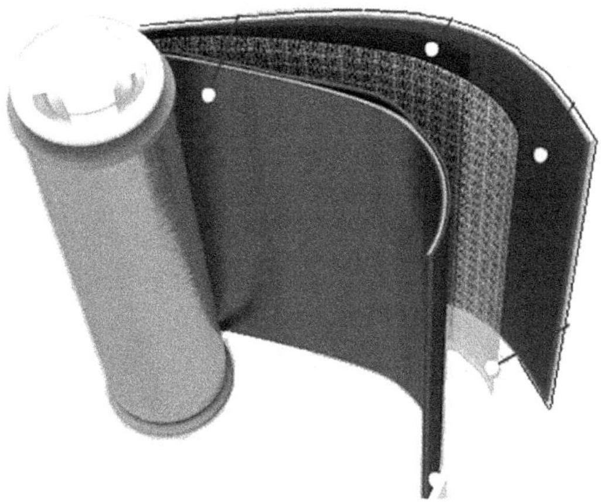

FIGURE 1.3 Internal construction of a lithium-ion battery.

materials like graphite or silicon-carbon composites, while lithium oxide composite is a common cathode material. The electrolyte, which facilitates ion transport between the anode and cathode, is usually prepared by dissolving a lithium salt in organic solvents such as ethylene carbonate, diethyl carbonate, or dimethyl carbonate [13]. The separator primarily regulates electron flow to minimize the likelihood of short circuits between the positive and negative electrodes [14]. The most common type of cylindrical cell is the 18650 (18 mm diameter, 65 mm length). Similar to these cells, we have 26650, 32650, and 42110 with different lengths and diameters compared to the 18650 cells. A typical cell has electrodes arranged in spiraled layers within a metal casing. These cells are used in battery modules and packs due to their advanced technology, high consistency, and robust mechanical stability [15]. Prismatic cells, on the other hand, offer better heat dissipation during discharge because of their larger surface area [16]. Rectangular cells provide ample space for integration in EVs. A pouch cell utilizes a layered structure of anodes, separators, and cathodes sealed within electrolyte-filled pouches. Their flexible design reduces weight, enhances energy density, and improves packaging efficiency, although it comes at the cost of increased mechanical vulnerability and the potential for swelling.

EV batteries are made up of thousands of small cells combined into a large battery pack. The heat generated by EV batteries increases with the discharge or acceleration of EVs. A compromise between affordability, energy density, storage capacity, and cost has to be reached while deciding on the battery for EVs. Typically, 60–75% of a battery's weight comes from the cells and their materials, while the remaining 25–40% consists of the metal casing, cables, and systems for thermal and battery management (TMS and BMS). In addition to lithium, EV batteries also contain cobalt and manganese. For example, a typical battery for an EV contains approximately 8 kg of lithium, 14 kg of cobalt, and 20 kg of manganese, with these quantities varying depending on the battery size, as shown in Figure 1.4. A Tesla Model S battery, for instance, contains approximately 62.6 kg of lithium [2].

The battery's casing and additional components can significantly increase the pack's total weight, impacting the vehicle's overall handling. To address these challenges, manufacturers are focusing on designing lighter, more efficient battery packs. Concurrently, research aims to enhance the energy density of lithium-ion batteries (LIB), with new structures and manufacturing techniques projected to realize energy densities of around 350 Wh/kg. This progress is expected to extend EV ranges while further reducing battery weight.

The size and weight of an EV battery are driven by its dimensions, capacity, and range. Larger batteries can store more energy but also weigh more, reducing the overall range of EVs. However,

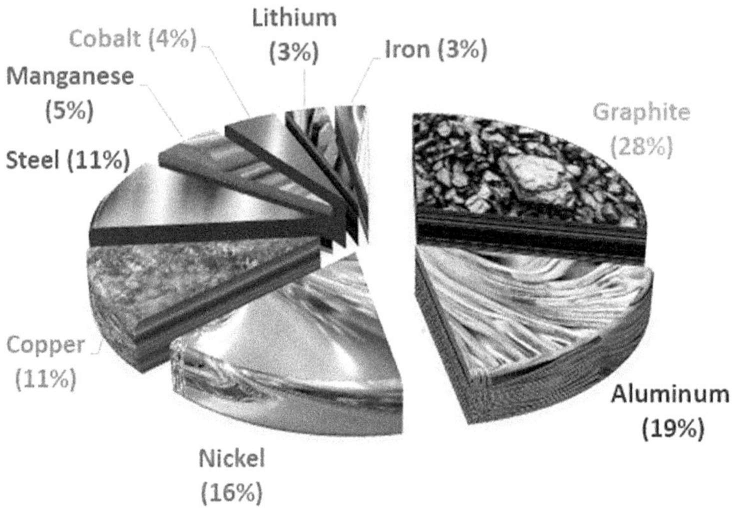

FIGURE 1.4 Elemental composition of the NMC-type battery pack.

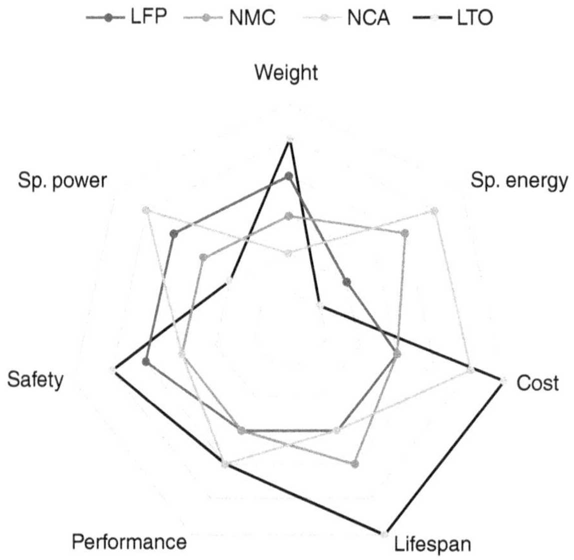

FIGURE 1.5 A graph showing a comparison of LFP, NMC, NCA, and LTO battery technologies.

advancements in battery technology are enabling reductions in weight while enhancing energy storage. Manufacturers are incorporating lighter materials and creating more efficient battery designs to decrease the weight of both the battery and the vehicle. Battery energy density and cost are key considerations in meeting mass-reduction goals and material qualification criteria. A few widely used forms of battery technology include, lithium iron phosphate (LFP), lithium nickel manganese cobalt oxide (NMC), lithium nickel cobalt aluminum oxide (NCA), and lithium titanate (LTO). Figure 1.5 compares these types of batteries on selection criteria.

From Figure 1.5, it can be seen that the NMC and NCA battery technologies have high specific energy and power and low weight. LTO has the highest lifespan, safety, weight, and cost. LFP has by far the most suitable properties, including moderate specific energy, weight, heat tolerance, and low cost. It was one of the most common batteries used in EV cars a few years ago. Currently, the EV car industry is using LFP batteries due to their availability and cost, battery life of about 4000 charge cycles, charging can be done for 100%, fast charging is possible safely, and the temperature range of LFP is better in warm weather conditions.

The framework to streamline the advancement in cutting-edge materials research for energy-storage applications is reported to be performed using methods like material modeling, AI integration, material design conceptualization, and practical experimental validation [17]. Material modeling begins with density functional theory (DFT) calculations, performed to determine the optimal structure or configuration of materials (e.g., ion arrangements on electrode surfaces) that minimizes interaction energy. This optimization leads to enhanced battery performance, such as enhanced storage capability and optimized open-circuit voltage (OCV) for the cathode. However, DFT calculations are computationally intensive and time-consuming. AI offers a solution by developing models that generalize DFT findings, thereby accelerating the material selection process. For instance, in evaluating the potential of a two-dimensional (2D) material like vanadium disulfide (VS_2) as a potential anode material in sodium-ion batteries (SIBs), the method of ab initio random structure searching is applied to identify the minimal-energy configurations for sodium interaction and adsorption phases on VS_2. These configurations are critical for predicting battery parameters like specific capacity and OCV. AI models trained on DFT data can extend this approach to screen a wide range of 2D materials, such as transition metal di-chalcogenides (TMDs), MXenes, and transition metal phosphides, for SIB applications. In such AI models, the input data might include material properties, while the outputs are the predicted capacity and OCV. By mapping these relationships, AI-driven optimization models can identify materials with the highest capacity and OCV. Additionally, this methodology can be expanded to predict properties of van der Waals heterostructures. In this case, input parameters for AI models could include structural details, for instance, the lattice average distances between atoms of sodium or their distances from the VS_2 surface. The principal outputs of the AI algorithm are battery capacity and OCV, enabling precise predictions. Continuing the material design, the physical battery is fabricated and tested. High energy density, lightweight battery with stable operation, and thermal are is the ultimate aims of lightweighting the battery cells.

Lightweighting batteries focuses on reducing their weight while preserving or enhancing performance, safety, and reliability. As illustrated in Figure 1.3, the casing protects the inner active layers of the battery cell and mitigates mechanical and environmental stresses. In 18650 batteries, steel is commonly used for the casing. To reduce weight, alternative materials including alloys of aluminum and magnesium, or carbon fiber composites can replace steel in certain applications, provided their thermal properties remain comparable to those of steel. High-strength, lightweight polymers, including poly-ether-ether-ketone (PEEK) and polycarbonate blends, are also viable options.

Material optimization techniques can minimize unnecessary mass while maintaining structural integrity. Topology optimization algorithms, which allow for multi-objective designs that account for thermal performance, are particularly effective. Additionally, three-dimensional (3D) printing enables the fabrication of intricate, lightweight geometries produced by these optimization routines, addressing manufacturing challenges associated with complex designs. A simpler approach involves incorporating honeycomb structures or ribbed patterns to enhance rigidity while reducing overall weight [18,19].

Battery form factor and packaging of a battery serve as a key factor in determining its overall weight [20–22]. Common battery form factors include cylindrical, prismatic, thin-film, and pouch cells. Pouch cells and prismatic designs offer advantages in lightweighting due to their lighter packaging compared to cylindrical cells. Another approach to reducing battery pack weight is the use of integrated structural designs, where the battery pack serves dual functions: providing energy storage and acting as a structural component of the device. This dual-purpose design eliminates the need for additional support structures [23]. Similarly, modular configurations designed for easy stacking can reduce redundant casing or support materials, further contributing to weight reduction.

Battery pack design contributes significantly to the safety, thermal management, and overall efficiency of EVs. The functions of battery casing include structural support, thermal management, and safety. A typical battery pack exploded view showing a number of components made from materials including batteries, plastics, metal connectors, printed circuit boards, integrated circuit chips, and sensors can be observed in Figure 1.6. By balancing material selection, thermal management, and structural design, EV manufacturers aim to enhance safety, performance, and efficiency while keeping costs competitive. Several strategies have been utilized in all these cases, including material substitution, design optimization, cell configuration, and use of composites in structural components.

Material substitution includes the replacement of steel structural components with aluminum or composites for battery pack enclosures. Lightweight thermal interface materials, particularly thermally conductive adhesives (TCAs), are used for heat management [24]. TCAs consist of a polymer matrix—usually epoxy—mixed with a thermally conductive filler. They are typically applied in liquid form and harden during curing to create a durable bond. While TCAs share some characteristics with thermal gels, they are generally stiffer and offer stronger adhesion between surfaces [24]. Design optimization of battery packs includes the implementation of structural adhesives instead of mechanical fasteners. Integrated designs for EV battery systems merge cooling plates with structural components, offering enhanced efficiency and compactness. These designs utilize multi-functional materials that not only provide mechanical support but also ensure effective thermal management for battery cells. Cell configuration changes, such as use of energy-dense NMC or NCA cells to reduce the number of cells required, lower overall weight. Use of prismatic or pouch cells instead of cylindrical cells makes the pack lighter in some designs. The other common strategy is the use of composites instead of metals in structural components, for example, sheet molding compound (SMC) and carbon fiber-reinforced polymer (CFRP) for enclosures, which reduce weight and improve stiffness. The combination of these elements provides the EV battery pack manufacturers the means to reduce overall system weight, improve energy density, and simplify assembly, contributing to improved performance and cost efficiency in EVs. Case studies showing these strategies incorporated in production vehicles include Tesla Model 3, which uses aluminum enclosures with liquid cooling and structural integration for lightweighting. BMW i3 utilizes CFRP for enclosures, combining lightweighting with high strength and safety. Nissan Leaf employs steel covers with air cooling for cost efficiency in mass-market applications.

A common battery cell assembly unit housing covers, encloses, and protects the battery cells and modules from environmental and physical damage. The battery housing primarily consists of a single-piece tray paired with a housing cover. This is complemented by a battery frame that functions as a crash frame and a lower protective cover integrated with a separate cooling system. The modules are arranged in alignment with the crash structure, which also defines the load path lines. The excess heat produced during the charging and discharging cycle needs to be safely disposed of. The battery cover needs to be sufficiently strong to prevent battery damage during car crashes. Aluminum alloys, high-strength steel, composites, and thermoplastics are the materials of choice for the battery casing. Recently, aluminum foam sandwich panels also have been researched for this purpose [26]. The battery pack is made sturdy to support its own weight using an underbody support made of steel or aluminum, and each cell is placed in a plastic support ring for crash energy absorption and maintaining the contact distance. The initial design approach can be based on statistical optimization of the base design, with the classical design of experiments to solve the structural integrity and thermal stability. It is reported that this approach was followed to reduce the weight from 7.68 kg to 1.48 kg, about an 80% reduction [27]. However, this approach does not consider the underlying dynamics of the vehicle and the modal analysis for structural assessment of the battery support.

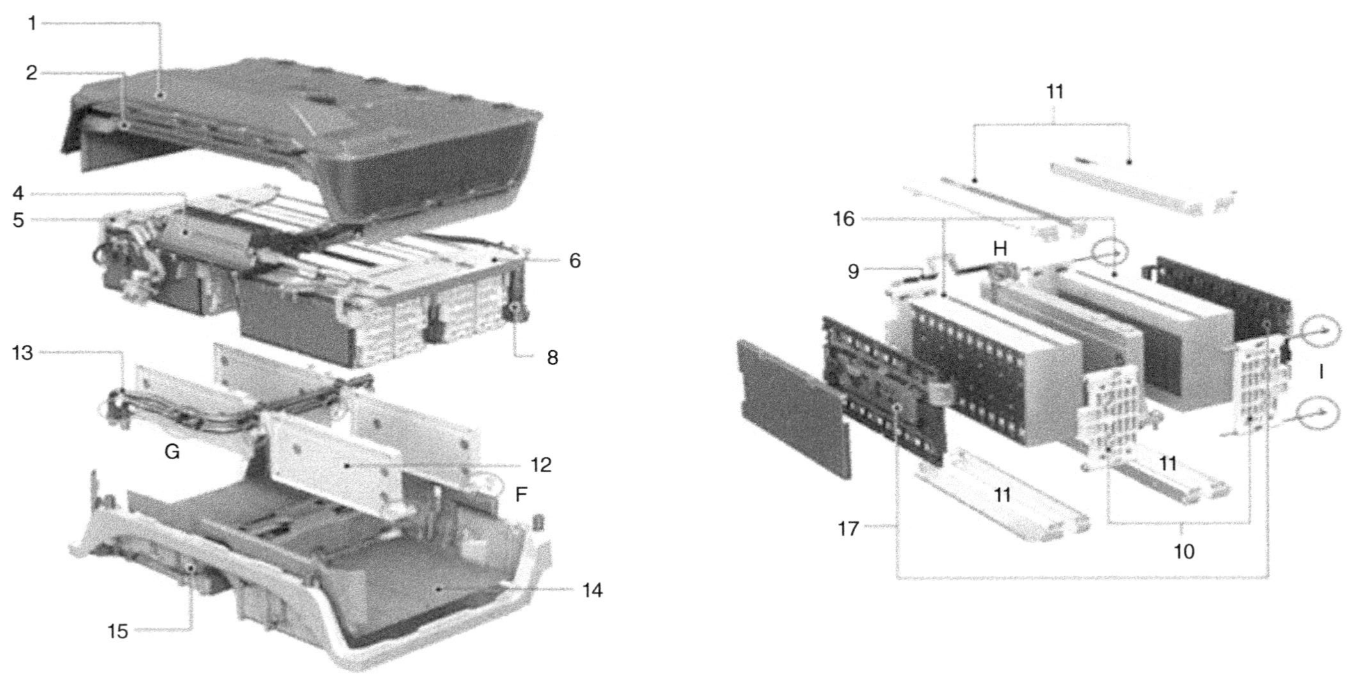

1. Upper housing shell
2. Upper insulator
4. Battery junction box
5. High voltage cables
6. Top transverse cover
8. Battery management controller (BMC)
9. Module connector
10. Side module junction
11. Top and bottom fastener of module
12. Cooling plate
13. Cooling pipe
14. Lower insulation
15. Lower housing shell
16. Modules
17. Cell management controller (CMC)

FIGURE 1.6 Exploded view of the battery pack of the Audi A3 sportback e-tron vehicle (Adapted from Ref. [25].)

The modules are provided with a steel housing cover and additional supporting panels to make the assembly rigid even during vehicle movement [18]. A comparison of dynamic modeling and simulation results of the base design and improved design indicates the battery tray weight was reduced by 5 kg, with a total battery pack weight reduction of 33% [18]. A design for a manufacturability study is presented, which caters to weight reduction of the battery tray bottom plate, and cross beam design optimization along with the optimization of crash safety parameters [28]. Multiple design iterations with simulations to corroborate the design changes are required to ensure weight reduction, crash safety parameters, and materials chosen before a final design recommendation can be suggested [28]. The modal analysis for a battery honeycomb pack also shows a higher frequency, indicating better performance compared to the base design. Currently, LIBs utilizing LFP and ternary electrochemical systems such as NMC and NCA are the main technological pathways for power system design. Ternary power batteries offer greater energy and power densities, which heighten the potential for thermal runaway due to their crystal structure and the chemical reactions involved in lithium-ion insertion and extraction under high-temperature conditions [29].

The use of cellular metallic and plastic structures in an EV car battery pack is reported in the literature to exhibit many advantages including weight reduction, energy absorption, vibration damping, material savings, and corrosion resistance [31]. Cellular structures provide high strength and stiffness while significantly reducing material usage, contributing to lower overall vehicle weight. This improves actual miles per charge in EVs. The unique geometry of cellular metallic materials enhances their ability to absorb impact energy, improving crashworthiness and passenger safety in the event of a collision. Similar to the crash test, these materials can reduce NVH levels and improve the safety of the battery connections. These structures optimize material placement, providing strength and rigidity precisely where needed without adding unnecessary bulk. Metallic cellular materials often have excellent thermal conductivity, making them suitable for heat dissipation in automotive components like battery system enclosures in EVs.

1.1.4 Cooling of the Lithium Battery Pack

The internal heat generation pattern is different for each of the battery geometries. Theoretical models with an electrochemical process along with some of the cooling strategies are reported in the literature [32–34]. Thermal runaway mitigation includes a variety of approaches including active and passive cooling. Considering commonly available battery geometries including 18650 batteries, a combination of radial cooling on the cylindrical surface and electrode cap cooling is required [32]. The jelly roll and battery end caps generally contribute to the uneven cooling of batteries, and the problem worsens during deep discharge cycles. Thermal management of batteries utilizes both active and passive cooling methods, such as air cooling, liquid cooling, phase change material cooling, and heat pipe (HP) cooling. Combination approaches are also routinely used for high discharge and high energy density batteries.

1.1.5 Air Cooling

This is inexpensive and occupies minimal space due to its simplified design with fewer components. Its maintenance and installation costs are low, and it is compatible with various batteries without concerns like liquid leakage. However, its efficiency is lower compared to other cooling systems. This is due to the lower thermophysical property values of air, for instance, heat transfer coefficient, thermal conductivity, and specific heat capacity. The power demand for a fan is significant, and the battery pack is vulnerable to external pollutants and dust carried by the air, potentially degrading its performance. Additionally, its performance heavily depends on ambient air temperature and may lead to uneven temperature distribution.

The open cell aluminum foam material is also used to prepare heat sink. It is reported that air cooling along with metal foam is more effective with metal foam in contact with the cell [35]. Harish et al. reported a numerical study to check air cooling of pouch batteries using aluminum foams [36]. The pore size and the porosity in the material are the primary factors influencing the heat transfer coefficient [36]. The results showed that the temperature rise is reduced by 45% with aluminum foam cooling compared to regular air cooling.

1.1.6 Liquid Cooling

Among all the battery-cooling solutions, liquid cooling is the most efficient owing to the superior thermal and thermophysical properties of common coolants used. Liquids are more effective at absorbing heat due to their higher heat transfer rate, greater mass flow rate, and superior specific heat capacity. Common cooling mediums like ethylene glycol, water, acetone, and oil offer better cooling performance and promote more uniform temperature distribution. If polar liquid such as water is used as a cooling medium, direct- and indirect-type cooling are employed to reduce the risk of a short circuit. However, this system features a more complex design, resulting in higher installation costs compared to air cooling. It also requires more components, leading to increased space requirements and added weight. Additionally, there is a risk of leakage and a shorter system lifespan.

1.1.7 Evaporative-Type Cooling

Evaporative-type cooling, including spray- or mist-type cooling, methods are a combination of air- and liquid-type cooling methods [37,38]. Evaporative-type cooling is a direct type of cooling involving a spray nozzle to produce spray of liquid droplets, and the battery is directly exposed to the coolant droplets blown over the batteries. Thus, for cooling the battery lateral surface, a fan may also be used to blow the mist-laden air. The cooling system is more sensitive to the injection pressure compared to air velocity for these systems. The maximum area of cooling is over the lateral surface of the cell, however, the high-temperature zones are located at the top and bottom electrode surfaces because of the electrical contacts [39].

1.1.8 Phase Change Materials

Phase change material (PCM) systems require minimal components, making them lightweight and compact compared to liquid and air cooling. These systems have a moderate initial cost and low maintenance. They consume no power, are more eco-friendly, and offer greater efficiency when measured against passive air cooling at low discharge rates of about 1°C with sustainable properties. Additionally, they have close to zero maintenance costs. However, PCMs have low thermal conductivity, and temperature regulation becomes challenging once the material fully transitions between phases, disrupting continuous operation. There is also a risk of leakage and the potential for super-cooling. Certain materials such as metallic nanoparticles, graphite, and carbon-based high thermal conductive additives are mixed with the PCM to increase the thermal conductivity and boost the heat transfer rates. However, only PCM-based cooling may not be sufficient for high discharge rate scenarios. A combination of PCM and air cooling is reported to be more effective for discharge rates higher than 2°C [40].

1.1.9 Heat Pipe Cooling

Cooling using HPs is a proven thermal management device, routinely used in the case of compact electronic devices such as performance laptops. The HP being a passive device reduces maximum temperature rise without requiring power consumption. Comparing HPs with PCM cooling, HPs having higher thermal conductivity, higher efficiency in heat transfer with no moving parts, and operate silently and reliably. This cooling system is lightweight, compact, and features a flexible design compared to other types of cooling methods. However, the HP has a limited contact area, resulting in lower efficiency and capacity, making it less effective in heating the battery. Additionally, it comes with high initial and operational costs, along with a risk of leakage.

1.1.10 Testing of Batteries and Lightweighting

The testing of these batteries is reported to be performed using the accelerated rate calorimentries (ARC) method [41]. The thermal runaway test is commonly employed to analyze the thermal runaway characteristics of a complete cell. ARC tests are performed using the heat-wait-seek-exotherm method. An adiabatic boundary is created around the cell during the exotherm phase allowing for precise measurement of heat generation during thermal runaway. The voltage and impedance (measured at 1 kHz) are recorded during the ARC test. Solid electrolyte interphase (SEI) is a thin, protective layer that forms on the anode surface at the time of initial charge/discharge cycles of a LIB. This layer is indispensable in the performance, safety, and lifespan of the battery. The mechanism of thermal runaway involves damage to the SEI, evaporation of electrolytes due to high temperatures, and a redox reaction that takes place rapidly increasing the core temperatures leading to thermal runaway [41].

Among the cooling methods discussed, the most lightweight option is air cooling with passive cooling. This method is not suited for the high discharge rates of modern battery usage. The most efficient is the liquid cooling technique. But this method is one of the most heavy, while a loop HP was found to be 12% lighter [42]. However, the use of fins, vortex generators, or heat spreader devices will increase the weight of the HP apparatus. The PCM-based hybrid cooling method uses metallic plates with flow passages for efficient heat dissipation.

The design and development of a battery pack is a complicated task involving structural integrity, thermal stability, and sufficient NVH isolation. Simulation methods used for battery EVs include multi-objective optimization including structural analysis, frequency response analysis, and thermal simulation. Results from a multi-physics electrochemical-thermally and electrically coupled model of battery packs have been reported in the literature [43]. Essentially, it captures the fundamental processes that take place when vehicles function in practical, everyday conditions. The range of parameters that can be optimized include anode and cathode thickness, porosity of the cell, cell gap, module construction parameters, thermal parameters of the module, module assembly, and pack structural analysis. AI, such as genetic programming, is utilized for the optimization of multi-objective problems such as redesign of the battery pack [44]. It is reported that a non-dominated sorting genetic algorithm performs better than base genetic programming. The electrochemical-thermally coupled multi-objective problem includes factors such as temperature, current, and porosities in the cathode and anodes. It provides information regarding the specific energy and state of charge (SOC) for the battery pack. Individual temperature variations and power imbalance are found between battery packs at various discharge rates [43].

1.1.11 AI and 3D Printing

AI-based design directions including 3D printed modules with integrated thermal management and structural stability are being tested [44]. Sensors to estimate the SOC, state of health (SOH), state of power (SOP), and remaining useful life (RUL) provide the measure of systematic improvements due to these design changes. AI and machine learning (ML) approaches to monitor and interpret the sensor data and fault finding have been reported in the literature [45]. These methods supersede the classical approaches such as Fourier transforms (FT), Pearson correlation coefficients (PCC), and discrete cosine transforms (DCT). The AI-based methods are more efficient in classifications

TABLE 1.1
AI and ML Algorithms used for EV Fault Finding

Function	Signal	AI algorithm
Bearing vibration	Vibration, stator current, AE	Convolution neural network (CNN), support vector machine (SVM), random forest (RF), gray relational analysis (GRA), deep neural network (DNN), ventral cochlear nucleus neural network (VCN)
Rotor	Motor current, stator current, vibration, thermal images, displacement	Support vector data description–SVM (SVDD-SVM), deep belief network (DBN), multilayer perceptron (MLP)
Stator	Vibration, stator current	Probabilistic neural network (PNN), CNN, ANN, SVM, error correction neural network (ENN)
Battery fault detection	Voltage and current	K-means, generalized regression neural network, SVM, RF
Open circuit	Voltage, current, three-phase currents	SVM, K-means, CNN, ANN, DNN, backpropagation neural network (BPNN)
Battery voltage prediction	Voltage	LSTM
Surface temperature prediction	Temperature, thermal images	BPNN, LSTM
SOH	Voltage, current, and temperature	Sparse Bayesian predictive modeling (SBPM), DNN, SVM, CNN,
SOC	Voltage, current, and temperature	Sparse learning machine (SLM), radial basis function neural network, long short-term memory (LSTM), SVM
RUL	Voltage, current, and temperature	Dynamic gray wolf optimizer–extreme learning machine (DGWO-ELM), LSTM, LSTM-GPR (Gaussian process regression)

and predictions, and a trained network requires less resources compared to classical methods [45]. Table 1.1 shows the ML methods employed in EV powertrain fault finding using AI and ML approaches in a brief consolidated manner [45].

From Table 1.1, it can be observed that a variety of network algorithms are developed to improve the fault finding and signal processing in EV cars and EV batteries. These methods are also employed to improve the design and maintenance of EV batteries and EV cars. Several components including the chassis, steering system, suspension system, braking system, tires, wheels, glass windows, lights and indicators, heating, ventilation, and air conditioning (HVAC) system, safety features, infotainment, controls, locks, and a few accessories are common to both internal combustion engine and EVs. In the subsequent sections, methods to lightweight these common systems are discussed.

1.2 EXTERIORS, CLOSURES, AND FENDERS

Exterior components are common to both EVs and ICE vehicles and include the hood cover, fender, roof cover, trunk lid, door covers, and body side panels, which are generally termed as closures and fenders. These components, shown in Figure 1.7, are traditionally made of steel, and this is the next component that can be considered for lightweighting. The common procedures of manufacturing these components include metal press molding and the joining process. Sheet metal is converted into the required covers for the geometries that are fixed to the metal door, hood, and trunk frames and body frame. The door, hood, and trunk frames are connected to the body frame with hinges and metal joining processes. The weight of these sheet metal components can be reduced by substitution with a lighter material such as aluminum or composites.

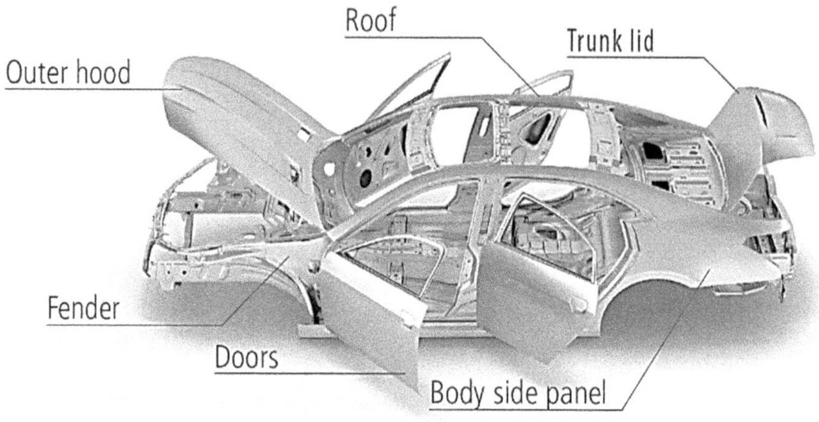

FIGURE 1.7 Components of the automobile that are part of exteriors, closures, and fenders. (Adapted from Ref. [46].)

Lightweight alternatives such as aluminum alloys, high-strength steel, and ultra-high-strength steel should also be conducive for manufacturability while maintaining sufficient strength for passenger safety. The materials of choice for these components require good physical properties such as modulus of elasticity and tensile strength. They have to be durable with good corrosion resistance, with high thermal performance and wear resistance. These materials should be easy to mold, support similar and dissimilar joining processes, and possess paint-retaining property for aesthetics. Further, it should be environmentally friendly and sustainable. This multi-criteria problem is not easily solved, with a variety of components and processes that needs to be analyzed [47]. Case study evidence and part-wise analysis for material substitution are needed for lightweighting the automobile.

Advanced materials such as composites have advantages including lower weight, strength and durability, corrosion resistance, and moldable to satisfy aerodynamic design. In general, however, design iterations are easier with metal forming compared to composite molding. Aluminum and steel have similar manufacturing flexibility, with aluminum parts being 50% lighter than steel but more expensive, and are usually used in premium and performance EVs. Steel, high-strength steel (HSS), and HSS (UHSS) are used in mass market EV cars. Aluminum is naturally corrosion resistant and needs minimal surface treatment and coatings. Steel covers need surface treatments and coatings to keep them free of corrosion. Both aluminum and steel components are in general more durable and easier to integrate to the body due to easy joining processes. They are easier to paint and several coatings are available for these materials to make them both aesthetic and durable.

In addition to these metals and alloys, composites are prevalent in the manufacture of high-performance road-legal cars called hypercars or supercars. These composites include CFRP, glass fiber-reinforced plastic (GFRP), natural fiber-reinforced plastic (NFRP), thermoplastic composites, and SMC. The primary methods of manufacture of these composites include hand layup, injection molding, spray layup, compression molding, and resin transfer molding. The suitability of these techniques depends on the type of matrix and reinforcement. The matrix used for composites used in exterior panels uses polymers that exhibit ductile property. As a base matrix, it should possess properties comparable to metals, such as impact strength, stiffness under loading, and resistance to environmental factors. Reinforcement further enhances the strength, rigidity, and ability to withstand strain due to regular and impact loading conditions. Synthetic fibers have high impact strength and mechanical strength. However, these fibers are more expensive and not recyclable. Natural fibers are cheaper compared to synthetic fibers but have low impact strength and the mechanical properties depend on the condition of the fibers.

Bere et al. proposed a multi-layered CFRP composite with stacking sequence to maximize crash protection compared to its competitors [48]. It has also been mentioned that vacuum bagging and curing of the composite using autoclaving is a common method for the fabrication of thin sheet-type composite parts. The model CFRP first developed reduced the weight of the hood by 85%, and helped in the energy cost saving and service of the automobile. It can further be improved by using a balanced stacking sequence of layers, and using metal foams or sandwich structures can further reduce the weight by 22% [48]. The lightweight hoods thus developed are reported to have close to 53% mass reduction and 42% weight reduction. The lightweighting of the hood using CFRP can also be applied to cost reduction in interior parts. The types of fibers used and the laying are critical aspects of the mechanical strength of these composites. The same technique can be used for fabrication of other components, such as fenders, side panels, roof panels, trunk lids, and rear fenders. However, the higher manufacturing cost of CFRP, GFRP, and NFRP, along with the complexity in joining these parts, restricts their wide applications [49].

One of the most difficult operations in the fabrication of a car body is the dissimilar metal joining process [49]. Welding techniques such as metal inert gas (MIG), tungsten inert gas (TIG), resistance spot welding, laser welding, element welding, friction stir welding, and cold metal transfer (CMT) welding are used to join steel–aluminum, aluminum–magnesium, and steel–magnesium. The alloying filler layer is used for steel–aluminum and steel–magnesium weldment, in MIG, TIG, CMT, and laser welding techniques. Intermetallic compounds (IMCs) are formed near the weld region, having both beneficial and detrimental effects depending upon the welding conditions. Steel–aluminum weld produces $FeAl_3$, $FeAl_5$ brittle IMCs depending upon the thermodynamics of reaction, copper–aluminum weld has $CuAl_2$ and Cu_4Al_9 IMCs, while nickel–aluminum IMCs are of the form NiAl and Ni_3Al. In general, IMCs can be controlled by optimization the welding, use of interlayers and fillers, and post-weld treatments. The automotive industry, however, prefers resistance spot welding and electric arc welding for dissimilar materials integration due to established infrastructures because they are therefore convenient and economical.

1.2.1 Cellular Materials

Cellular structures are lightweight architectures composed of repeating units or cells, such as honeycomb patterns, lattice frameworks, or foam-like structures. Figure 1.8 shows the unit cell structures of common cellular materials. These designs are particularly advantageous in EV components due to their ability to enhance the strength-to-weight ratio while maintaining structural integrity. Additionally, they offer multifunctional benefits, including thermal management, noise and vibration reduction, and crash safety, making them integral to modern EV design. By employing cellular structures, manufacturers can achieve lightweight, efficient, and safe designs that align with the industry's goals of sustainability and performance optimization.

Cellular structures can be optimized using topology optimization algorithms to minimize weight without compromising mechanical strength. These algorithms distribute material efficiently, targeting specific load paths for optimal performance. Common unit cell geometries include hexagonal honeycombs, octet trusses, gyroids, and Voronoi patterns. These geometries allow customization of mechanical properties such as stiffness, energy absorption, and thermal expansion compensation. Cellular structures are highly scalable and can be designed for applications ranging from microscopic components to macroscopic vehicle frames. They can also be embedded into larger components, enhancing the structural and functional performance of EV systems.

Applications in EVs include chassis and body panels to provide lightweight, load-bearing structures and battery enclosures to facilitate thermal management and structural support. Battery

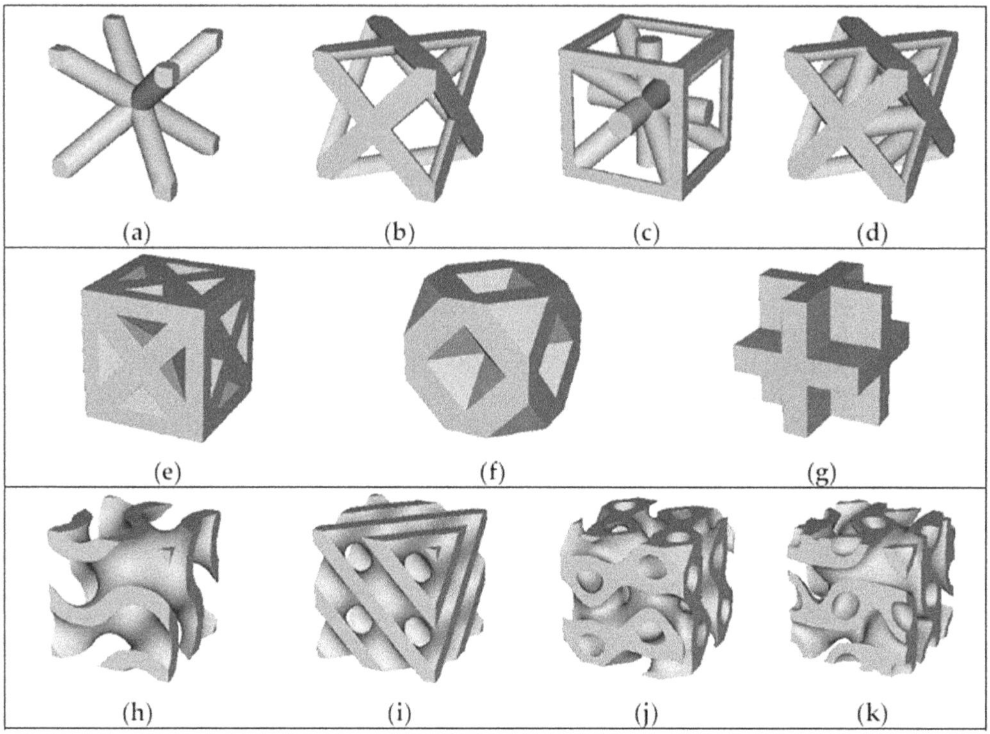

FIGURE 1.8 Images of unit cells of cellular materials: open cell beam-based lattice structures (top row), closed-cell foam-based structures (middle row), and triply periodic minimal surface (TPMS) structures (bottom row). (Adapted with permission from Ref. [50].)

cooling systems use these type of materials to improve heat dissipation with high surface-area-to-volume ratios. Interior components gain enhanced efficiency in seats, door panels, dashboards, and flooring. Manufacturing of cellular structures uses some of the state-of-the-art technologies such as additive manufacturing (AM), brazing, and electron beam melting (EBM). The manufacturability of cellular structures depends on the material, geometric complexity, and scale of the design. Recent advances in manufacturing technologies, particularly AM, have enabled the fabrication of complex cellular geometries with tailored meso-structures [51]. AM techniques such as selective laser melting (SLM) and other technologies such as EBM have significantly advanced the production of honeycomb patterns and lattice framework cellular structures. For instance, microscopic cellular structures have been successfully fabricated using the EBM process [52]. Simple honeycomb geometries are easily manufactured via 3D printing [53]. Complex octet truss structures have been created using axial lattice extrusion with carbon composites [50]. A 20 × 20 × 40 mm parallelepiped structure made from aluminum alloy $AlSi_{10}Mg$ was fabricated using an SLM machine [54]. Traditional high-precision manufacturing methods such as cellulose nanocrystal (CNC) milling and water jet cutting are also used for CFRP and hollow cylindrical tube-based octet lattice structures [55]. Other approaches such as titanium building blocks with notches for braze connections have been reported for small mechanical test structures [56].

A geometrically complicated gyroid lattice structure is a type of TPMS characterized by zero mean curvature and minimized local surface area, as shown in Figure 1.9. These structures are mathematically defined by sine and cosine functions, and feature smooth, continuous surfaces that partition space into two congruent, interwoven regions.

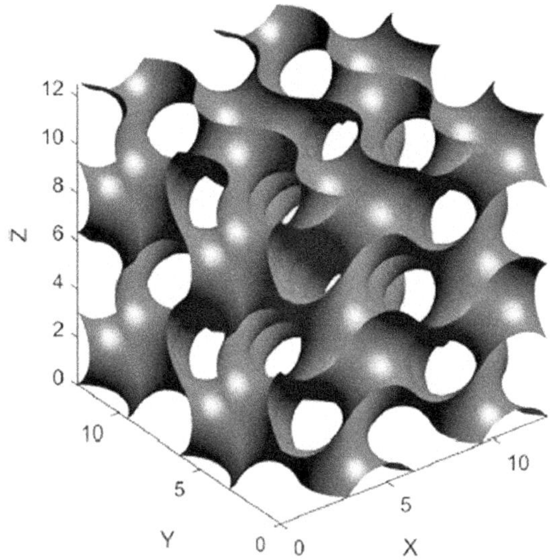

FIGURE 1.9 A gyroid shape used for lightweight structural material. (Adapted from Ref. [57].)

The mathematical function to produce gyroid geometry is as shown in equation (1.1).

$$\Phi(x, y, z) = \cos(\omega x)\sin(\omega y) + \cos(\omega y)\sin(wz) + \cos(\omega z)\sin(\omega x) - c$$

where ω are the TPMS function periodicities, and is given by:

$$\omega = \frac{2\pi}{l_i} \quad \text{(Eq. 1.1)}$$

The pores in the TPMS structure are controlled with the factor c and l_i is the length of the unit cell in x, y, and z directions. It is the factor that also controls the shape of the surfaces as well as the topology of the TPMS structures. Gyroid lattices are best fabricated using AM techniques due to their geometric complexity. Gyroid structures have been manufactured from 304 stainless steel using the SLM method [57]. Post-fabrication treatments, such as heat treatment in a nitrogen-protected tube furnace at 300°C for 2 hours followed by furnace cooling, are essential to improve material properties.

By leveraging advanced manufacturing technologies and carefully selecting materials, cellular structures can be effectively integrated into EV designs, contributing to improved performance, efficiency, and sustainability.

1.2.2 Body in White (BiW)

BIW refers to the stage in vehicle manufacturing where the car body's structural frame has been assembled but has not yet been painted or equipped with mechanical components, interior fittings, or trim. The BIW consists of welded sheet metal components that form the skeleton of the vehicle, including elements like the roof, floor, side panels, and pillars [58]. A representative image of an automotive body structure is shown in Figure 1.10.

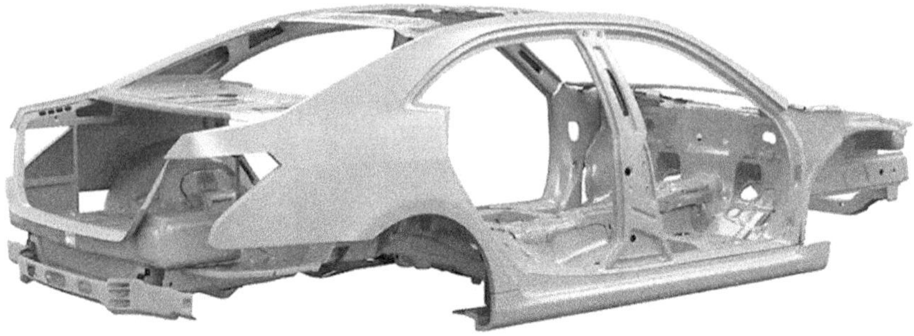

FIGURE 1.10 A representative image of an automotive body structure.

The primary purpose of the BIW is to provide a rigid and lightweight structure to ensure safety, aerodynamics, and support for other components. This phase is critical in determining the vehicle's crashworthiness, structural integrity, and overall weight. A variety of materials including aluminum, magnesium, HSS [59], advanced HSS (AHSS), Mg and Ti alloys of steel [60], multi-material [61], carbon fiber, and foam are used to make the BiW of EVs lightweight. Zhang et al. developed an aluminum foam double tube structure for the threshold of the car [30].

The application of aluminum in the automotive industry has increasingly shifted toward EV body structures and chassis, as well as components specific to EVs, such as battery trays and electric drive systems [62]. Currently, these components are predominantly fabricated from sheet and extruded aluminum products, with limited use of castings. The castings employed are often structural, necessitating advanced manufacturing processes and the use of high-quality aluminum alloys, sourced either from primary production or clean secondary scrap.

A recent development in this field is the adoption of giga-castings (also referred to as "mega-castings" by certain original equipment manufacturers). These large structural castings consolidate multiple components into a single, ultra-large casting [62]. This innovation has the potential to transform automotive manufacturing by altering the overall utilization of aluminum and the distribution of product forms—castings, sheet, and extrusions—in light vehicle production. Specifically, giga-castings may drive significant growth in the use of castings while moderating the demand for sheet and extruded aluminum products.

1.3 INTERIOR PLASTIC COMPONENTS

1.3.1 Fire Retardation in Polymers

Passenger vehicular interior cabin components can be considered to be predominantly plastic based. However they consists of metals and alloys as support materials. Crucial innovations in polymer science focusing on high strength, light weight, durability, and sustainability have paved the way for the design, development, and production of EVs with an effective range of above 500 km. The major weight in an EV cabin is due to polymers in car interiors, passenger seating and harnesses, and electrical components. As such plastics are low density and lightweight further reductions in weight are not easy to accomplish. Reports of reputed car companies such as Mercedes reported reduced car overall weight by 20%, by using jute fibers [63], further natural fibers were incorporated into the production of components for Evobus, Setra Top-class vehicles, and Travego buses. In 1999, the company used a composite material consisting of 65% flax, hemp, and sisal fibers combined with 35% Baypreg, a semi-rigid elastomer developed by Bayer. Similarly, Ford Montero reported the extensive use of kenaf fiber-based composites in car interior applications as early as 1996 [63]. The plastic parts used in automobiles are required to possess properties such as chemical resistance, thermal stability, shatterproof, UV and weather resistance, high optical appearance, surface finish, insulation against external weather, shock absorption, and water impermeability.

Additionally, these parts are prone to fire hazards and in particular pose a greater threat in EVs. The risk of fires due to LIBs is high in EV automobiles, due to the high driving performance demands, speed of charging, and unavoidable accidents. A number of fires in EV cars have been reported in the literature [64]. In the event of fire in passenger cars, EVs produce more heat compared to ICE vehicles [64] . EV car fires are reported to produce toxic chemicals such as hydrogen fluoride (HF) from burning LIBs. LIB fires are difficult to suppress due to the risk of re-ignition, and much water needs to be used to keep the batteries at temperatures below the ignition temperature [64]. Specific to EVs, UNECE R100 imposes a number of fire safety tests for EV cars. A total of about 100 kg of different types of plastics are used in external and internal plastic components. These plastics have to pass mandatory fire tests.

The traditional flame-retardant materials developed are halogen based including methyl [2,2,2-trifluoroethyl] carbonate (FEMC), chloroparaffins, tetrabromobisphenol A (TBBA), and dede cachloropentacyclooctadecadiene (cechlorane) [65]. Additional research for mitigating environmental effects resulted in the following phosphorus-based flame retardants being developed: dihydro-oxa-phosphaphenantrene-oxide (DOPO), ammonium polyphosphate (APP), and resorcinol bis [diphenyl phosphate] (RDP) [65]. Many bio-carbon- and bio-graphite-based reactive extrusion systems with in situ compatibilization-based composites have the advantages of flame-retardant properties along with lightweight, superior performance, durability, and reduced density [65].

Materials such as natural fiber composites, graphene-based plastics, environmentally friendly materials such as polylactic acid (PLA), polyhydroxyalkanoates (PHA), and bio-based polyurethane (Bio-PU) are being used as car interior materials. In particular, natural fiber composites with reinforcements such as flax fiber, jute, bamboo, kenaf, abaca, sisal, hemp, caryota, banana fiber, coconut fiber, and hay fiber, are used in a variety of car components [63]. Some of the applications proposed using natural fibers include flax fibers in door panel covers, instrument panel housing, covered inserts, arm rests, seat components, hemp fibers in door panels, seat back panels, dashboard covers, Abaca and banana fibers in floor panels and body panels, jute and coir in car seat covers, back panels, door panels, seat mattresses, sound proofing, and insulation, and similarly, cotton fibers are used for sound proofing and insulation, wool for upholstery and seat covers, wood for dashboards, internal door handles, floor panels, and arm rests, foamed wood for seat panels, and other fibers in seat back cushions, inserts, and interior insert parts.

The major obstacles in using natural fibers in car interior components include moisture absorption and bacterial infestation, odor, inconsistent quality, lower thermal stability, limited compatibility with polymers, premature biodegrading, low durability and fire resistance, cost, supply chain issues, and low thermal stability during manufacturing with thermoplastic materials. Overcoming these challenges often involves modifications such as chemical treatments, hybrid composites, or coatings to enhance the durability, performance, and compatibility of natural fibers in automotive applications. Natural fibers contain non-cellulosic components such as hemicellulose, lignin, pectin, and wax, on the outer surface of fibers, which negatively impact their mechanical properties. These substances hinder interfacial adhesion between the matrix and the reinforcing fibers. To enhance bonding with various matrices, it is essential to modify the surface properties of natural fibers. Surface modifications can be achieved through two primary approaches. Physical treatments, such as laser processing, argon treatment, heat application, and plasma treatment, are effective in improving adhesion at the fiber–matrix interface. However, with physical treatment being more expensive than chemical methods, industry prefers chemical treatment of fibers. Some of the chemical treatments include alkaline treatment, silane treatment, acetylation, maleated coupling, benzoylation treatment, and enzyme treatment.

The alkaline treatment method involves immersing natural fibers in a 2–6% sodium hydroxide (NaOH) solution to remove non-cellulosic substances such as lignin, pectin, hemicellulose, wax, and oils from the fiber's outer cell wall. This process modifies the fiber's cellulosic molecular structure, enhancing its compatibility as a reinforcement material for thermosets and thermoplastics. After the fibers are treated in the alkaline solution for a specified duration, they are rinsed with 0.2 N HCl to neutralize any residual alkalinity, followed by thorough washing with water and subsequent drying for further processing. Silane coupling agents act as surface coatings, filling micropores and voids on the fiber surface, thereby forming interlocked coatings. Silane-treated natural fiber-reinforced composites (NFRCs) exhibit superior tensile properties compared to those treated with alkali alone. The silane treatment is done with the aim to modify the surface and make the fibers water resistant. Common silanes such as amino-propyl trimethoxysilane and vinyl-trimethoxysilane are often used to modify fibers. The process involves mixing an ethanol–water solution with the silane compound (ratio 2:3) and adjusting the pH to 4 using acetic acid. After 1 hour, fibers are immersed in the solution for 3 hours, followed by drying at 65°C for 8 hours. Jute polyester composites had silane treatment enhanced flexural strength and modulus by 20% and 8%, respectively, compared to alkali treatment alone [66]. Jute epoxy composites had tensile strength and modulus increased by 12% and 7%, respectively, with silane treatment versus alkali treatment [66].

Acetylation is a chemical treatment method for natural fibers that replaces their hydroxyl (OH–) groups with acetyl groups, increasing their hydrophobicity. This process aims to reduce the hydrophilic nature of fibers, thereby improving the dimensional stability of NFRCs. This treatment involves grafting the cellulose structure of the fibers with acetyl groups, either with or without the use of a catalyst. Typically, acetic anhydride and acetic acid are used, though they may not react efficiently with the fibers under standard conditions. The process begins with immersing the fibers in acetic anhydride, followed by acetic acid, for 1–4 hours at elevated temperatures. In a specific case, sisal fibers pretreated with an alkaline solution (5% and 10% NaOH for 2 hours at 280°C) underwent acetylation, which significantly enhanced the fiber–matrix interfacial bonding in the resulting composites. These composites are aimed for sustainability along with lightweight properties in EVs.

1.4 DISCUSSION

Lightweighting in EVs is a multidisciplinary effort that leverages advanced materials, innovative design, and state-of-the-art manufacturing technologies. By reducing vehicle mass, these strategies enhance energy efficiency, extend driving range, lower production costs, and contribute to environmental sustainability. In conclusion, these are the brief key approaches that promise a transformative impact on the EV industry of the future.

1.4.1 Solid-State Batteries

A solid-state battery with lithium sulfide, lithium polymer, and lithium oxide chemistry is touted to be better than a LIB, with solid materials as the electrolyte instead of a liquid or gel. These solid materials also function as the separator, resulting in fewer distinct components in the battery's construction. Additionally, the carbon anode is omitted, allowing solid-state batteries to achieve a much higher energy density. The charging of these batteries is thermodynamically more reversible, and this leads to enhanced lifespan [67]. Furthermore they charge at a faster rate, since they do not need lithium to diffuse into carbon particles, unlike traditional LIBs. Solid-state batteries are expected to offer improved safety and stability while being more cost-effective than LIBs. However, solid-state batteries face several challenges, particularly in the formation and maintenance of electrical interfaces during fabrication and operation. For instance, while sulfides are more deformable than oxides and can be effectively processed into thin films using cold compression, it is crucial to preserve these interfaces during volumetric expansion caused by charging cycles. Solid-state batteries are capable of rapid charge/discharge rates exceeding 10°C, operate effectively at extreme temperatures (above 80°C or below 10°C), and offer enhanced safety [68]. However, the high- and low-voltage reactivity of sulfides poses stability issues. Research into hybrid battery technologies, which integrate the benefits of polymer ion, oxide, and sulfide batteries, is ongoing. Efforts are also focused on hybrid chemistry, and scaling solid-state batteries for applications in automotive industry.

1.4.2 Structural Batteries

Incorporating batteries into a vehicle's structural framework offers a dual benefit of weight reduction and improved space efficiency. These integrated batteries, known as structural batteries, combine carbon fiber and battery components within the framework's geometry. Additionally, alternative coupling mechanisms, such as polymer binders, stainless steel meshes, interconnected joints inspired by biological systems, aluminum sheets, tubular structures, and interlocking polymer rivets, have been explored in the literature. Evaluations of the multifunctional performance of structural batteries consistently highlight their ability to extend the range and autonomy of EVs without increasing overall mass. Moreover, for EVs with equivalent battery capacity, a reduction in total mass has been observed [69].

1.4.3 Design Innovations Supported by Numerical Simulations and Artificial Intelligence (AI)

AI is driving the advancements in optimization of component designs, assembly of vehicles, service, and maintenance in the charging infrastructure network. AI-driven tools predict wear and optimize the lifecycle of lightweight components. The integration of functional large aluminum casting in EVs has created a maximum strength-to-weight ratio. The AI is trained on the driving styles of the individual user to optimize the power distribution during vehicular use. Virtual replicas of system components, often called the digital twins, are being used to optimize the designs, and generative AI is used for design variations to speed up the design iterations and achieve a competitive edge. Digital twins are used to simulate and optimize the performance of lightweight components in real-world conditions, while AI-driven advanced extension of digital twins, referred to as digital triplets, incorporate more comprehensive data layers, including predictive analytics, to refine lightweighting strategies further [70]. Structural integration at battery pack level, cell level, and micro-level component integration is aimed with the thin-film battery coating for structural members of EVs [71].

1.4.4 Advanced Materials Research

High-performance Alloys: Research into aluminum–lithium, magnesium, and ultra-high-strength steel alloys is enabling the production of lightweight yet robust vehicle structures currently. Several strength-enhancing processes are used to produce high-strength steels such as lightweight steels with induced plasticity (L-IP) and twinning-induced plasticity (TWIP) steels [72]. Titanium alloys (Ti-6Al-4V) have been explored for reducing bending and cycling deformation. Zinc and nickel plating and surface treatments are researched for solid solution strengthening and second-phase strengthening. Future chassis designs may incorporate multi-material configurations, combining cellular materials in energy-absorbing zones with fiber-reinforced polymers to enhance chemical and corrosion resistance. Inspired by natural structures like honeycombs and bones, bionic designs are being used to achieve superior lightweighting with strategizing and minimizing material usage. Cellular structures such as foams, honeycombs, octet trusses, gyroids, and Voronoi patterns will likely be preferred for their lightweight characteristics. Additionally, car door structures requiring similar impact strength could benefit from origami-inspired panels, which offer superior energy absorption capabilities.

Materials such as CFRPs, glass-fiber composites, and bio-based composites provide substantial weight reductions, particularly in structural applications. These composites are already utilized in the automotive sector due to their favorable results in impact and ballistic testing. Incorporating materials such as carbon nanotubes, graphene, and nanoclay enhances mechanical strength and thermal conductivity while minimizing weight. These emerging nanocomposite material options are being explored for use in bumpers, fixtures, and closures. However, the mechanical properties and crash performance of these innovative materials must be thoroughly validated through predictive modeling and testing prior to large-scale adoption.

1.4.5 Advanced Manufacturing Techniques

The hard-to-machine and brittle lightweight materials need advanced manufacturing techniques to process and integrate various material types. Some of these approaches, such as hot stamping, hydroforming, and laser-assisted machining, aim to enhance the body structure's bending strength, torsional strength, roll strength, and collision energy absorption while keeping the complex structural geometries. High-strength complex components are produced by combining shaping and heat treatment through the press hardening process. Adiabatic shear cutting is employed as a post-treatment method for hardened steels, while sheet forming processes with high local deformation rates benefit from elevated temperatures to reduce the required cutting force.

Laser-assisted machining (LAM) technology is widely utilized in precision and ultra-precision machining processes. Key metrics for assessing and optimizing machining quality include surface integrity, cutting force, tool life, and material removal rate. Laser irradiation facilitates energy transfer, where surface particles absorb photon energy and undergo rapid localized heating. The workpiece gradually reaches thermal equilibrium through internal heat transfer within the material. Workpiece temperature is controlled depending on its brittle-to-ductile transition behavior. Accurate prediction and control of the workpiece temperature during machining are critical [73].

1.4.6 Additive Manufacturing (3D Printing)

Traditional manufacturing techniques rely on the production and subsequent assembly of mating parts. Three-dimensional (3D) printing technology has the capability of producing functional mating components that can potentially reduce the number of assembly steps and components to be assembled. It enables intricate geometries that use less material while maintaining strength and functionality. This aspect of design will be critical to leap to realize the full potential of techniques of Industry 5.0. AM reduces waste compared to traditional subtractive processes, making it ideal for lightweight production. Research into fabrication of 3D-printed carbon fiber composite objects with embedded multi-layer circuitry which can directly support the kind of capacitive sensing has been reported [74]. Multi-material electronically functional embedded sensors in carbon composite structures can be utilized both for lightweight and active monitoring. Hybrid printing combines materials to produce parts optimized for weight and performance in specific applications. Advanced multi-material joining technologies, including adhesive bonding, friction stir welding, and laser welding help in multi-material integration. High-performance adhesives join dissimilar materials, such as metals and composites, effectively eliminating the use of mechanical fasteners like rivets. Also, the joints produced are strong, lightweight, with minimal material distortion. Strategic use of lightweight and high-strength materials in different sections of the vehicle optimizes vehicle performance and range.

Polymer nanocomposites and advanced coatings with wear resistance and thermal stability reduce degradation of plastics and enhance performance. The demand for polymer nanocomposites with specific properties is increasing in the EV car manufacturing industry [75]. Considerable research into nanostructured surfaces has produced material properties, such as hardness and friction, for lightweight components. These surface coatings are often applied to window glass and painted surfaces to achieve de-icing and defrosting. Fundamental structural elements, such as nanowires, nanoparticles, or nanotubes, are organized in one-dimensional, 2D, or 3D configurations to create nanostructured surfaces. This straightforward approach is efficient, cost-effective, and easy to implement, making it well-suited for large-scale production.

There are some key challenges in the lightweighting endeavors of the EV car industry. Extreme design iterations with rapid design changes make spare part availability and reusability difficult. Specific specialized designs may lead to end-of-life wastage. Structural batteries pose a major challenge. The battery failure or damage to a structural battery may necessitate replacement of the entire component leading to the mounting cost of repairs. Innovations in material recycling need to be expedited to ensure that newly used lightweighting materials, such as composites and high-strength alloys, are recovered sustainably.

1.5 SUMMARY

With the integration of these diverse strategies, the EV industry is poised to achieve significant advancements in lightweighting. Technology innovations in allied fields such as space engineering, ballistics, and aircraft design and manufacturing are utilized to improve crash safety, performance, and sustainability. Combining materials innovation, AI-assisted design, cutting-edge manufacturing, and sustainability practices ensures EVs are not only lighter but also more efficient, safe, cost-effective, and eco-friendly. This holistic approach aligns with the broader goals of reducing energy consumption and greenhouse gas emissions in transportation, driving a future of sustainable mobility.

REFERENCES

[1] Z. Liu *et al.*, "Comparing total cost of ownership of battery electric vehicles and internal combustion engine vehicles," *Energy Policy*, vol. 158, pp. 1–41, 2021, doi: 10.1016/j.enpol.2021.112564.

[2] D. Berjoza and I. Jurgena, "Effects of change in the weight of electric vehicles on their performance characteristics," *Agron. Res.*, vol. 15, no. Special Issue 1, pp. 952–963, 2017.

[3] G. S. Daehn, K. E. Daehn, and O. Kuttner, "Environmentally responsible lightweight passenger vehicle design and manufacturing," *Automot. Innov.*, vol. 6, no. 3, pp. 300–310, 2023, doi: 10.1007/s42154-023-00241-4.

[4] B. Wang, Z. Zhang, G. Xu, X. Zeng, W. Hu, and K. Matsubae, "Wrought and cast aluminum flows in China in the context of electric vehicle diffusion and automotive lightweighting," *Resour. Conserv. Recycl.*, vol. 191, no. July 2022, p. 106877, 2023, doi: 10.1016/j.resconrec.2023.106877.

[5] A. Mayyas, M. Omar, M. Hayajneh, and A. R. Mayyas, "Vehicle's lightweight design vs. electrification from life cycle assessment perspective," *J. Clean. Prod.*, vol. 167, pp. 687–701, 2017, doi: 10.1016/j.jclepro.2017.08.145.

[6] F. Ballo, P. Stabile, M. Gobbi, and G. Mastinu, "A lightweight ultra-efficient electric vehicle multi-physics modeling and driving strategy optimization," *IEEE Trans. Veh. Technol.*, vol. 71, no. 8, pp. 8089–8103, 2022, doi: 10.1109/TVT.2022.3172174.

[7] N. V. Martyushev, B. V. Malozyomov, S. N. Sorokova, E. A. Efremenkov, and M. Qi, "Mathematical modeling of the state of the battery of Cargo electric ehicles," *Mathematics*, vol. 11, no. 3, pp. 536–555, 2023, doi: 10.3390/math11030536.

[8] Z. Pusztai, P. Kőrös, F. Szauter, and F. Friedler, "Vehicle model-based driving strategy optimization for lightweight vehicle," *Energies*, vol. 15, no. 10, pp. 3631–3651, 2022, doi: 10.3390/en15103631.

[9] T. Wang, R. Dong, S. Zhang, and D. Qin, "Research on lightweight design of automobile collision safety structure based on multiple materials," *J. Phys. Conf. Ser.*, vol. 1670, no. 1, pp. 1–10, 2020, doi: 10.1088/1742-6596/1670/1/012004.

[10] F. Del Pero, L. Berzi, A. Antonacci, and M. Delogu, "Automotive lightweight design: Simulation modeling of mass-related consumption for electric vehicles," *Machines*, vol. 8, no. 3, 2020, doi: 10.3390/machines8030051.
[11] A. Dér, A. Kaluza, L. Reimer, C. Herrmann, and S. Thiede, "Integration of energy oriented manufacturing simulation into the life cycle evaluation of lightweight body parts," *Int. J. Precis. Eng. Manuf. – Green Technol.*, vol. 9, no. 3, pp. 899–918, 2022, doi: 10.1007/s40684-021-00412-w.
[12] C. Bailo, S. Modi, M. Schultz, T. Fiorelli, B. Smith, and N. Snell, "Vehicle mass reduction roadmap study 2025-2035," 2020. [Online]. Available: www.cargroup.org/publication/vehicle-mass-reduction-roadmap-study-2025-2035/.
[13] J. G. Zhang, W. Xu, J. Xiao, X. Cao, and J. Liu, "Lithium metal anodes with nonaqueous electrolytes," *Chem. Rev.*, vol. 120, no. 24, pp. 13312–13348, 2020, doi: 10.1021/acs.chemrev.0c00275.
[14] W. Zhang, Z. Tu, J. Qian, S. Choudhury, L. A. Archer, and Y. Lu, "Design principles of functional polymer separators for high-energy, metal-based batteries," *Small*, vol. 14, no. 11, pp. 1–21, 2018, doi: 10.1002/smll.201703001.
[15] T. Thomas and T. Thomas, "& self-healing: DRDO In Leh, Army holds continuing medical education (CME) programme," *The Economic Times*, Chennai, pp. 1–2, Jul. 04, 2019.
[16] X. Hua *et al.*, "The prismatic surface cell cooling coefficient: A novel cell design optimisation tool & thermal parameterization method for a 3D discretised electro-thermal equivalent-circuit model," *eTransportation*, vol. 7, pp. 1–150, 2021, doi: 10.1016/j.etran.2020.100099.
[17] W. Li *et al.*, "Intelligent optimization methodology of battery pack for electric vehicles: A multidisciplinary perspective," *Int. J. Energy Res.*, vol. 44, no. 12, pp. 9686–9706, 2020, doi: 10.1002/er.5600.
[18] A. Dhoke and A. Dalavi, "Optimal design of honeycomb battery pack enclosure for electric vehicle," *Int. J. Simul. Multi. Design Optim.*, vol. 23, pp. 1–12, 2024.
[19] A. Atiqah, M. N. M. Ansari, and L. Premkumar, "Impact and hardness properties of honeycomb natural fibre reinforced epoxy composites," *Mater. Today Proc.*, vol. 29, no. November 2018, pp. 138–142, 2019, doi: 10.1016/j.matpr.2020.05.645.
[20] S. Baazouzi, N. Feistel, J. Wanner, I. Landwehr, A. Fill, and K. P. Birke, "Design, properties, and manufacturing of cylindrical li-ion battery cells—A generic overview," *Batteries*, vol. 9, no. 6, pp. 1–20, 2023, doi: 10.3390/batteries9060309.
[21] H. Löbberding *et al.*, "From cell to battery system in BEVs: Analysis of system packing efficiency and cell types," *World Electr. Veh. J.*, vol. 11, no. 4, pp. 1–15, 2020, doi: 10.3390/wevj11040077.
[22] E. Sahraei, M. M. Keshavarzi, X. Zhang, and B. Lai, "Mechanical properties of prismatic li-ion batteries—electrodes, cells, and stacks," *J. Electrochem. Energy Convers. Storage*, vol. 19, no. 4, pp. 1–13, 2022, doi: 10.1115/1.4054823.
[23] S. Arora, A. Kapoor, and W. Shen, "Application of robust design methodology to battery packs for electric vehicles: Identification of critical technical requirements for modular architecture," *Batteries*, vol. 4, no. 3, pp. 1–25 2018, doi: 10.3390/batteries4030030.
[24] R. V. Maddila and S. Rostami, *Investigation of the effect of thermal interface materials on the cooling of battery cells: A comparative study on the heat transfer properties*, Department of Mechanics and Maritime Sciences Chalmers University of Technology Gothenburg, Sweden, Master's Thesis, pp. 1–2000, 2023. www.chalmers.se
[25] M. Alipanah, A. K. Saha, E. Vahidi, and H. Jin, "Value recovery from spent lithium-ion batteries: A review on technologies, environmental impacts, economics, and supply chain," *Clean Technol. Recycl.*, vol. 1, no. 2, pp. 152-184, 2021. doi: 10.3934/ctr.2021008
[26] C. Breuning, P. Hommel, D. Roth, and M. Kreimeyer, "Investigation of the use of aluminum foam sandwich in battery housings for electric vehicles," *DS 130: Proceedings of Nord Design 2024, Reykjavik, Iceland, 12th-14th August 2024*, pp.226–235.
[27] T. Ganguli, "Lightweight battery enclosure design," *Met. Form.*, vol. 44, pp. 465–474, 2024, doi: 10.21741/9781644903254-50.
[28] A. Bala and M. C. Kamaraju, "Design and optimization of battery housing in electric cars," Masters thesis, Chalmers University Of Technology Gothenburg, Sweden, pp. 1–114, 2020. www.chalmers.se
[29] S. Landini, J. Leworthy, and T. S. O'Donovan, "A review of phase change materials for the thermal management and isothermalisation of lithium-ion cells," *J. Energy Storage*, vol. 25, pp. 1–25, 2019, doi: 10.1016/j.est.2019.100887.

[30] J. Zhang, Y. An, and H. Ma, "Research progress in the preparation of aluminum foam composite structures," *Metals (Basel).*, vol. 12, no. 12, pp. 1–28, 2022, doi: 10.3390/met12122047.

[31] A. I. Komara, B. Budiwantoro, and R. Setiawan, "Cellular structure design and manufacturability for electric vehicle: A review," *Int. J. Sustain. Transp. Technol.*, vol. 5, no. 2, pp. 70–79, 2022, doi: 10.31427/ijstt.2022.5.2.5.

[32] D. Dubey, A. Mishra, S. Ghosh, M. V. Reddy, and R. Pandey, "Geometry-influenced cooling performance of lithium-ion battery," *Appl. Therm. Eng.*, vol. 230, pp. 1–39, 2023, doi: 10.1016/j.applthermaleng.2023.120723.

[33] S. Shahid and M. Agelin-Chaab, "A review of thermal runaway prevention and mitigation strategies for lithium-ion batteries," *Energy Convers. Manag. X*, vol. 16, no. July, p. 100310, 2022, doi: 10.1016/j.ecmx.2022.100310.

[34] A. H. Mohammed *et al.*, "Dual-purpose cooling plate for thermal management of prismatic lithium-ion batteries during normal operation and thermal runaway," *Appl. Therm. Eng.*, vol. 160, pp. 1–25, 2019, doi: 10.1016/j.applthermaleng.2019.114106.

[35] H. Behi *et al.*, "Aluminum heat sink assisted air-cooling thermal management system for high current applications in electric vehicles," *2020 AEIT Int. Conf. Electr. Electron. Technol. Automotive, AEIT Automot.* 2020, no. November, 2020, doi: 10.23919/aeitautomotive50086.2020.9307413.

[36] S. Harish, S. Jeyanthi, M. Abinash, R. Suresh Kumar, S. Agarwal, and G. Venkatachalam, "Study of thermal behavior of Aluminium foams in batteries of electric vehicles," *IOP Conf. Ser. Mater. Sci. Eng.*, vol. 1128, no. 1, p. 012042, 2021, doi: 10.1088/1757-899x/1128/1/012042.

[37] L. H. Saw *et al.*, "Novel thermal management system using mist cooling for lithium-ion battery packs," *Appl. Energy*, vol. 223, pp. 146–158, 2018, doi: 10.1016/j.apenergy.2018.04.042.

[38] L. H. Saw, Y. J. King, M. C. Yew, T. Ching Ng, W. T. Chong, and N. A. Pambudi, "Feasibility study of mist cooling for lithium-ion battery," *Energy Proc.*, vol. 142, pp. 2592–2597, 2017, doi: 10.1016/j.egypro.2017.12.197.

[39] A. Mohanan and K. Chidambaram, "Multi-strategical thermal management approach for lithium-ion batteries: Combining forced convection, mist cooling, air flow improvisers and additives," *World Electr. Veh. J.*, vol. 15, no. 5, pp. 1–20, 2024, doi: 10.3390/wevj15050213.

[40] Z. Ling, F. Wang, X. Fang, X. Gao, and Z. Zhang, "A hybrid thermal management system for lithium ion batteries combining phase change materials with forced-air cooling," *Appl. Energy*, vol. 148, pp. 403–409, 2015, doi: 10.1016/j.apenergy.2015.03.080.

[41] X. Feng *et al.*, "Key characteristics for thermal runaway of Li-ion batteries," *Energy Proc.*, vol. 158, pp. 4684–4689, 2019, doi: 10.1016/j.egypro.2019.01.736.

[42] H. Fayaz *et al.*, "Optimization of thermal and structural design in lithium-ion batteries to obtain energy efficient Battery Thermal Management System (BTMS): A critical review," *Arch. Comput. Methods Eng.*, vol. 29, no. 1, pp.129-194, 2022.

[43] M. Astaneh, J. Andric, L. Löfdahl, and P. Stopp, "Multiphysics simulation optimization framework for lithium-ion battery pack design for electric vehicle applications," *Energy*, vol. 239, pp. 1–22, 2022, doi: 10.1016/j.energy.2021.122092.

[44] X. Cui, B. Panda, C. M. M. Chin, N. Sakundarini, C. T. Wang, and K. Pareek, "An application of evolutionary computation algorithm in multidisciplinary design optimization of battery packs for electric vehicle," *Energy Storage*, vol. 2, no. 3, pp. 1–18, 2020, doi: 10.1002/est2.158.

[45] X. Zhang, Y. Hu, J. Deng, H. Xu, and H. Wen, "Feature engineering and artificial intelligence-supported approaches used for electric powertrain fault diagnosis: A review," *IEEE Access*, vol. 10, pp. 29069–29088, 2022, doi: 10.1109/ACCESS.2022.3157820.

[46] N. Aleris, "Automotive lightweighting with aluminum closures," *Industry Today*, 2018.

[47] Y. Pu, F. Ma, L. Han, and G. Wang, "Material selection of green design processes for car body via considering environment property," *Math. Probl. Eng.*, vol. 2020, pp. 1–15, 2020, doi: 10.1155/2020/9241634.

[48] P. Bere, M. Dudescu, C. Neamțu, and C. Cocian, "Design, manufacturing and test of CFRP front hood concepts for a light-weight vehicle," *Polymers (Basel).*, vol. 13, no. 9, pp. 1–17, 2021, doi: 10.3390/polym13091374.

[49] M. Abdul Karim and Y.-D. Park, "A review on welding of dissimilar metals in car body manufacturing," *J. Weld. Join.*, vol. 38, no. 1, pp. 8–23, 2020, doi: 10.5781/jwj.2020.38.1.1.

[50] P. Poddar, M. Olles, and D. Cormier, "Mechanical response of carbon composite octet truss structures produced via axial lattice extrusion," *Polymers (Basel).*, vol. 14, no. 17, pp. 1–22, 2022, doi: 10.3390/polym14173553.

[51] J. Arul, *Method to design and fabricate an octahedral-tetrahedral spaceframe from repurposed scaffolding,* 2023, Available: https://dspace.mit.edu/handle/1721.1/151253%0Ahttps://dspace.mit.edu/bitstream/handle/1721.1/151253/arul-jarul-sm-idm-2023-thesis.pdf?sequence=1&isAllowed=y.

[52] N. Jorapur, R. West, C. B. Williams, and A. Druschitz, "Design of fiber-reinforced cellular structures with tensegrity behavior manufactured using 3D printed sand molds," *Solid Free. Fabr. 2016 Proc. 27th Annu. Int. Solid Free. Fabr. Symp. – An Addit. Manuf. Conf. SFF 2016*, pp. 2043–2059, 2016.

[53] B. G. Compton and J. A. Lewis, "3D-printing of lightweight cellular composites," *Adv. Mater.*, vol. 26, no. 34, pp. 5930–5935, 2014, doi: 10.1002/adma.201401804.

[54] C. G. Ferro, S. Varetti, and P. Maggiore, "Experimental evaluation of mechanical compression properties of aluminum alloy lattice trusses for anti-ice system applications," *Machines*, vol. 12, no. 6, pp. 1–16, 2024, doi: 10.3390/machines12060404.

[55] X. Liu, V. Alizadeh, and C. J. Hansen, "The compressive response of octet lattice structures with carbon fiber composite hollow struts," *Compos. Struct.,* vol. 239, no. January, p. 111999, 2020, doi: 10.1016/j.compstruct.2020.111999.

[56] M. R. O. Masta, L. Dong, L. St-pierre, H. N. G. Wadley, and V. S. Deshpande, "The fracture toughness of octet-truss lattices," *J. Mech. Phys. Solids*, vol. *98*, pp.271–289, 2017.

[57] H. Ramos, R. Santiago, S. Soe, P. Theobald, and M. Alves, "Response of gyroid lattice structures to impact loads," *Int. J. Impact Eng.*, vol. 164, no. July 2021, p. 104202, 2022, doi: 10.1016/j.ijimpeng.2022.104202.

[58] S. Sandberg and M. Näsström, "A proposed method to preserve knowledge and information by use of knowledge enabled engineering," *2007 Proc. ASME Int. Des. Eng. Tech. Conf. Comput. Inf. Eng. Conf. DETC2007*, vol. 2 PART A, no. May, pp. 207–212, 2008, doi: 10.1115/DETC2007-35188.

[59] J. Enriquez, "Body in white architecture for an electric vehicle concept," Masters thesis, Halmers University Of Technology Gothenburg, Sweden 2016 Department of Product- and Production Development Division of Design & Human Factors, pp. 1–145, 2016.

[60] G. H. Okoth, R. Ndeda, P. Raghupatruni, and E. O. Olakanmi, "Trends in lightweighting for automotive applications: A case study," *J. Sustain. Res. Eng.*, vol. 8, no. 1, pp. 50–67, 2024.

[61] A. M. Deshpande *et al.*, "Design and development of a multi-material, cost-competitive, lightweight mid-size sports utility vehicle's body-in-white," *Compos. Adv. Mater. Expo, CAMX 2022*, 2022, doi: 10.33599/nasampe/c.22.0148.

[62] A. Hartlieb and M. Hartlieb, "The impact of giga-castings on car manufacturing and aluminum content – Light metal age magazine," pp. 24–27, 2023, www.lightmetalage.com.

[63] V. Naik, M. Kumar, and V. Kaup, "A review on natural fiber composite materials in automotive applications," *Eng. Sci.*, vol. 18, pp. 1–10, 2022, doi: 10.30919/es8d589.

[64] P. Sun, R. Bisschop, H. Niu, and X. Huang, "A review of battery fires in electric vehicles," *Fire Technol.,* vol. 56, no. 4, pp.1361–1410, 2020.

[65] A. K. Mohanty *et al.*, "Sustainable composites for lightweight and flame retardant parts for electric vehicles to boost climate benefits: A perspective," *Compos. Part C Open Access*, vol. 12, no. July, p. 100380, 2023, doi: 10.1016/j.jcomc.2023.100380.

[66] N. Ramli *et al.*, "Natural fiber for green technology in automotive industry: A brief review," *IOP Conf. Ser. Mater. Sci. Eng.*, vol. 368, no. 1, pp. 1–7, 2018, doi: 10.1088/1757-899X/368/1/012012.

[67] J. Janek and W. G. Zeier, "Challenges in speeding up solid-state battery development," *Nat. Energy*, vol. 8, no. 3, pp. 230–240, 2023, doi: 10.1038/s41560-023-01208-9.

[68] A. Ulvestad, "A brief review of current lithium ion battery technology and potential solid state battery technologies," arXiv:1803.04317, 2018, [Online]. Available: http://arxiv.org/abs/1803.04317.

[69] G. Silva, T. Assis Dutra, J. Nunes-Pereira, and A. P. Silva, "Coupled and decoupled structural batteries: A comparative analysis," *J. Power Sources*, vol. 604, no. December 2023, p. 234392, 2024, doi: 10.1016/j.jpowsour.2024.234392.

[70] H. Alimam, G. Mazzuto, N. Tozzi , F. Emanuele Ciarapica, and M. Bevilacqua, "The resurrection of digital triplet: A cognitive pillar of human-machine integration at the dawn of industry 5.0," *J. King Saud Univ. – Comput. Inf. Sci.*, vol. 35, no. 10, p. 101846, 2023, doi: 10.1016/j.jksuci.2023.101846.

[71] T. Jin, G. Singer, K. Liang, and Y. Yang, "Structural batteries: Advances, challenges and perspectives," *Mater. Today*, vol. 62, pp. 151–167, 2023, doi: 10.1016/j.mattod.2022.12.001.
[72] A. Kalhor, J. Dykas, K. Rodak, and A. Grajcar, "Materials and constructional design for electric vehicles: A review," *Adv. Sci. Technol. Res. J.*, vol. 19, no. 1, pp. 178–196, 2025, doi: 10.12913/22998624/195457.
[73] K. You, G. Yan, X. Luo, M. D. Gilchrist, and F. Fang, "Advances in laser assisted machining of hard and brittle materials," *J. Manuf. Process.*, vol. 58, no. September 2019, pp. 677–692, 2020, doi: 10.1016/j.jmapro.2020.08.034.
[74] S. Swaminathan, K. B. Ozutemiz, C. Majidi, and S. E. Hudson, "Fiber wire: Embedding electronic function into 3D printed mechanically strong, lightweight carbon fiber composite objects," *Conf. Hum. Factors Comput. Syst. – Proc.*, vol. 2019-Jan, no.1, pp. 1–11, 2019, doi: 10.1145/3290605.3300797.
[75] V. Shah, J. Bhaliya, G. M. Patel, and K. Deshmukh, "Advances in polymeric nanocomposites for automotive applications: A review," *Polym. Adv. Technol.*, vol. 33, no. 10, pp. 3023–3048, 2022, doi: 10.1002/pat.5771.

2 Thermal and Mechanical Advancements in Epoxy-Enhanced Bamboo Nanocomposites

A Comprehensive Review on Synthesis, Structure, and Emerging Applications

Yogendra Kumar Verma, Amit Kumar Singh, Sandeep Kumar Singh, Paswan M.K., Prashanth S., and Subhankur Dutta

2.1 INTRODUCTION

Bamboo-based nanocomposites have gained significant attention as a promising class of sustainable materials due to the increasing demand for eco-friendly alternatives to synthetic composites. Bamboo fibers, known for their exceptional mechanical properties, renewability, and availability, make them ideal candidates for reinforcing advanced polymer composites [1]. When integrated with epoxy resins, bamboo nanocomposites exhibit enhanced mechanical strength, thermal stability, and durability, making them suitable for structural and functional applications [2]. The incorporation of nanofillers, such as nanoclay, carbon nanotubes, and metal oxides, further improves these properties by refining the microstructure, increasing interfacial bonding, and enhancing the dispersion of reinforcements, thereby elevating the performance of bamboo–epoxy composites [3]. Bamboo's fast growth and unique mechanical and thermal properties make it an attractive option for reinforcing composite materials. Epoxy resins, which are known for their superior adhesion, low shrinkage, and excellent mechanical properties, provide the ideal matrix for bamboo fibers, resulting in bamboo–epoxy nanocomposites with improved performance for a range of industrial applications [4]. The incorporation of bamboo with epoxy resins in the form of nanocomposites has demonstrated notable enhancements in mechanical strength, thermal stability, and morphological characteristics. These improvements have significantly broadened the scope of bamboo-based materials, offering potential for diverse applications across industries. With the added benefits of durability and versatility, these nanocomposites are becoming increasingly valuable for structural and functional uses in advanced material technologies [5]. For example, the combination of bamboo with kenaf fibers and nanoclay in hybrid nanocomposites significantly enhances thermal resistance and flame retardancy. The synergistic effect of these components, particularly the improvement

DOI: 10.1201/9781003596097-2

in char formation and barrier properties due to nanoclay, reduces heat and oxygen penetration. This, combined with the structural reinforcement provided by bamboo and kenaf fibers, results in improved thermal stability and lower flammability, which are essential attributes for lightweight and thermally resilient applications [6–8]. In addition to enhancing mechanical properties, bamboo cellulose contributes to the eco-friendly nature of composite materials. Its interaction with poly(2-hydroxyethyl methacrylate) (PHEMA) not only reinforces the matrix but also supports sustainable material development, particularly in biomedical applications [9,10]. Moreover, the cellulose, hemicellulose, and lignin content in bamboo plays a crucial role in improving the mechanical strength, thermal stability, and functionality of epoxy-based nanocomposites. Advancements in synthesis techniques, such as chemical treatments, surface functionalization, and high-energy milling, have successfully addressed challenges like fiber agglomeration and poor matrix adhesion. These techniques enable the customization of bamboo nanocomposites for specific applications, ranging from automotive parts to energy-storage devices and biomedical implants [11]. As the demand for renewable materials increases, bamboo nanocomposites are positioned as essential contributors to sustainable engineering and design, particularly as part of the global push toward circular economies. The incorporation of nanoclays, nanoparticles, and cellulose nanofibers (CNFs) into bamboo–epoxy composites enhances their mechanical and thermal properties, including tensile strength, flexural strength, and impact resistance. The improved interfacial bonding and uniform dispersion of these fillers within the matrix contribute to the overall performance. Additionally, the use of CNFs enhances environmental sustainability, making these nanocomposites suitable for high-performance applications in industries such as automotive, construction, and electronics [12,13]. Morphological analysis reveals reduced void content and better fiber–matrix adhesion, which directly contributes to the improved mechanical behavior. Thermal analysis further supports these findings, showing enhanced thermal stability and reduced degradation rates, thus making these hybrid composites suitable for high-performance and durable applications in various industries [14–16]. Moreover, the integration of nanocellulose crystals (NCCs) significantly enhances the material's structural integrity, while also promoting biodegradability, making these composites an attractive option for environmentally friendly and sustainable solutions, particularly in tissue engineering [17]. The optimized formulations of these nanocomposites have shown superior performance compared to non-optimized versions, highlighting their potential for eco-friendly, high-performance applications in packaging and structural materials. This demonstrates the effectiveness of statistical modeling in designing nanocomposites with tailored properties to meet specific performance criteria [18]. In addition, the incorporation of electrospun nanocellulose into the epoxy matrix not only boosts the thermal stability but also improves tensile strength, modulus, and impact resistance. These findings further reinforce the potential of nanocellulose-based composites in industries requiring high-performance materials, positioning them as a sustainable alternative to traditional reinforcement agents. Electrospun nanocellulose can thus play a key role in advancing the design of high-performance, environmentally friendly nanocomposites for a wide range of applications [19]. Recent advancements in lightweight materials and their impact on electric vehicles (EVs) are depicted in Table 2.1.

This review aims to provide a comprehensive overview of the recent advancements in the synthesis, structure, and applications of epoxy-enhanced bamboo nanocomposites, focusing on their thermal and mechanical properties. It aims to highlight the methods for preparing bamboo–epoxy composites, including mechanical milling, chemical treatments, and the incorporation of nanomaterials. The review also explores the structural characterization techniques used to understand the composites' morphology and bonding, as well as their thermal stability and viscoelastic properties. Furthermore, it examines the emerging applications of these composites in industries such as automotive, aerospace, and construction, while addressing challenges and future directions in their development [20–23].

TABLE 2.1
Recent Advancements in Lightweight Materials and Their Impacts on EVs

Aspect	Advancement	Impact on EVs
Material synthesis	Development of epoxy-enhanced bamboo nanocomposites with optimized properties	Reduced weight of EV components, improving efficiency and extending battery life
Mechanical properties	Improved tensile strength, flexural strength, and impact resistance through nanoclay and fiber treatments	Enhanced structural integrity, contributing to improved safety and durability in EVs
Thermal properties	Enhanced thermal stability with bamboo nanocomposites treated with various additives	Better heat resistance in high-performance EVs, ensuring longevity of vehicle components
Environmental benefits	Use of eco-friendly bamboo fibers and reduced use of petroleum-based materials	Supports sustainability goals by reducing the carbon footprint of EV production
Cost efficiency	Lower material costs due to the abundant availability of bamboo and ease of composite processing	Reduced manufacturing costs for EVs, promoting mass adoption of EVs
Durability and reliability	Bamboo-based composites exhibit resistance to wear, UV degradation, and environmental factors	Improved lifespan and reduced maintenance costs for EV components, enhancing overall reliability
Energy storage Potential	Integration of bamboo nanocomposites in supercapacitors for enhanced energy storage	Contributes to higher energy-storage capacity and faster charging times for EV batteries
reduction of carbon footprint	Bamboo composites sequester carbon and reduce reliance on synthetic polymers	Helps EV manufacturers meet green energy standards and lower environmental impact during production
Emerging applications in EVs	Development of interior and exterior panels, lightweight structural elements, and battery casings	Increased use of bamboo composites in EV interiors and structures, reducing overall vehicle weight

2.2 MATERIALS AND METHODS

2.2.1 Materials

Chimono bamboo, which was gathered from the Nagaland Bamboo Resource Centre, India, was mechanochemically processed into micro-sized chips, successfully eliminating lignin and hemicellulose for increased usefulness. In order to speed up the curing process, the matrix system used Araldite® LY556, a low-viscosity, anhydride-cured epoxy resin, combined with Aradur® 906 anhydride hardener and DY070 imidazole accelerator, all of which were purchased from Funique Composites Private Limited, Delhi, India.

2.2.2 Methodology

2.2.2.1 Green and Sustainable Techniques for the Preparation of Bamboo Nanocellulose

Green and sustainable production techniques for preparing bamboo nanocellulose involve eco-friendly processes emphasizing minimal environmental impact and efficient resource utilization. One common method is alkaline pre-treatment, which removes lignin and hemicellulose, followed by mechanical processes such as high-pressure homogenization, ultrasonication, or ball milling to isolate nanocellulose. These techniques enhance fiber separation and reduce energy consumption. Enzymatic hydrolysis using cellulose enzymes offers a biodegradable and non-toxic alternative, aiding in cellulose depolymerization under mild conditions. TEMPO-mediated oxidation is another widely used green approach that selectively oxidizes hydroxyl groups on cellulose to carboxyl groups, enabling better dispersion of nanocellulose in aqueous media. Recently, ionic liquids and deep eutectic solvents have been explored as sustainable options to dissolve cellulose without harsh chemicals, ensuring reduced environmental toxicity. These methods prioritize renewable solvents, energy-efficient processes, and high cellulose yield, making them integral to sustainable nanocellulose production [24].

2.2.2.2 Preparation of Bamboo Nanocellulose from Extracted Bamboo Fibers via Ball Milling

The mechanochemical method of extracting bamboo fibers and bamboo nanocellulose from extracted bamboo (chimono bamboo) is shown in Figure 2.1. The bamboo columns are first prepared by stripping them of their waxy inner and hard outer coatings, then turning them into chips. To aid in fiber extraction, these chips are soaked in a NaOH solution and heated to 70°C for 12 hours. A mechanical stirrer running at 1400 rpm provides agitation. After being properly cleaned with distilled water, the extracted fibers are baked for 12 hours at 70°C to cure them. Before undergoing a series of chemical treatments, including an alkaline treatment with 5 wt.% NaOH at 90°C for 12 hours, an acid treatment with HCl at 80°C for 3 hours, and a second alkaline treatment with 5 wt.% NaOH at 80°C for 5 hours, the fibers are first cleaned and dried in an oven at 70°C for 24 hours. After each treatment, the fibers are rinsed, dried in a vacuum oven at 70°C for 24 hours, and ground using a ball mill at 500 rpm for 4 hours. Nano-sized particles are then separated using a sieve shaker, completing the extraction and processing of bamboo nanocellulose.

2.2.3 Chemical Processing (Acid Hydrolysis, Enzymatic Treatments)

The preparation of bamboo nanocellulose often involves chemical processing methods such as acid hydrolysis and enzymatic treatments. Acid hydrolysis uses strong acids, typically sulfuric acid, to break down amorphous regions of cellulose while preserving the crystalline structure, producing nanocrystals with a uniform size. In a typical method, bamboo fibers are treated with 64% sulfuric acid at 45°C under continuous stirring for 1–2 hours, followed by quenching in cold water, centrifugation, and dialysis to neutralize and remove excess acid, yielding nanocellulose with high crystallinity and thermal stability. Enzymatic treatments, on the other hand, involve cellulose enzymes to selectively hydrolyze amorphous cellulose. In this process, bamboo fibers are pre-treated with alkali

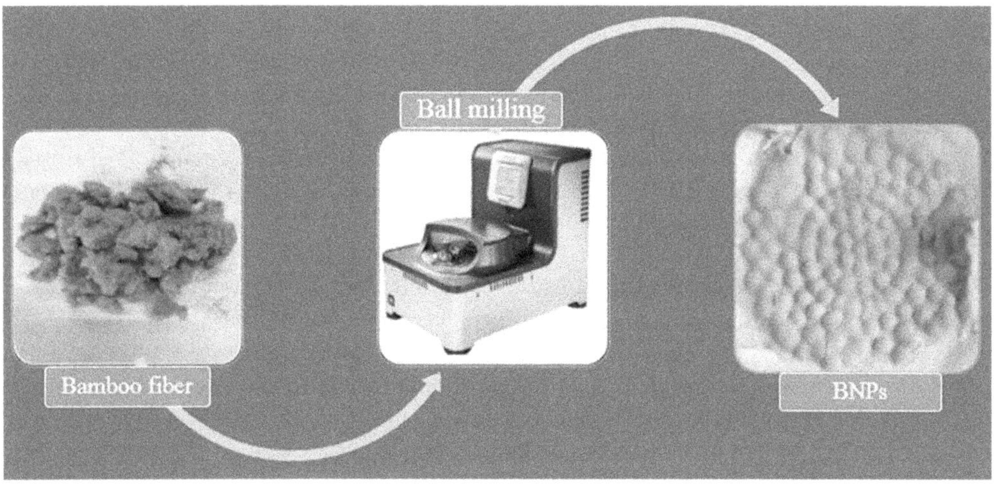

FIGURE 2.1 Preparation of bamboo nanocellulose.

TABLE 2.2
Comparison of Methods

Aspect	Acid Hydrolysis	Enzymatic Treatment
Product type	CNCs (high crystallinity)	CNFs (longer fibrils)
Processing time	Short (1–2 hours)	Long (24–48 hours)
Chemical use	High (concentrated acids)	Minimal (enzymes, mild buffers)
Environmental impact	Generates acid waste	Eco-friendly

to remove lignin and hemicellulose, followed by incubation with cellulose (10–20 U/g) at 50°C and pH 4.8 for 24–48 hours under gentle agitation, resulting in nanocellulose with enhanced biocompatibility and improved dispersion in aqueous systems. Both methods are widely used depending on the intended application, as acid hydrolysis yields nanocrystals with high mechanical properties, while enzymatic treatments provide nanofibrils with higher aspect ratios and eco-friendliness [25,26]. A comparison of these two methods is depicted in Table 2.2.

2.2.4 Chemical Modifications

Surface functionalization methods like phosphorylation, carboxymethylation, oxidation, and sulfonation offer a versatile approach to modify nanocellulose, enhancing its ionic characteristics and enabling customization of its properties for diverse applications. Pre-treatment with sulfuric (H_2SO_4) and phosphoric acid (H_3PO_4) facilitates the extraction of nanostructured cellulose, modifying its surface by introducing charges that improve its dispersion and stability in aqueous solutions during the isolation process. These functionalization routes play a critical role in optimizing the physicochemical properties of nanocellulose for advanced material applications.

2.2.4.1 Phosphorylation-Based Functionalization of Nanocellulosic Materials

Phosphorylated nanocellulose, modified through the introduction of phosphate (PO_4^{3-}) ester groups, exhibits enhanced mechanical strength, thermal stability, and biocompatibility, making it highly suitable for applications in orthopedics, textiles, biomedical devices, adsorption systems, fuel cells, wastewater treatment, and ion exchange processes. This modification improves biocompatibility with calcium phosphate ($Ca(PO_4)_2$), facilitating hybrid structure formation, while also imparting flame-retardant properties through the catalytic action of phosphate and sulfate esters, which promote dehydration and char formation. Although nanocellulose exhibits reduced thermal stability due to its flammable nature, the crystalline regions are more thermally stable, whereas the amorphous parts remain highly reactive. Phosphorylated nanocellulose polymers can further enhance the performance of other polymeric chains, enabling large-scale applications. Common agents used in this functionalization include $POCl_3$, P_2O_5, $(NH_4)_2HPO_4$, H_3PO_4, and organophosphates, offering tailored modifications to meet specific application needs [27].

2.2.4.2 Acetylation-Based Functionalization of Nanocellulosic Surfaces

Acetylation is a widely used method to enhance the hydrophobicity of nanocellulose, where acetic anhydride ($C_4H_6O_3$) and dry acetic acid (CH_3COOH) react in the presence of H_2SO_4 or $HClO_4$ as catalysts, facilitating the substitution of hydroxyl groups with acetyl groups to reduce moisture absorption. Acetylation occurs through two primary mechanisms: the fibrous process, which employs a swelling diluent like toluene to produce insoluble, highly acetylated cellulosic material while maintaining its original morphology, and the homogeneous method, which operates without diluents and, when catalyzed by H_2SO_4 in CH_3COOH, results in soluble acetylated chains.

Surface-functionalized cellulosic nanowhiskers through acetylation with vinyl acetate ($C_4H_6O_2$) under mild conditions, preserving the crystallinity and dimensions of the material. However, prolonged reactions reduced crystallinity and size, and increased the degree of acetylation, indicating diffusion-controlled kinetics [28].

2.2.4.3 Functionalization of Nanocellulose through Amidation

Amidation is a common approach for modifying cellulosic surfaces, relying on a carbodiimide-mediated reaction that selectively interacts with the carboxyl groups of preoxidized nanocellulose, with EDAC [N-ethyl-N-(3-(dimethylamino)-propyl) carbodiimide hydrochloride] serving as the preferred coupling agent. The pH range for this reaction typically falls between 7 and 10. Ruiz-Palomero and colleagues demonstrated the successful covalent bonding of β-cyclodextrin ($(C_6H_{10}O_5)_2$) to amine-functionalized nanocellulose via amidation. This functionalized nanocellulose showed potential as an effective sorbent for danofloxacin ($C_{19}H_{20}FN_3O_3$), facilitating the fluorimetric determination of the antibiotic in milk, showcasing the versatility and application of amidation in enhancing nanocellulose functionality [29].

2.2.4.4 Functionalization of Nanocellulose through the Sulfonation Method

The sulfonation technique is commonly used to functionalize nanocellulosic materials by introducing anionic charges to their surfaces, particularly through the treatment of cellulose nanocrystals (CNCs) with concentrated sulfuric acid (H_2SO_4). The introduction of sulfate half-esters to cellulose nanocrystals results in their surface modification, allowing the CNCs to self-assemble into a chiral nematic phase at certain concentrations, while the negatively charged sulfate groups improve their water dispersion through enhanced electrostatic interactions. To improve the thermal stability of H_2SO_4-isolated nanocellulose, a neutralization process with NaOH is often employed. Hydrolysis with H_2SO_4 and HCl, followed by sonication, produces spherical cellulose nanocrystals with a low-density sulfate group and maximum thermal stability compared to those obtained by pure H_2SO_4 hydrolysis. Additionally, the use of $NaIO_4$ (periodate) and $NaHSO_3$ (bisulfite) in the functionalization of nano-fibrillated hardwood pulp has resulted in sulfonated NFCs with diameters ranging from 10 to 60 nm [30].

2.2.5 Physical Treatments (Plasma Treatment, UV-Curing)

Plasma treatment and UV-curing are highly effective surface modification techniques used to improve the properties of natural fibers like bamboo, enhancing their compatibility with polymer matrices in composite applications. Plasma treatment utilizes ionized gases to activate functional groups on the fiber surface, which increases surface roughness and alters surface energy, resulting in better adhesion to polymers; it is also an environmentally friendly method as it avoids the use of harmful chemicals, as shown in Figure 2.2. In contrast, UV-curing involves ultraviolet light to trigger photochemical reactions that promote the cross-linking of polymer coatings on the fiber, which leads to faster curing times and improved fiber properties such as increased hydrophobicity and enhanced mechanical strength. Both plasma treatment and UV-curing have demonstrated significant efficacy in improving the interfacial bonding between natural fibers and polymer matrices, thereby boosting the mechanical performance and durability of the resulting composite materials, making them suitable for various advanced applications in industries seeking eco-friendly yet high-performance materials [31,32].

2.2.6 Preparation of Bamboo Nanocomposite

Bamboo nanofillers–epoxy bio-composites were fabricated by first combining Araldite® LY 556 epoxy resin and Hardener HY 906 in a stoichiometric ratio of 100:95 wt, followed by thorough

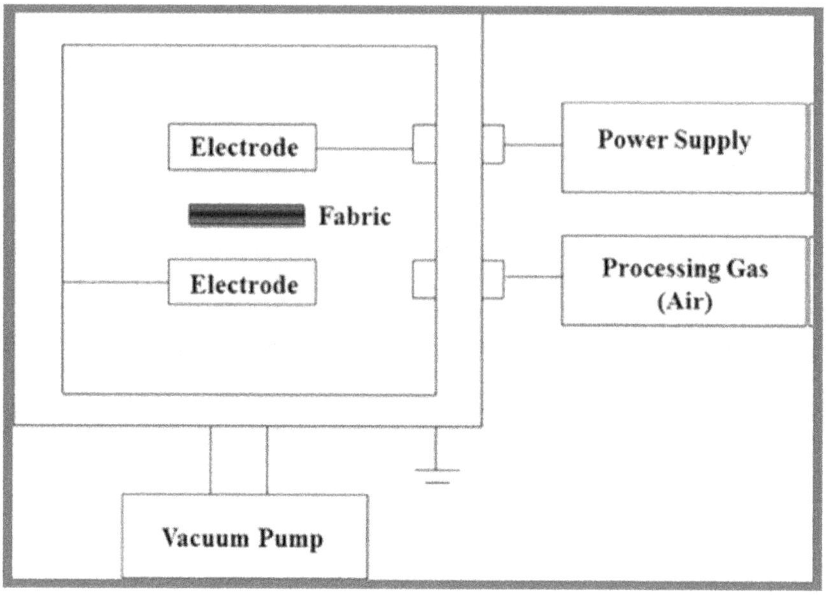

FIGURE 2.2 Plasma treatment of bamboo nanocellulose.

mixing. Bamboo nanofillers were then added at a specified weight ratio relative to the epoxy resin, and the mixture was blended for 30 minutes. Accelerator DY 070 was incorporated at 2 wt.% of the epoxy resin weight, and the blend was mixed for an additional 15 minutes. The slurry was placed in a vacuum chamber to remove air bubbles before being poured into molds. The curing process involved heating the mixture in a hot air oven at 120°C for 2 hours, followed by 160°C for 8 hours. Finally, the cured samples were polished using progressively finer grit emery paper to ensure uniform thickness, preparing them for the evaluation of tensile and fracture properties.

2.3 RESULTS AND DISCUSSION

2.3.1 XRD, FTIR, TGA, and DTA Analysis of Nanocellulose

The X-ray diffraction (XRD) pattern reveals a distinct peak around 2θ = 22°, indicating the crystalline nature of cellulose due to the presence of ordered cellulose I structures, while the broadness of the peak reflects the amorphous regions inherent to cellulose. The Fourier transform infrared spectroscopy (FTIR) spectrum confirms the functional groups present in bamboo nanocellulose, showing a strong O–H stretching peak at 3400 cm^{-1}, indicating hydrogen bonding, and characteristic peaks at 1720 cm^{-1} and 1640 cm^{-1} corresponding to C=O and C=C stretching vibrations, respectively, suggesting ester and conjugated functional groups. Peaks around 1100 cm^{-1} indicate C–O–C linkages, further confirming the cellulose framework. The thermogravimetric analysis (TGA) curve demonstrates the thermal stability of bamboo nanocellulose, with an initial weight loss around 50–100°C due to moisture evaporation, followed by significant decomposition around 300°C, corresponding to cellulose degradation. The differential thermal analysis (DTA) profile corroborates the TGA findings, highlighting exothermic peaks indicating decomposition stages at specific temperatures. Collectively, these analyses validate the crystalline structure, functional groups, and thermal behavior of bamboo nanocellulose, making it suitable for applications requiring stability and functionality [33].

2.3.1.1 Morphological Analysis (TEM, SEM)

Figure 2.3 shows scanning electron microscopy (SEM) and transmission electron microscopy (TEM) analyses of bamboo nanocellulose provide valuable insights into the structural and morphological characteristics of the material. The high-speed ball-milling process significantly reduces the particle size of bamboo fibers, resulting in finer particles that exhibit enhanced properties. This reduction is primarily due to the increased number of collisions and grinding actions within the mill, which progressively refines the fiber structure. SEM analysis reveals the surface morphology, showing a smoother and more uniform appearance of the milled bamboo nanocellulose. TEM analysis further delves into the internal structure, highlighting the nanofibrils with a precise and consistent diameter. These findings are essential for understanding the material's potential in advanced applications, including nanocomposites, where properties like surface area, reactivity, and ease of incorporation into other materials are critical for performance. The consistent particle size distribution and the presence of nanofibrils enhance the material's functionality, making it suitable for diverse industrial uses.

2.3.2 Modification Effects on Nanocellulose Crystallinity

The crystallinity of bamboo nanocellulose is influenced by multiple factors, including the lignocellulosic structure of bamboo and the type of chemical treatment applied. Studies have shown that nanocellulose extracted using H_2SO_4 typically exhibits lower crystallinity compared to that obtained with HCl, attributed to the more pronounced hydrolytic action of sulfuric acid. Additionally, the hydrolysis duration plays a crucial role; extended hydrolysis time enhances crystallinity by removing amorphous components. For bamboo nanocellulose, crystallinity values often surpass those of other lignocellulosic sources like banana, jute, or pineapple fibers, due to bamboo's inherently high cellulose content. For example, bamboo-derived nanocellulose often demonstrates crystallinity values in the range of 70–85%, similar to those observed in other high-cellulose sources such as sisal or cotton. However, variations can occur depending on the processing conditions and the extent of lignin and hemicellulose removal, highlighting the importance of optimized extraction techniques in achieving high crystallinity [34].

2.3.3 Modification and Its Impact on Bamboo Nanocellulose Mechanical Properties

The addition of bamboo nanocellulose as a filler in composite materials significantly enhances mechanical properties such as rigidity, toughness, strength, and barrier features, with potential improvements in flame retardancy. These enhancements are attributed to the high surface area of

FIGURE 2.3 Morphological analysis of bamboo nanocellulose: (a) SEM; (b) TEM.

bamboo nanocellulose, allowing substantial changes with minimal addition. While its strength is lower compared to carbon nanotubes, bamboo nanocellulose offers a cost-effective alternative, making it attractive for specialized applications. The strengthening ability of bamboo nanocellulose whiskers arises from their high surface area, combined with the material's eco-friendly and renewable nature, reinforcing its appeal for sustainable and advanced material applications [35].

2.3.4 Impact of Modification on Bamboo Nanocellulose Thermal Stability

A film forms when bamboo nanocellulose suspensions are dried, resulting in a material with a poor thermal expansion coefficient (TEC). This limitation arises from the high crystallinity and strong interactions within the nanocellulosic network. To reduce the TEC of polymer nanocomposite sheets, the inherently low TEC of bamboo nanocellulose can be leveraged if the network strength is sufficient to limit thermal expansion. However, the low thermal stability of bamboo nanocellulose can restrict its application and processing at high temperatures. This issue is particularly significant for bamboo-derived CNCs hydrolyzed using H_2SO_4, which display significantly reduced thermal stability compared to untreated bamboo or other forms of nanocellulose. The reduced stability is attributed to the presence of sulfate ester groups introduced during acid hydrolysis [36].

2.3.5 Mechanical and Thermal Behavior of Nanocomposites

2.3.5.1 Enhancement in Mechanical Strength

The mechanical performance of bamboo nanoparticle (BNP)–epoxy composites was evaluated by analyzing their tensile properties at different nanoparticle loadings, revealing how BNP content influences strength. Tensile strength measurements revealed a progressive enhancement at lower BNP loadings, starting from 68.28±1.79 MPa at 0.5% compared to the neat epoxy value of 60.34±1.27 MPa. The strength further increased to 79.98±1.81 MPa at 1% and peaked at 83.92±1.74 MPa with 1.5% BNP content, indicating an optimal dispersion and effective load transfer between the nanoparticles and the epoxy matrix. However, at 2% BNP loading, the tensile strength decreased to 70.24±1.80 MPa, likely due to particle agglomeration and a consequent rise in plastic deformation, which reduced the composite's ability to withstand tensile stresses, which are shown in Figure 2.4. This behavior suggests that while BNPs effectively reinforce the epoxy matrix at lower concentrations, excessive filler content can compromise mechanical properties by inducing strain-to-failure mechanisms. These findings underline the importance of optimizing filler content for achieving superior performance in BNP–epoxy composites for applications demanding high mechanical strength.

2.3.5.2 Thermal Stability of Bamboo Nanoparticle (BNP)–Epoxy Composites

The incorporation of BNPs into epoxy composites significantly enhances their thermal stability, as evidenced by dynamic thermogravimetric analysis (TGA), which reveals that the thermal degradation of BNP–epoxy composites initiates at higher temperatures compared to pure epoxy composites, with degradation temperatures for composites containing 1 wt.% and 1.5 wt.% BNPs increasing by 6.5°C and 5.3°C, respectively, and this improvement in thermal properties is attributed to the dispersal of BNPs that restricts the mobility of molecular chains, preventing further decomposition, while also highlighting the crucial role of interfacial adhesion between BNPs and the epoxy matrix in reducing the thermal motion of the epoxy molecules, thus improving the composite's stability, with the TGA curves displaying a typical three-stage decomposition process where moisture evaporation occurs below 150°C, decomposition of hemicelluloses and cleavage of glucosidic linkages in cellulose takes place between 250–450°C, and further degradation of products from the second stage into charred residue occurs above 450°C, in line with the thermal decomposition patterns of lignocellulosic materials, where hemicellulose decomposes between 150–350°C, cellulose between

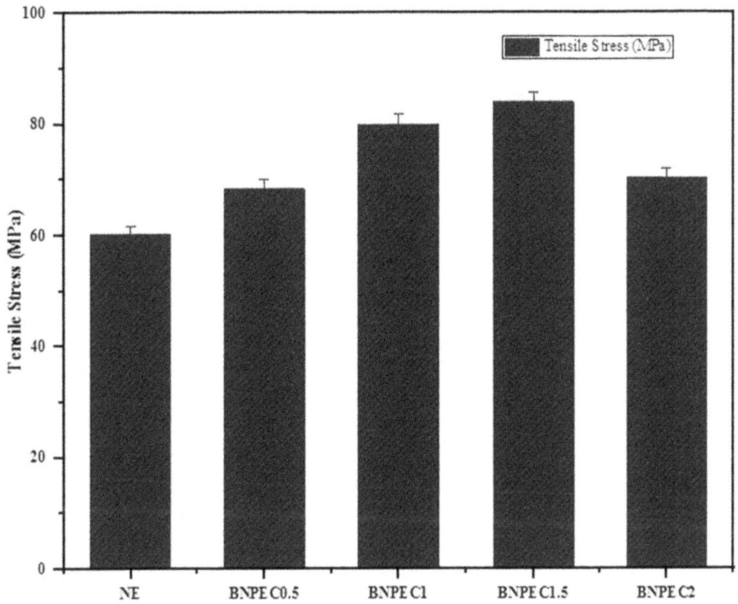

FIGURE 2.4 Tensile strength of bamboo nanocomposite.

275–350°C, and lignin between 250–450°C (Figure 2.5), ultimately demonstrating the effectiveness of BNPs in enhancing the thermal performance of epoxy composites [37].

2.3.6 Emerging Applications of Bamboo Nanoparticle (BNP)–Epoxy Composites

BNP–epoxy composites have gained significant attention due to their exceptional mechanical, thermal, and environmental properties, positioning them as promising materials for diverse applications. The incorporation of BNPs into epoxy matrices enhances tensile strength, thermal stability, and impact resistance while maintaining a lightweight profile, making these composites suitable for automotive, aerospace, and construction industries. Their bio-based origin and eco-friendly nature align with the growing demand for sustainable materials, while their improved viscoelastic properties broaden their utility in structural and load-bearing components. Furthermore, the enhanced interfacial adhesion between BNPs and epoxy resin contributes to superior durability and moisture resistance, enabling applications in marine environments and outdoor structures. Recent advancements in functionalization techniques have also paved the way for BNP–epoxy composites in electronic devices, where they exhibit potential as dielectric materials with excellent insulating properties. These emerging applications underline the versatility and transformative potential of BNP–epoxy composites across various industrial sectors [38].

2.4 CONCLUSIONS

This comprehensive review provides an in-depth analysis of recent advancements in the fabrication, characterization, and multifunctional applications of BNP-reinforced epoxy composites, highlighting their superior mechanical robustness and thermal resilience. The integration of BNPs into epoxy matrices has led to remarkable enhancements in tensile strength, thermal stability, and dynamic mechanical properties, positioning these composites as high-performance materials for a broad spectrum of industries, including automotive, aerospace, construction, and marine sectors.

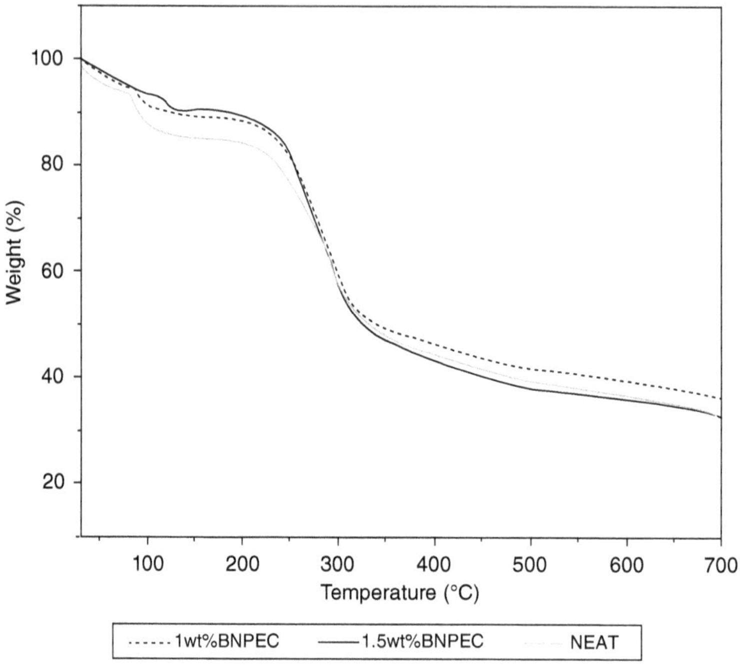

FIGURE 2.5 TGA analysis of BNPEC.

Structural analyses through XRD, FTIR, SEM, TEM, and TGA have provided detailed insights into the material's crystalline structure, interfacial adhesion, and thermal stability. Optimized BNP content has been shown to enhance performance, although excessive filler loading can lead to diminished mechanical properties due to agglomeration. The composites' bio-based origin and alignment with sustainability goals further elevate their potential for broader adoption. Emerging functionalization techniques have expanded their utility into electronic and dielectric applications, demonstrating their versatility and transformative potential in modern materials science. These findings underscore the importance of balancing performance with scalability and sustainability for the commercial success of BNP–epoxy composites.

REFERENCES

[1] Gao X, Zhu D, Fan S, et al (2022) Structural and mechanical properties of bamboo fiber bundle and fiber/bundle reinforced composites: A review. *J Mater Res Technol* 19:1162–1190. https://doi.org/10.1016/j.jmrt.2022.05.077

[2] Elsacker E, De Laet L, Peeters E (2022) Functional grading of mycelium materials with inorganic particles: The effect of nanoclay on the biological, chemical and mechanical properties. *Biomimetics* 57:1–23. https://doi.org/10.3390/biomimetics7020057

[3] Chin SC, Tee KF, Tong FS, et al (2020) Thermal and mechanical properties of bamboo fiber reinforced composites. *Mater Today Commun* 23:100876. https://doi.org/10.1016/j.mtcomm.2019.100876

[4] Composites F, Norizan MN, Norrrahim F, et al (2022) Mechanical performance evaluation of bamboo fibre reinforced polymer composites and its applications: A review mechanical performance evaluation of bamboo fibre reinforced polymer composites and its applications: A review. *Funct Composites Struct* 4(1), 1–36. https://doi.org/10.1088/2631-6331/ac5b1a

[5] Abdul Khalil HPS, Bhat IUH, Jawaid M, et al (2012) Bamboo fibre reinforced biocomposites: A review. *Mater Des* 42:353–368. https://doi.org/10.1016/j.matdes.2012.06.015

[6] Pang W, Ni Z, Chen G, et al (2015) Mechanical and thermal properties of graphene oxide/ultrahigh molecular weight polyethylene nanocomposites. *RSC Adv* 5:63063–63072. https://doi.org/10.1039/c5ra11826c

[7] Verma YK, Singh AK, Paswan MK, Sonker PK (2022) Surface functionalization of jute fibers in the presence of sodium hydroxide and graphene oxide in the preparation of bio composite. *IOP Conf Ser: Mater Sci Eng* 1259(1):12049.

[8] Kumar Verma Y, Kumar Singh A, Paswan MK, Kumar Sonker P (2023) Importance of functionalized jute fibers in the field of Nano cellulose for the preparation of biodegradable Nano composites. *Mater Today Proc* 80(xxxx):254–257. https://doi.org/10.1016/j.matpr.2023.01.031.

[9] Rao SS, Jeyapal SG, Rajiv S (2014) Composites: Part B Biodegradable electrospun nanocomposite fibers based on Poly (2-hydroxy ethyl methacrylate) and bamboo cellulose. *Compos PART B* 60:43–48. https://doi.org/10.1016/j.compositesb.2013.12.068

[10] Verma YK, Singh AK, Paswan MK, Vinay P (2024) Mechanical and tribological characterization of vacuum pressurized impregnation (VPI) treated bamboo fiber-reinforced polymer (FRP) composites with jute mat layer under fretting wear conditions. *Wood Mater Sci Eng*:1–13. Published online: 31st Dec 2024. https://doi.org/10.1080/17480272.2024.2386692

[11] Fan J, Yang J, Li H, et al (2021) Cryogenic mechanical properties of graphene oxide/epoxy nanocomposites: Influence of graphene oxide with different oxidation degrees. *Polym Test* 96:107074. https://doi.org/10.1016/j.polymertesting.2021.107074

[12] Yang X, Song X, Hu Z, et al (2022) Improving comprehensive properties of high-density polyethylene matrix composite by boron nitride/coconut shell carbon reinforcement. *Polym Test* 115:107728. https://doi.org/10.1016/j.polymertesting.2022.107728

[13] Chen D, Zhen W (2021) Performance, interfacial compatibility testing and rheonaut technology analysis for simultaneous rheology and FTIR of poly (lactic acid)/modified saponite nanocomposites. *Polym Test* 100:107232. https://doi.org/10.1016/j.polymertesting.2021.107232

[14] Azhary T, Wildan MW (2022) Mechanical, morphological, and thermal characteristics of epoxy/glass fiber/cellulose nanofiber hybrid composites. *Polym Test* 110:107560. https://doi.org/10.1016/j.polymertesting.2022.107560

[15] Verma YK, Singh AK, Paswan MK, Gurmaita PK (2023) A novel approach to extract and characterize vacuum-pressurized bamboo fibers using mechano-chemical techniques. *J Wood Chem Technol* 43(4):1–13. https://doi.org/10.1080/02773813.2023.2224301

[16] Verma YK, Singh AK, Paswan MK, Prashanth S (2024) Mechano-chemical extraction and characterization of tannic acid-functionalized bamboo fibers from vacuum-pressure impregnation (VPI) treated and untreated sources to enhance mechanical properties of fiber-reinforced polymer (FRP) laminated composites. *Wood Mater Sci Eng* 1–16. https://doi.org/10.1080/17480272.2024.2436517

[17] Oliveira PE, Petit-breuilh X, Díaz PE, Gacitúa W (2020) Nano-structures & Nano-objects manufacture of a bio-tissue based on nanocrystalline cellulose from chilean bamboo Chusquea quila and a polymer matrix using electrospinning. *Nano-Structures Nano-Objects* 23:100525. https://doi.org/10.1016/j.nanoso.2020.100525

[18] Adamu M, Rahman MR, Hamdan S (2019) Formulation optimization and characterization of bamboo/polyvinyl alcohol/clay nanocomposite by response surface methodology. *Compos Part B Eng* 176:107297. https://doi.org/10.1016/j.compositesb.2019.107297

[19] Gabr MH, Phong NT, Okubo K, et al (2014) Thermal and mechanical properties of electrospun nanocelullose reinforced epoxy nanocomposites. *Polym Test* 37:51–58. https://doi.org/10.1016/j.polymertesting.2014.04.01035.

[20] Yan L, Xu H (2025) Lightweight composite materials in automotive engineering: State-of-the-art and future trends. *Alexandria Eng J* 118(November 2024):1–10. https://doi.org/10.1016/j.aej.2024.12.002

[21] Phiri R, Mavinkere S, Siengchin S (2024) Heliyon advances in lightweight composite structures and manufacturing technologies: A comprehensive review. *Heliyon* 10(21):e39661. https://doi.org/10.1016/j.heliyon.2024.e39661

[22] Zhou X (2022) Lightweight materials in electric vehicles. *Automotive Manufact Mater* 1(1):28–34.

[23] Fekiač JJ, Krbata M, Kohutiar M, et al (2025) Comprehensive review: Optimization of epoxy composites, mechanical properties, & technological trends. *Polymers* 17(3), 271:1–39.

[24] Verma YK, Singh AK, Paswan MK (2023) Preparation of bamboo-based nano-cellulose by ball milling. In: Edited by Bhingole Pramod P., Ulkesh B. Desai Sunkulp Goel *Advances in Material Science and Metallurgy: Select Proceedings of ICFAMMT 2022*. Springer, pp. 89–98.

[25] Chen W, Yu H, Fan Z (2018) Nanocellulose: A promising nanomaterial for advanced electrochemical energy storage. *Chem Soc Rev*. 47:2837–2872. https://doi.org/10.1039/C7CS00790F

[26] Gomri C, Cretin M, Semsarilar M (2022) Recent progress on chemical modification of Cellulose Nanocrystal (CNC) and its application in nanocomposite films and membranes- A comprehensive review. *Carbohydrate Polym*. 294:119790.

[27] Gindl-altmutter W, Bacher M, Edler M, Griesser T (2017) Surface chemical functionalization of cellulose nanocrystals by 3-aminopropyltriethoxysilane. *Int J Biol Macromol*. 106:1288–1296. https://doi.org/10.1016/j.ijbiomac.2017.08.136

[28] Wohlert J, Chen P, Berglund LA, Re G Lo (2024) Acetylation of nanocellulose: Miscibility and reinforcement mechanisms in polymer nanocomposites. *ACS Nano* 18(3):1882–1891. https://doi.org/10.1021/acsnano.3c04872

[29] Thakur V, Guleria A, Kumar S, et al (2021) Recent advances in nanocellulose processing, functionalization and applications: A review. *Mater Adv* 2:1872–1895. https://doi.org/10.1039/d1ma00049g

[30] Onyango MS (2021) Recent advances in the synthesis of nanocellulose functionalized – Hybrid membranes and application in water quality improvement. *Processes* 9(4):611.

[31] John MJ, Thomas S (2008) Biofibres and biocomposites. *Carbohydrate Polym*. 71:343–364. https://doi.org/10.1016/j.carbpol.2007.05.040

[32] Seisa K, Ude AU (2022) Surface treatments of natural fibres in fibre reinforced composites: A review. 151. https://doi.org/10.2478/ftee-2022-0011

[33] Kumar Y, Kumar A, Paswan MK, Kumar P (2024) Preparation and characterization of bamboo based nanocellulose by ball milling and used as a filler for preparation of nanocomposite. *Polymer (Guildf)* 308:127396. https://doi.org/10.1016/j.polymer.2024.127396

[34] Mariano M, El Kissi N, Dufresne A (2014) Cellulose nanocrystals and related nanocomposites: Review of some properties and challenges. *J Polym Sci Part B: Polym Phys* 52(12):791–806. https://doi.org/10.1002/polb.23490

[35] Elgharbawy AA, Alam MZ, Moniruzzaman M, Goto M (2016) *Ionic Liquid Pretreatment as Emerging Approaches for Enhanced Enzymatic Hydrolysis of Lignocellulosic Biomass*. Elsevier B.V.

[36] Kisin ER, Murray AR, Sargent L, et al (2011) Genotoxicity of carbon nano fi bers: Are they potentially more or less dangerous than carbon nanotubes or asbestos? *Toxicol Appl Pharmacol* 252:1–10. https://doi.org/10.1016/j.taap.2011.02.001

[37] Kumar V, Kumar R (2012) Improved mechanical and thermal properties of bamboo – Epoxy nanocomposites. *Polym Compos* 33(3):362–370. https://doi.org/10.1002/pc

[38] Verma RK, Kesarwani S, Davim JP (2023) *Polymer nanocomposites: Fabrication to applications,* 1st edition. CRC Press, Taylor & Francis.

3 Recent Developments in Battery Materials and Their Management System for Electric Vehicle Applications

A Perspective

Pritam Bhat and Mahesh K. Varpe

3.1 INTRODUCTION

The main source of global air pollution and CO_2 is fossil fuel combustion. Burning fossil fuels (coal, diesel, petrol, oil, and natural gas) for power, heating, transportation, and industry causes most air pollution worldwide, generating around 85% of airborne toxic particulate pollution and almost all sulfur dioxide and nitrogen oxide emissions. Fossil fuels are predominantly composed of carbon and hydrogen.[1] The combustion of fossil fuels results in the reaction of oxygen and carbon to generate CO_2, while oxygen also reacts with hydrogen to produce water. Energy is derived from the heat generated by these reactions. The composition of carbon and hydrogen in fuel influences CO_2 emissions and heat generation. Natural gas, primarily composed of methane (CH_4), possesses a high hydrogen content and produces lower CO_2 emissions per unit of heat compared to other fossil fuels. Burning fossil fuels like petrol and diesel produces greenhouse gases (GHGs) such as carbon dioxide. The buildup of carbon dioxide (CO_2) and other GHGs like methane (CH_4), nitrous oxide (N_2O), and hydrofluorocarbons (HFCs) is causing the Earth's atmosphere to warm, resulting in changes to the climate. Transportation emits 28% of US GHGs, making it the greatest contributor. [2] In 2022, fossil fuel burning accounted for 74% of US anthropogenic GHG emissions. [3] The world is thus trying to move toward a sustainable future by adopting various technologies in energy production and transportation that reduce the dependence on polluting fossil-based conventional energy resources. Electric vehicles (EVs) are becoming more popular, which is a huge step forward in the direction of more ecologically friendly and sustainable future mobility. These vehicles can overcome the environmental issues that relate to conventional internal combustion engine (ICE)-based automobiles. The energy storage system (ESS) forms a major sub-system of EVs that powers its drivetrain. The ESS stores the electric energy and supplies it to electric motors to propel the vehicle. Lithium-based secondary (rechargeable) batteries are preferred for EVs due to their higher Coulombic efficiency (CE), longer lifespan, reduced self-discharge rates, and high energy density, as seen in Figure 3.1. Lithium ions move through the electrolyte between the positive and negative electrodes of Li-ion batteries (LIBs), releasing and storing energy. As seen in Figure 3.2, a lithium cell has electrodes, an electrolyte, and a separator, just like a regular battery. At the beginning of the charging and discharging cycles, lithium is found in the cathode as metal oxides and oscillates because of chemical processes. Graphite, silicon, or lithiated titanium oxides

DOI: 10.1201/9781003596097-3

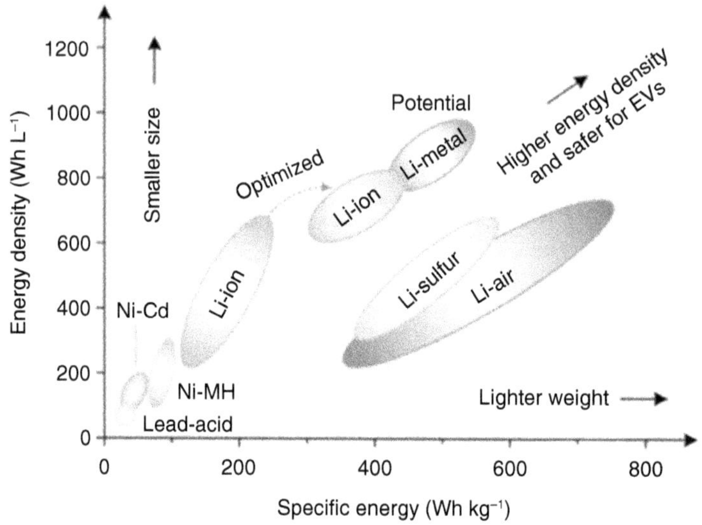

FIGURE 3.1 Cell level-specific energy and energy density of different secondary batteries [7].

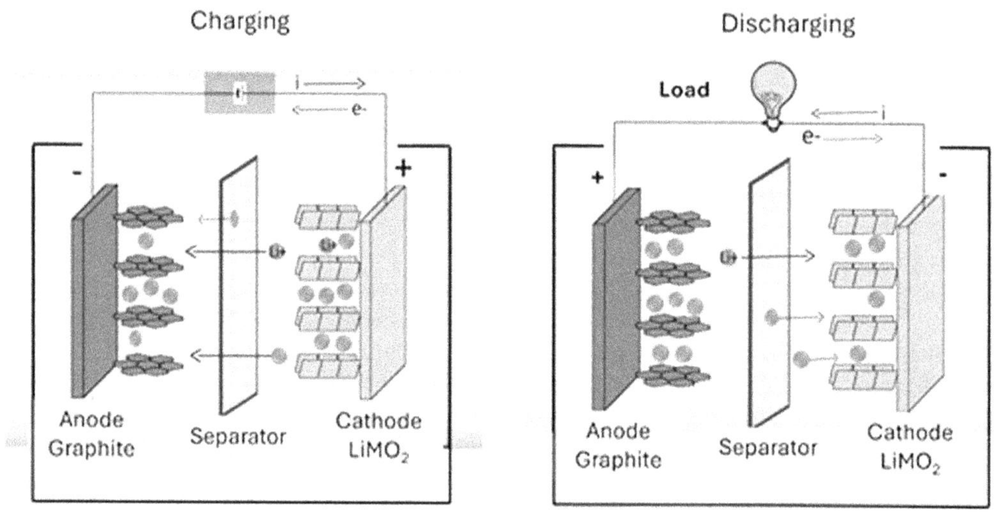

FIGURE 3.2 LIB working principle.

are commonly used as anode materials, while lithiated metal oxides or phosphates are commonly used as cathode materials. The chemical reactions of charge and discharge occur in accordance with equations (3.1)–(3.6). The mix of anode, cathode, and electrolyte materials (cell chemistry) results in a variety of LIB types. Different cell chemistries have a major impact on lithium-ion cell performance. The core components of LIB and their functions are outlined in Table 3.1. [4,5] LIBs come in a variety of form factors, such as prismatic cells, cylindrical cells, button cells, and pouch cells, in addition to different battery chemistries. LIBs in cylindrical and prismatic form factors are preferred for present-day EVs due to their manufacturing, packaging, and maintenance advantages. Table 3.2 presents the properties of commonly used metals in LIB technology. [6] NCA, LFP, and NMC types of LIB are commonly used in EVs. This chapter intends to present a summary of recent advancements in battery materials and the methods utilised for their thermal control, emphasising the prospects and challenges for future EV applications.

TABLE 3.1
Li-Ion Battery Key Components and Their Characteristics

Component	Key materials/ technologies	Main functions	Limitations	Recent developments
Cathode	High-capacity layered Li-/Mn-rich oxides, lithium nickel manganese cobalt oxide (NMC), lithium cobalt oxide (LCO), lithium nickel cobalt aluminum oxide (NCA), lithium manganese oxide (LMO), lithium iron phosphate (LFP), high-voltage spinels	Provides energy storage, contributes to capacity and voltage, influences battery stability and lifespan	Capacity fading due to Mn^{3+} instability, resource limitations, high cost, safety concerns	Doping strategies, advanced coatings, high-nickel NMC, cobalt-free materials, zirconium doping, and Mn-rich cathodes
Anode	Graphite, silicon composites, transition metal oxides (TMOs), alloys, lithium metal anodes	Influences charge/discharge C-rates, stores lithium ions, impacts battery capacity	Limited capacity, potential for dendrite formation, safety concerns, volume expansion	Silicon anodes, hybrid anodes, Li-metal anodes, improved solid electrolyte interphase (SEI) formation, TMOs, and alloy-based anodes
Electrolyte	Liquid electrolytes (e.g., $LiPF_6$ in organic solvents), solid-state electrolytes, additives	Facilitates ion transport between cathode and anode, impacts safety and performance	Flammability, thermal stability, ion conductivity, electrolyte instability	Solid-state electrolytes, gel electrolytes, non-flammable electrolytes, electrolyte additives for stability
Separator	Polyethylene (PE), polypropylene (PP), ceramic-coated separators, γ-alumina separators	Prevents short circuits, maintains structural integrity, influences battery safety and longevity	Thermal stability, mechanical integrity, risk of internal short circuits due to melting or shrinking	Enhanced thermal stability, thinner separators, ceramic or composite separators, electrode-coated separators
Current collectors	Aluminum for cathode, copper for anode	Conducts electricity between external circuit and electrode, distributes current uniformly	Higher metal cost, corrosion, compatibility with electrode materials	Development of lightweight, corrosion-resistant materials, coatings for current collectors
Binder	Polyvinylidene fluoride (PVDF), carboxymethyl cellulose (CMC), styrene- butadiene rubber (SBR)	Maintains mechanical integrity of the electrode, ensures adhesion between materials	Electrochemical stability, flexibility to accommodate volume changes, processability	Development of flexible and durable binders, improved SEI formation, eco-friendly binders

Chemical reactions taking place during charging of the LIB are as follows (Eqn. 3.1–3.3):
At the cathode:

$$LiMO_2 \rightarrow Li^+ + MO_2 + e^- \quad (3.1)$$

TABLE 3.2
Properties of Commonly Used Metals in LIB Technologies

Material	Characteristics	Advantages	Limitations
Lithium	Low standard reduction potential (–3.05 V) and high capacity (3860 mAhg^{-1}) and is the core electrolyte component	High energy density, lightweight, excellent electrochemical characteristics, facilitates high-capacity batteries	Highly reactive; requires protection from air and water, needs secure and tamper-resistant sealing
Nickel	High energy density, improves battery performance, commonly used in NMC cathodes having nickel content in range of 33% to 90%	Increases energy density and battery efficiency, provides extended driving range of EVs, lower cost compared to some alternatives	Stability issues at high temperatures, higher cost than some other metals
Manganese	Has high energy capacity, improves safety, reduces risk of thermal runaway, enhances structural stability	Enhances safety and stability, good rate capability, improved battery life by reducing capacity fading effect	Relatively low specific energy and specific strength
Cobalt	Enhances thermal stability, improves low-temperature performance, has high energy density, and optimises rate performance	Improves thermal stability and low-temperature performance, increases energy density and rate performance	High cost, environmental and ethical concerns related to mining
Aluminum	Provides better stability in NCA chemistries, extended service life of LIB, and high specific power	Provides structural stability and battery life, lightweight and cost-effective	Generally used in small quantities of around 5–10%

At the anode:

$$C_6 + Li^+ + e^- \rightarrow LiC_6 \tag{3.2}$$

Cell reaction:

$$LiMO_2 + C_6 \rightarrow MO_2 + LiC_6 \tag{3.3}$$

The chemical reactions taking place during discharging the LIB are as follows (Eqn. 3.4–3.6):
At the cathode:

$$MO_2 + Li^+ + e^- \rightarrow LiMO_2 \tag{3.4}$$

At the anode:

$$LiC_6 \rightarrow C_6 + Li^+ + e^- \tag{3.5}$$

Cell reaction:

$$MO_2 + LiC_6 \rightarrow LiMO_2 + C_6 \tag{3.6}$$

3.2 NEED FOR IMPROVEMENTS IN EXISTING LIB MATERIALS

Batteries in electric cars must satisfy numerous stringent standards, such as high energy density, extended cycle life, safety, and environmental sustainability. LIB technology is used as it satisfies the requirements of EVs. As conventional LIBs continue to encounter issues related to material scarcity, limitations in energy density, cost, and thermal instability, new research in enriching existing LIB raw materials such as anode, cathode, and separator (electrolyte) to increase its performance have been reported in the literature.

3.2.1 Anode Materials

Advanced LIB anode materials are needed for next-generation energy storage. Many researchers have suggested that silicon-based anodes can boost LIB energy density because of their large theoretical capacity. Due to volume expansion during lithiation, silicon anodes degrade mechanically and lose capacity, making their practical usage difficult. Composite materials, carbon coatings, and nanostructured designs solve these problems. Nanostructured LIB anode materials are making progress in present-day applications. Nanostructured materials made of metal oxides, metal sulphides/nitrides/phosphides, carbon, and metal/semiconductors have high surface area, electric and ionic conductivity, and low diffusion distance, making them ideal for LIB applications. Hossain et al. reported that the anode, which includes a copper collector, typically accounts for 15–30% of the battery's weight. [8]

Recent advancements include the development of physical vapour deposited (PVD) silicon/carbon-based anode materials for LIB. The investigation of PVD methods has shown promise in resolving significant issues with silicon anodes, potentially improving LIB stability and performance. By improving electrical conductivity and prolonging cycle life, PVD deposition of Si/C films successfully addresses problems related to volume expansion and degradation. [9] Although Si/C thin-film electrodes have made significant strides in improving their electrochemical efficiency, enduring issues need the creation of reliable multilayer and nanocomposite Si/C electrode materials for use as anodes in LIB. The volume increase experienced by silicon-based electrodes during cycling poses a substantial difficulty. Mechanical deterioration brought on by the expansion eventually causes electrode damage and a drop in battery capacity. Although Si/C thin-film electrodes have made significant strides in improving their electrochemical efficiency, enduring issues need the creation of reliable multilayer and nanocomposite Si/C electrode materials for use as anodes in LIB. The volume increase experienced by silicon-based electrodes during cycling poses a substantial difficulty. Mechanical deterioration brought on by the expansion eventually causes electrode damage and a drop in battery capacity. [10] PVD techniques must be used to design electrode architectures to mitigate the dangers associated with Si/C-based thin-film anodes. Long-term battery performance depends on interface compatibility between multilayer Si/C electrodes and other battery components. Interfaces are essential for electron conduction, ion transport, and mechanical integrity preservation. By improving energy density, cycle life, and safety in lithium batteries, the creation of Si/C thin-film anodes using PVD techniques holds great promise for accelerating the mainstream use of electric cars and massive energy storage systems. Furthermore, studies show that adding graphene and other conductive additions enhances the stability and functionality of silicon anodes. Even with these developments, there are still many obstacles to overcome, especially in achieving long-term cycle stability, scalability, and cost-effectiveness. Table 3.3 presents a comprehensive list of materials utilised as anodes in LIB along with descriptions, benefits, and drawbacks, as well as current advancements in their properties.

TABLE 3.3
Summary of Anode Materials Used in LIB

Anode materials	Description	Advantages	Common issues	Recent advancements/ improvements	Remarks
1. Intercalation/ de-intercalation materials A. Carbon-based materials (soft and hard carbon)	One of the most used anode materials in LIB due to their thermal chemical and electrochemical stability, availability, low cost and good lithium intercalation and de-intercalation reversibility. Soft carbon (graphite), hard carbon (non-graphitisable), porous carbon, carbon nanotubes, nanofibres, and graphene are some of the popular carbon-based materials researched for use in LIBs	Graphite has a peak theoretical capacity of around 4200 $mAhg^{-1}$ and undergoes a minimum volume change of around 12% during lithium insertion helping in preserving structural integrity across multiple charge/ discharge cycles. Hard carbon accommodates lithium ions in voids helping in lithiation/de-lithiation process which has high reversible capacity	Graphite has low capacity and undergoes high-voltage hysteresis during the de-lithiation process. Hard carbon has low initial CE and low tap density	Hard carbon capacities of around 200–600 $mAhg^{-1}$ were reported. The limitations of hard carbon can be overcome by metal coating such as using a thin layer of soft carbon, surface oxidation, and fluorination	
Carbon nanotubes (CNT)	CNTs are highly ordered carbon nanostructures obtained by self-assembly of unidirectional growth process. Classified into single and multiwalled CNTs	Exhibited reversible capacity of around 1116 $mAhg^{-1}$ in LiC_2 stoichiometry due to intercalation of lithium into stable sites that are located inside the nanotubes as well as pseudo-graphite layers	CNTs contain large structure defects and high-voltage hysteresis, hindering the improvement of CE	Modification of morphological features such as tube diameter, porosity, wall thickness, and shapes to overcome the limitations. Researchers are also exploring doping CNTs with nano-structured materials such as Si, Ge, Sn, Sn-Sb, metal oxides (M_xO_y; M= Fe,Mn,Ni,Mo, Cu,Cr etc.)	CNT mass production and cost is posing a challenge for their widespread adoption in LIB applications

TABLE 3.3 (Continued)
Summary of Anode Materials Used in LIB

Anode materials	Description	Advantages	Common issues	Recent advancements/ improvements	Remarks
Graphene	This is a honeycomb structure of sp_2 carbons that are bonded into 2D sheets having nanometre thickness	Multiple graphene layer sheets have a theoretical capacity of around 780–1116 $mAhg^{-1}$ due to the presence of edges and other defects creating additional active sites for lithium storage	It has poor electronic conductivity due to the defects	Doped hierarchically porous graphene (DHPG) electrodes were developed and investigated to have a long cycle life of around 3000 cycles at current density of 5 Ag^{-1}	Recent research suggests graphene in conjunction with metal oxides/ phosphides can be a suitable candidate for use as anode in LIB
B. Titanium-based oxides		These materials are inexpensive, less toxic, with low volume changes of 2–3% and have a good cycle life	Low electronic conductivity and theoretical capacity in the range of 175–330 $mAhg^{-1}$	Nanostructured titanium oxides have higher rate capability and higher cycle life as found by various researchers and thus many studies have been conducted on different nanostructured allotropic forms and spinel $Li_4Ti_5O_{12}$ (LTO) and titanium dioxide (TiO_2, titania)	
Spinel $Li_4Ti_5O_{12}$ (LTO)		LTO has a better reversible capacity at high operating potential of 1.55 V vs. Li/Li^+, thus guaranteeing better safety. It also mitigates the formation of SEI and dendrite development – a typical issue in carbon-based anodes	Reduced Li-ion diffusion due to poor electronic conductivity and low theoretical capacity of 175 $mAhg^{-1}$	Electronic conductivity can be enhanced by surface treatments. Li-ion diffusion can be increased by employing nano-crystalline LTO having particle size of 20–50 nm	Nanosizing, surface treatments and doping can enhance the theoretical performance of titanium-based oxides

(continued)

TABLE 3.3 (Continued)
Summary of Anode Materials Used in LIB

Anode materials	Description	Advantages	Common issues	Recent advancements/ improvements	Remarks
2. Alloy/de-alloy materials	Materials such as silicon (Si), silicon monoxide (SiO), germanium (Ge), and tin oxide (SnO_2) can easily react with lithium according to alloy/de-alloy method	Specific theoretical capacity ranges from 783 $mAhg^{-1}$ for tin oxide to as high as 4211 $mAhg^{-1}$ for silicon	These materials exhibit very high-volume expansion/ contraction leading to poor cyclability and large irreversibility in initial cycles	Nano-scaling the particles and composite fabrication involving active and inactive material can overcome the common issues	
3. Conversion materials	Anodes based on materials such as iron oxides, cobalt oxides, metal phosphides (MP_x), metal sulphides (MS_x), and nitrides (MN_x) exhibit high reversible capacities of around 500–1000 $mAhg^{-1}$ due to the participation of a large number of electrons arising out of conversion reactions	Low cost, nontoxic, and abundantly available in nature	Iron oxides exhibit poor cyclic life, low electrical conductivity and poor diffusion of Li ions due to high volume expansion	Recent development of carbon coating and nanosized iron oxide particles (haematite and magnetite) have reported improvements in the electrical conductivity	

3.2.2 Cathode Materials

In LIB chemistry, the cathode material is the main and active source of lithium ions. [11] A comparison of LIB materials is shown in Table 3.4, with a focus on the potential uses of different cathode materials. Although lithium has the highest specific capacity of these materials, the higher risk of cell explosiveness and dendritic formation limits its use. [12]

Lithium-ion batteries use lithium-layered cathode materials such as LCO, LMO, LFP, NCA, and NMC. Along with a number of lithium-layered metal oxides, the most commonly utilised cathode materials are LCO, LMO, and LFP. [13,14] Regarding energy density, thermal stability, and cost-effectiveness, many cathode materials display unique properties. [15,16] NMC performs better in a few areas, including price, power, energy density, durability, efficiency, and safety. As improvements in its design features are documented, the NMC battery is becoming more and more well-known in the worldwide LIB industry because of its many benefits. [15] During charge and discharge cycles, cathode materials that use lithium intercalation can effectively and reversibly absorb and release lithium ions inside their crystal structure. [17] For LIBs to operate effectively, safely, and sustainably, cathode materials are essential. The development of safer, more effective, and greener LIBs is being propelled by advancements in cathode materials. To satisfy the growing demands of many applications, recent research focuses on enhancing energy density, cycle life, and safety.

3.2.3 Binder, Electrolyte, and Separator in LIB

For battery anode materials, the choice of binders is crucial since it guarantees the electrodes' mechanical soundness and electrochemical stability. Throughout battery operation, an efficient binder keeps the current collector, conductive additives, and active material particles cohesive, maintaining electrical contact. [18] For battery capacity and cycle life to be negatively impacted by degradation and side reactions, the binder should have electrochemical stability. [19] A solid electrolyte interphase (SEI) layer is created by an efficient binder, improving cycle stability and durability. To avoid cracking or losing its adherence, the binder needs to be flexible enough to accept the anode's volume expansion during lithiation. [20] The binder plays a crucial role in the longevity and efficacy of batteries due to its ease of production, environmental impact, and adherence throughout many charge/discharge cycles. [21] Water-soluble polymers, including styrene-butadiene rubber (SBR), polyacrylic acid (PAA), alginates (ALG), and carboxymethyl cellulose (CMC), are common binders in anode applications.

Positive lithium ions in batteries are obtained from dissolved lithium salts in the electrolyte. Electrons transfer between the negative and positive current collectors through the electrolyte and separator. [22] Lithium exists as conductive salts (electrolytes) such as lithium hexafluorophosphate ($LiPF_6$), lithium bis(trifluoromethylsulfonyl)imide ($LiC_2F_6NO_4S_2$), lithium tetrafluoroborate ($LiBF_4$), lithium triflate [$Li(CF_3SO_3)$], lithium perchlorate ($LiClO_4$), and lithium hexafluoroarsenate ($LiAsF_6$), comprising 1.2–2.0% of the battery's mass. Crystalline salts exhibit solubility in low-viscosity, highly conductive solvents such as linear carbonates [diethyl carbonate (DEC), dimethyl carbonate (DMC), ethyl methyl carbonate (EMC)] and cyclic carbonates [ethylene carbonate (EC), propylene carbonate (PC), γ-butyrolactone] within liquid electrolytes. [23]

In LIBs, the separator, typically a thin microporous polymer membrane, enables ionic transport and serves as an electrolyte reservoir while isolating the negative and positive electrodes. [24] This isolation facilitates lithium ion exchange while preventing the movement of electrons, thereby avoiding internal short-circuits. [13,14] Barrier materials, including semi-crystalline polyolefins, non-woven fabric mats, inorganic composites, and their modified variants, are selected for their optimal balance of permeability, mechanical strength, and electrochemical inertness to prevent electrode particle migration and maintain sufficient electrolyte retention. [25]

3.3 OTHER EMERGING BATTERY TECHNOLOGIES FOR EV APPLICATIONS

3.3.1 Solid-State Batteries

Solid-state batteries (SSBs) represent a significant advancement for EVs, offering enhanced energy density and improved safety over traditional LIBs. Recent advancements in solid-state electrolyte materials exhibiting high Li+ conductivity (>10^{-3} Scm^{-1}) have propelled the progress of SSBs, positioning them as viable alternatives to liquid electrolytes. [26] The fundamental operating principle of SSBs and LIBs is identical. In contrast, SSBs employ a solid material between the electrodes, while conventional LIBs utilise a liquid electrolyte. The potential of SSBs is rooted in their solid-state architecture, which may allow for the utilisation of high-capacity electrode materials that extend beyond traditional LIBs.

Anode materials in SSBs: Lithium metal anodes and alloy anodes (metals/materials that alloy with lithium) are the two main types of anode materials that have garnered a lot of attention to be used as

TABLE 3.4
Summary of Cathode Materials used in LIB

Cathode materials	Description	Advantages	Common issues	Recent advancements/improvements	Remarks
Lithium manganese (LMO) spinels	The research and development of Li-Mn-based oxides resulted in the identification and improvement of high-voltage spinel cathodes such as $LiMnO_3$, $nLiNi_{1/3}Co_{1/3}Mn_{1/3}O_{2(1-n)}$ $LiNi_{1/2}Mn_{3/2}O_4$	It is readily available, non-toxic, cost-effective, and has desirable electrochemical properties	Their capacity diminishes with frequent cycling, attributed to the instability of electrochemically active Mn^{3+} ions resulting from manganese dissolution at elevated temperatures above 55 °C, a phenomenon associated with the Jahn–Teller effect. The application of acid-resistant coatings and the doping of $LiMn_2O_4$ with specific ions, including Al, Co, Cr, and Fe, can enhance structural stability	Over the years the discovery of the combination of different composition has resulted in ionic conductivity, good rate capability, high thermal tolerance, and a consistent voltage profile of approximately 4 V, enhancing safety and suitability for high discharge applications, including EVs and power tools	Li-Mn-based oxides can serve as a better alternative to expensive and toxic Li-Co and difficult to produce Li-Ni-based oxides having better diffusion coefficients in the range of 10^{-6} to 10^{-14} sq cm s^{-1}
Lithium metal oxides	Lithiated nickel and cobalt oxides such as NMC, LCO, and NCA are best suited Li-ion battery cathode materials due to their structural and thermal stability	LCO has layered structure and high density of 4.2 g cm^{-3} but can provide theoretical capacity of around 140–150 $mAhg^{-1}$	Synthesis and production of lithiated nickel-cobalt oxides are expensive due to resource limitations in addition to capacity loss due to degradation of cobalt over time. LCO has limited cycle life of 500–1000 cycles due to its thermal instability at higher operating temperatures or overcharging conditions. Cobalt is toxic in nature and thus its mining and usage have ethical and environmental concerns	Solid materials with diverse compositions are being researched to reduce the cobalt composition and increase nickel and other doping elements composition. For instance, NMC ternary battery material having a chemical composition of lithium-nickel-cobalt-aluminum oxide is represented by $LiNi_xMn_yCo_{1-x-y}O_2$. Recent advances in NCA-type cathode materials have focused on increasing nickel constant while reducing cobalt along with doping, surface coatings, and nano-structuring techniques to enhance the performance thermal stability and cycle life. Researchers have studied characteristics of NCA doped with elements like magnesium,	NMC 811 ($LiNi_{0.8}Mn_{0.1}Co_{0.1}O_2$) has gained significant attention recently as a viable substitute to NMC as it can deliver a discharge capacity of 190–200 m Ah g^{-1}, making it a promising material for high-capacity LIB. However, adding more nickel than the optimum content can result in capacity fading and thermal and structural instability

				titanium, or zirconium which suppressed phase transitions and high structural integrity was observed at high-voltage conditions. Also, increasing nickel content in NMC-type cathodes (NMC 424,532,622,712,811, etc.) is being researched extensively to enhance the electrochemical performance	
Olivines (olivine crystal structures/ polyanion oxides)	Lithium iron phosphate (LFP) represented by $LiFePO_4$ has an olivine crystal structure in which the oxygen atoms are arranged in a hexagonal close-packed lattice. Phosphate anions fill the tetrahedral voids, while iron and lithium cations occupy octahedral holes	The polyanion oxides exhibit minimum capacity degradation over time and are also environmentally safe	Olivines have a very low energy density compared to other cathode materials such as LMO and NCA	Other isomorphous elements within the olivine family of metals (Ni, Mn, Co, or their combination) are being explored as a substitute to the LFP cathode. $LiCoPO_4$ and $LiNiPO_4$ were found to have higher operating potentials	Alternative olivian-structured materials come with issues involving low electronic conductivity, poor Li^+ diffusion, limited stability of electrolytes at high voltage and drop in charging voltage below 4 V that needs to be addressed

active anode material in SSBs. Lithium metal anodes have garnered significant research and development focus.

Lithium metal exhibits the largest theoretical charge storage capacity (3861 mAh g^{-1}) among all anode choices and is considered to maintain more stable surfaces when interacting with certain solid-state electrolytes compared to liquid electrolytes. Lithium metal anodes paired with high-capacity intercalation cathodes can attain stack-level energy density surpassing 1100 Wh L^{-1} and specific energy approaching 400 Wh kg^{-1}. There is a growing emphasis on investigating alternative high-capacity anode materials for SSBs. Alloy anodes containing lithium demonstrate advantageous electrochemical characteristics in SSBs. Alloy anode materials, including silicon, tin, and aluminum, have enhanced lithium storage capacity per unit volume relative to conventional graphite anodes. [27] Nonetheless, alloy-based cells demonstrate reduced specific energies, as depicted in Figure 3.3. Alloy anodes provide benefits over lithium by alleviating various degradation mechanisms associated with lithium metal in solid-state batteries, including dendrite formation that results in short-circuiting. Conversely, substantial volumetric and structural changes in alloy materials during charging and discharging may result in material degradation. Table 3.5 enumerates the electrochemical, volumetric, and transport properties of several alloy materials. In the last 15 years, extensive research has concentrated on replacing graphite anodes in liquid-electrolyte lithium-ion batteries

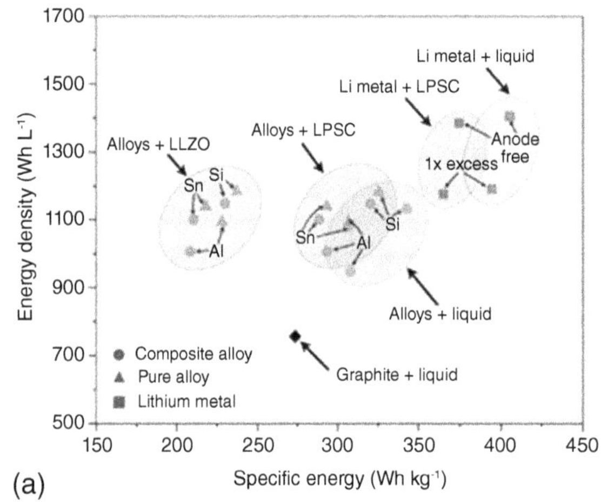

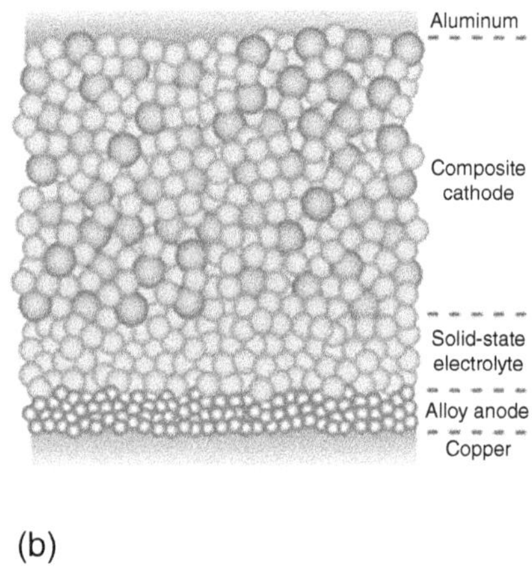

FIGURE 3.3 (A) Energy density comparison of SSBs having alloy anodes and lithium metal anodes and (B) schematic representation of SSBs with typical material composition. (Adapted from Ref. 28.)

TABLE 3.5
Specific Capacity, Volume Change, and Other Electrochemical Properties of Lithium Alloy Materials

Alloy	Most lithiated phase	Specific capacity (mAhg^{-1})	Average potential vs. Li/Li^+	Volume change (%)	Lithium diffusion co-efficient (cm^2 s^{-1})
Si	$Li_{15}Si_4$	3579	0.4	320	10^{-11}–10^{-13}
Sn	$Li_{4.4}Sn$	993	0.50	260	10^{-14}–10^{-16}
Al	LiAl	990	0.38	96	10^{-9}–10^{-11}
Sb	Li_3Sb	660	0.95	200	10^{-8}–10^{-9}
In	$Li_{13}In_3$	1012	0.3	105	~10^{-8}
Mg	Li_3Mg	3350	0.03	100	10^{-7}
Zn	LiZn	410	0.38	98	10^{-8}–10^{-10}

with alloy materials, especially silicon. This encompasses comprehensive investigations at both the research and industrial levels focused on clarifying reaction mechanisms and constructing intricate material architectures to guarantee long-term stability. Recent commercial progress, including the incorporation of silicon into commercial anodes for lithium-ion batteries LIBs, has yielded a 20% enhancement in energy density relative to comparable graphite cells.

Solid electrolytes: The predominant varieties of solid electrolytes are solid polymer electrolytes (SPEs) and solid inorganic electrolytes (SIEs). SIEs consist of crystalline or amorphous materials, such as ceramics, oxides, sulphides, or phosphates. These materials exhibit a structured and rigid configuration that enhances the efficient transport of ions. SIEs exhibit elevated ionic conductivity, facilitating swift and efficient ion transport inside the material. This feature is crucial for enhancing the performance of SSBs. In contrast, SPEs are polymers that have been included or infused with ionic salts, exhibiting flexibility and amorphous characteristics. [29] This versatility improves processing capacity and compatibility across various battery configurations. Although both SPEs and SIEs have unique benefits, SIEs generally demonstrate enhanced ionic conductivity relative to SPEs. The increased ionic conductivity of SIEs enhances charge and discharge rates in SSBs, hence boosting their energy transfer efficiency. SIEs exhibit an improved electrochemical stability window, facilitating operation at elevated voltages without incurring harmful side reactions. SIEs demonstrate exceptional mechanical stability owing to their rigid crystalline or glassy structure. This stability prevents the formation of internal short circuits within the battery, hence ensuring safety and longevity. Ultimately, SIEs demonstrate enhanced thermal stability, enabling SSBs to function at elevated temperatures without degrading performance. The improved thermal stability reduces the likelihood of thermal runaway incidents and enables the implementation of superior battery-cooling methods to prevent fire or explosion. A succinct summary of several types of SIEs and their characteristics is provided in Figure 3.4.

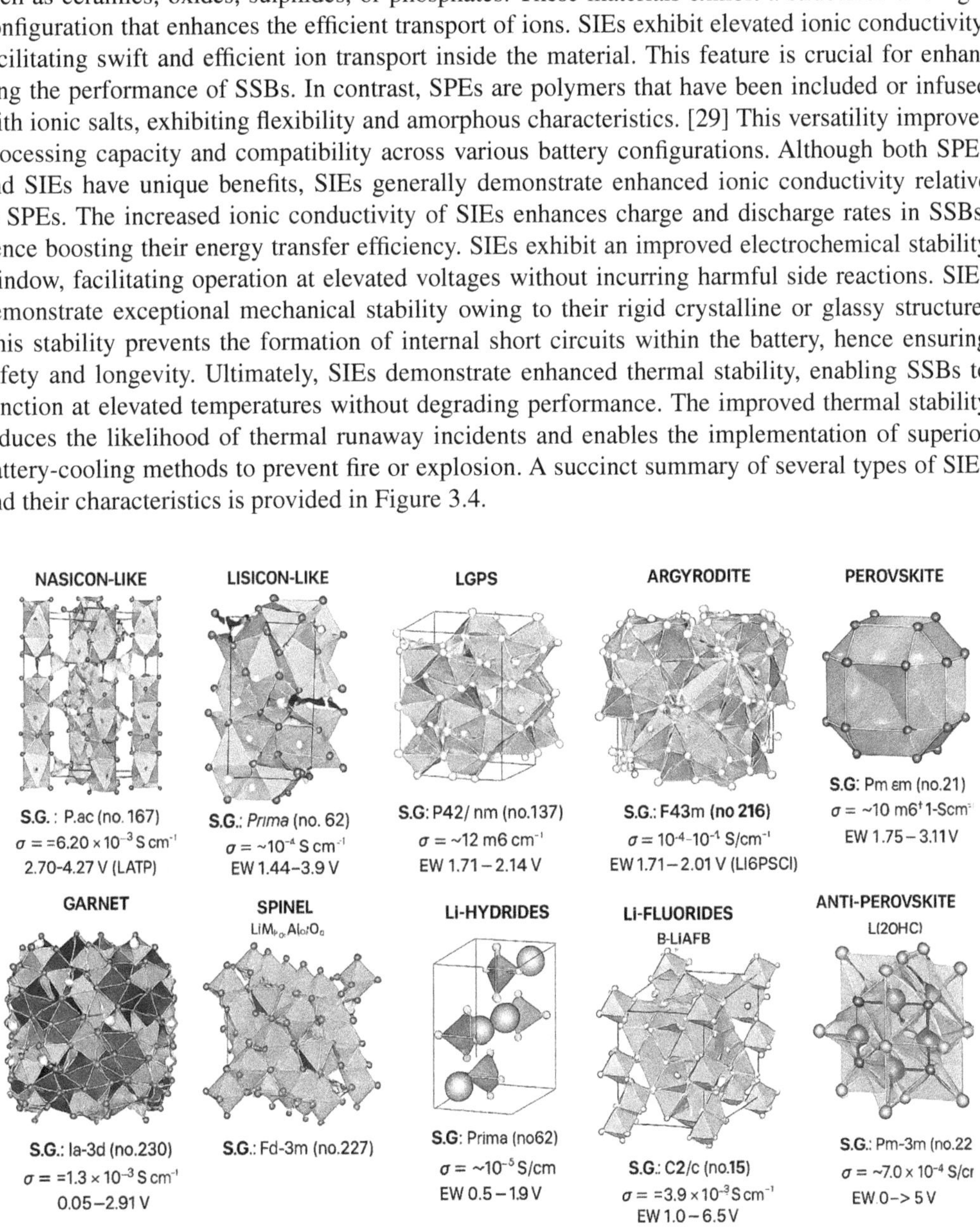

FIGURE 3.4 The different types of SIEs with their electrochemical structure and ion transport characteristics. (Adapted from Ref. 26.)

Table 3.6 provides a brief classification, composition, and ionic conductivity of various types of SIEs. [26] Garnet-type SIEs demonstrate notable characteristics, including elevated ionic conductivity, a broad electrochemical stability range, and exceptional chemical stability. The Li superionic conductor (LISICON) and Thio-LISICON possess a g-Li_3PO_4 framework, in addition to garnet-type materials characterised by a nominal $Li_5La_3M_2O_{12}$ composition. These materials demonstrate elevated Li-ion conductivity. The perovskite-type lithium lanthanum titanate (LLT) has garnered attention for its exceptional ionic conductivity and its ability to maintain stability in the presence of lithium metal.

Glass-based SIEs demonstrate remarkable chemical stability and compatibility with a range of electrode materials. LiPON-type SIEs have demonstrated effective applications and compatibility in thin-film SSBs. The NASICON-type superionic sodium conductor LAGP [$Li_{1.5}Al_{0.5}Ge_{1.5}(PO_4)_3$] demonstrates impressive ionic conductivity along with remarkable stability.

3.3.2 Other-Ion Batteries

Silicon (Si) has emerged as a viable next-generation anode material for alkali metal (Li, Na, and K) ion batteries (AIBs) owing to its high theoretical capacity and operational voltage. Sodium and potassium ion batteries (SIBs and PIBs) are being investigated as alternatives to lithium-ion

TABLE 3.6
Classification of Solid Inorganic Electrolytes (SIEs) with Their Various Compositions

Solid inorganic electrolyte	Composition	Ionic conductivity (Scm^{-1})
Garnet type	$Li_5La_3M_2O_{12}$	~10^{-6}
LISICON/ThioLISICON type	$Li_7La_3Zr_2O_{12}$	6×10^{-4}
	$Li_{6.75}La_3Zr_{1.75}Nb_{0.25}O_{12}$	4×10^{-5}
	$Li_7La_3M_2O_{12}$	8×10^{-5}
	Li_4GeS_4	3.6×10^{-4}
	Li_4SnS_4	1×10^{-4}
	$Li_{14}ZnGe_4O_{16}$	1.25×10^{-8}
	$Li_{10}GeP_2S_{12}$	1×10^{-2}
Perovskite-type	$Li_{1.5}La_{1.5}MO_6$	…
	$La_{0.9}Sr_{0.1}Ga_{0.8}Mg_{0.2}O_3$-δ	8×10^{-2}
	$CaTiO_3$	1.7×10^{-3}
	$BaZrO_3$	5.5×10^{-3}
Glass type	LLTO	10^{-3}
LiPON type	Li_2S–SiS_2 Li_2S– P_2S_5	6.9×10^{-4}
	Li_2O–B_2O_3 Li_3PS_4	1.2×10^{-2}
	$Li_{10}GeP_2S_{12}$	1×10^{-7}
	$Li_7La_3Zr_2O_{12}$ Na_2S-P_2S_5	2.2×10^{-3}
	$Li_{0.99}PO_{2.55}N_{0.30}$	10^{-2}
	$Li_{0.95}PO_{3.00}N_{0.60}$	3.6×10^{-4} 10^{-5}
		3×10^{-7}
		6.6×10^{-7}
NASICON type	Li_2PO_2N	6.51 ±
	$Li_{13 \cdot 33}Al_{05}Ti_{1 \cdot 73}(PO_4)_3$ $Li_{17}La_3Zr_2(PO_4)_3$	0.36×10^{-7}
	$Li_{174 \cdot 8}Al_{0 \cdot 8}Ge_{1 \cdot 71}(PO_4)_3$	10^{-4} to 10^{-3}
	$Na_3Zr_2Si_2PO_{12}$	2.2×10^{-3}
		5.87×10^{-4}
		10^{-4}

batteries (LIBs) due to their enhanced availability, reduced cost, and comparable chemical properties to lithium. Nevertheless, the conventional carbon-based anode is approaching its theoretical capacity, constraining enhancements in energy density for material-based electrode design. The silicon anode exhibits a significant theoretical capacity of 954 mAh g^{-1} as NaSi in sodium-ion batteries and 995 mAh g^{-1} as KSi in potassium-ion batteries, surpassing that of conventional anodes. [20] Silicon anodes exhibit limitations. The substantial volumetric expansion during charging and discharging may result in particle fragmentation, uncontrolled proliferation of the SIE, and diminished CE. Research efforts have concentrated on enhancing silicon-based anode materials through nano-structuring, elemental doping, surface modification, composites, and alloying with metals, metal oxides, and carbon-based substances to augment the performance of aluminum-ion batteries. These strategies have proven beneficial in reducing volume expansion and enhancing performance. Table 3.7 presents a summary of the electrochemical performance of silicon-based anodes for aluminum-ion batteries. [30–32] Silicon-based anodes provide a promising option for next-generation aluminum-ion batteries owing to their elevated theoretical capacity and reduced cost.

A crucial element in the advancement of Si-based anodes for practical applications is attaining a balance between superior performance and cycle stability. Increasing the silicon content in the anode may improve energy density, but it simultaneously results in reduced cycle stability and safety. Various issues, including irreversible capacity deterioration during discharge and unstable SEI development, contribute to the inadequate CE of silicon-based anodes, particularly the initial CE, which often remains below 80%. [33] Future research should focus on the design and synthesis of innovative binders and electrolyte additives that improve the mechanical durability and adherence of the anode to resolve these challenges. This will result in the stabilisation of the SEI layer and an improvement in CE. The development of suitable cathode materials for silicon anodes is another area that necessitates attention. The advancement of cathode materials has lagged that of silicon anode materials. Identifying cathode materials that can achieve the high capacity and cycling stability of silicon anodes is essential for enhancing the efficiency and longevity of lithium, sodium, and potassium ion batteries. The safety of silicon-based anodes in AIBs is a major problem, mostly due to the substantial volume expansion of silicon during charge and discharge cycles. The expansion induces instability in the SEI layer and leads to electrolyte degradation, heightening the probability of lithium dendrite formation. Dendrites pose a significant safety risk, as they can induce short circuits and potentially result in cell explosions. [34] The costs associated with traditional preparation methods substantially impede the practical implementation of Si-based anodes in energy storage and diverse technologies, including additive manufacturing, fluorescence probes, and solar

TABLE 3.7
Electrochemical Properties of Si-Based Anodes for SIB and PIB

Type of battery	Anode materials	Initial capacity @ current density/cycles	Capacity retention @ current density/cycles
SIBs	Si nanoparticles	410 mAh g^{-1}	92% @100
	Mesoporous	400 mAh g^{-1}	390 mAh g^{-1}@100
	silicon microspheres a-Si	390 mA h g^{-1}	85% @2000
	coated c-Si	300 mAh g^{-1}@100 mA g^{-1}	98% @200 mA g^{-1} /100
	Rolled up a-Si	378 mAh g^{-1} @	71% @800 mA g^{-1}/2000
	C-Si nanocomposite	100 mA g^{-1} 681.6 mAh g^{-1}@	
	Silicon coated with reduced graphene oxide	100 mA g^{-1}	
PIBs	c-silicene	180.1 mAh g^{-1} @10 mA g^{-1}	100% @3000
	Amorphous Si and graphene composite	510 mAh g^{-1}	

Source: Ref. [35].

cells. The development of cost-effective, efficient, and uncomplicated silicon material production methods is essential. This involves improving existing methods to reduce costs or developing whole new production systems. Realising this objective will improve the economic feasibility of Si-based anodes and advance multiple scientific and technological domains.

3.4 BATTERY MANAGEMENT SYSTEM

The battery management system (BMS) is an advanced electronic system tasked with overseeing, controlling, and enhancing the performance of the battery pack. A well-engineered and efficient BMS is crucial for ensuring the battery's safety, durability, and reliable power output. BMS comprises of methods and technologies specifically engineered to enhance the efficiency of battery packs by monitoring various battery characteristics such as voltage, current, and temperature. The BMS monitors the voltage levels of lithium cells and prevents overcharging and undercharging. To enhance battery performance and longevity, the BMS should implement cell balancing strategies through charge/voltage equalization. The BMS monitors the cell's working temperature at various stages, regulates a power converter, and manages the battery cell to ensure its health and safe functionality despite thermal conditions. The cell protection mechanism safeguards the cell from short circuits, overloads, and current/voltage stress over time. [36] In the EV system, the battery management system evaluates and quantifies the energy storage distribution processes and anomalies. The main functions of BMS include monitoring of current and voltage in the LIB cell; estimation and protection for LIB charge/discharge control; cell equalization; management of temperature, power, and heat; data storage and acquisition; communication and networking; and fault evaluation and diagnostics, as shown in Figure 3.5. [37] The BMS monitors the battery pack measurements to optimise the efficiency of the battery pack by accessing the various battery parameters using intelligent algorithms and control strategies to mitigate any adverse battery conditions during operation of the EV. The different BMS functions are detailed in Figure 3.5 and below.

3.4.1 Types of BMS

3.4.1.1 *Centralised BMS*

A centralised battery management system architecture features a single control unit that manages each individual cell. The central BMS within the battery pack is interconnected with all other battery packs, creating a unified network. The centralised nature of this system poses challenges for

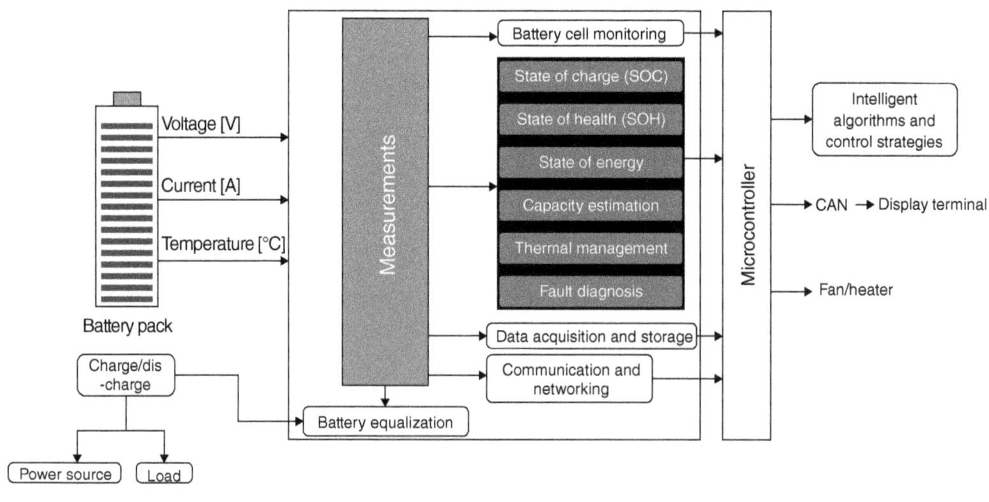

FIGURE 3.5 Block diagram of a BMS in an EV.

maintenance and troubleshooting. Furthermore, an increased risk factor exists; a malfunction in the central BMS compromises the integrity of the entire system.

3.4.1.2 *Distributed BMS*

A distributed battery management system architecture utilises multiple control units to oversee the battery cells. The control board contains all electronic components and is situated on the cells or modules. It reduces wiring and cable clutter, thereby simplifying troubleshooting and maintenance. Although the initial investment is greater, the distributed BMS design offers improved reliability, flexibility, and customised management systems for EV batteries.

3.4.1.3 *Modular BMS*

A modular battery management system architecture organises identical battery cells, each responsible for a segment of the battery pack, and assigns them a control unit. The control unit subsequently communicates with the master control unit to execute the functionalities. This modular BMS structure achieves a balance between centralised and distributed BMS, optimising system structure, complexity, maintenance, and performance.

3.4.1.4 *Cloud-Based BMS*

Cloud-based BMSs are becoming more popular, facilitating remote monitoring, data storage, and performance analysis of batteries. Utilising cloud computing and Internet of Things (IoT) technologies enables remote access and analysis of BMS data, hence enhancing predictive maintenance, fleet management, and performance optimisation across various battery systems.

3.4.1.5 *Wireless BMS*

Conventional BMSs depend on electrical connections linking the battery modules to the central control unit. Wireless BMSs seek to eradicate physical connections, facilitating a more adaptable and economical BMS. Wireless communication protocols such as Bluetooth, ZigBee, or proprietary technologies are employed in wireless BMS to convey data signals between battery modules and the BMS controller. This method diminishes wiring intricacy and cable disarray, hence enhancing maintenance and scalability at significantly reduced expense.

3.4.1.6 *Machine Learning (ML)- and Artificial Intelligence (AI)-based BMS*

The implementation of ML and AI algorithms into BMS is increasing in recent EV development. These advanced techniques can enhance several aspects of battery management, such as:

- Estimation of state of charge (SOC) and state of health (SOH): Machine learning models can enhance the accuracy of SOC and SOH estimations by analysing past data and adjusting to battery degradation and usage trends.
- Predictive maintenance: AI algorithms can evaluate battery data to forecast prospective failures or performance decline, facilitating proactive maintenance and minimising possible downtime.
- Thermal management optimisation: ML models can enhance thermal management techniques by analysing usage patterns, environmental conditions, and battery parameters, hence enhancing efficiency and increasing battery lifespan.

3.4.2 Li-Ion Battery Cells/Pack Monitoring

Real-time surveillance is a crucial component of the battery management system in EVs. The BMS collects data on critical factors such as cell voltage, temperature, and current to continuously evaluate battery health. The cell-monitoring outcomes inform the device's operational efficacy in energy management, power distribution, and safety. LIBs provide a stable voltage and current throughout the discharge period. Monitoring these factors facilitates safe operation and prevents overcharging or excessive discharge. BMS algorithms systematically manage these parameters to avert overcharging, excessive discharge rates, or hazardous thermal runaway incidents that may jeopardise battery integrity and provide safety risks. The estimation of parameters such as SOC, SOH, and battery capacity helps protect the LIB by controlling the charge/discharge rates. The SOC indicates the percentage of remaining battery capacity, offering essential information regarding available energy reserves. A precise SOC estimate is crucial for efficient power management and averting over-discharge situations that may irreparably harm the battery. SOC estimate mitigates overcharging and deep draining, hence preserving the battery's integrity and lifespan. Also, this information is essential for determining vehicle range and strategically stopping at charging stations, so guaranteeing a seamless and dependable electric driving experience. The SOH is a diagnostic metric that assesses a battery's overall condition and deterioration during its operational lifespan. This metric considers elements such as cycle ageing, temperature variations, and impedance alterations, facilitating predictive maintenance and prompt battery replacements. The BMS in EVs oversees critical characteristics such as voltage, current, and temperature, facilitating precise SOC assessments that indicate the battery's remaining charge.

3.4.3 Cell Equalization

LIB cells are arranged in series and parallel to obtain a large battery pack of the required kWh for operating an EV. Individual cells often experience SOC imbalance because of various internal or external factors. The cell imbalance can be attributed to a manufacturing defect, self-discharge rate, internal impedance, and charge storage volume. The uneven distribution serves as an external factor contributing to the rise in temperature within an LIB, influenced by different self-discharge rates during the charge and discharge cycles in an imbalanced cell string. Thus, a BMS for EVs plays a crucial role in ensuring that the charge levels of individual cells are balanced, thereby preventing complications associated with overcharging or undercharging. This function guarantees that all cells maintain consistent voltage levels throughout the battery pack, thereby improving the performance and lifespan of EV batteries. The cell balancing is mainly classified into two primary types: active balancing and passive balancing, determined by their energy storage elements, usage, and energy-balancing techniques.

3.4.3.1 *Active cell balancing*

Active cell balancing is more effective than passive cell balancing as it facilitates the transfer of additional energy among LIB cells through the utilisation of capacitors, converters, transformers, and inductors, rather than employing shunt resistors. Excess energy cells transfer energy from regions of higher energy to those of lower energy, according to their energy carrier. Thus, cell balance is achieved without energy loss. The topology can be employed independently of the chemical properties of all cells in the technology. High efficiency and rapid balancing speed are significant advantages; however, they are accompanied by implementation complexity and elevated costs. Active cell balancing is categorised into three distinct types based on the active elements: capacitors, converters, and inductors or transformers. [38]

3.4.3.2 *Passive cell balancing*

Shunt resistors are necessary for balancing passive cells, as they mitigate excess energy by dissipating it as heat. In dissipative cell balance, the topology minimises excess power until the power levels of the lower and higher cells are equalised. This topology provides cost-effectiveness, simplicity, and a compact form factor. Disadvantages encompass heat dissipation, energy loss, and prolonged cell balancing. [39]

3.4.4 Thermal management

The BMS in EVs plays a crucial role in thermal management. Thermal management refers to the process of controlling the temperature of a system to ensure optimal performance and reliability. Regardless of the battery chemistry, one of the most essential aspects of battery performance is thermal management. Inadequate temperature management can result in shortened battery lifespan, decreased performance, and severe malfunctions, including thermal runaway. To guarantee safe and efficient operation, it is essential to pair advanced materials with cutting-edge thermal management systems. This involves the use of various techniques and materials to dissipate heat and maintain desired thermal conditions. The BMS continuously monitors battery pack temperatures to prevent overheating and maintain an optimal operating temperature range. The system manages cooling and heating within the battery pack, safeguarding battery integrity and ensuring safety by preventing thermal runaway, which is essential for improving battery life and overall performance in EVs.

A LIB consists of an electrode, electrolyte, and separator encased within a shell. The functional temperature range for safe operation and optimal performance of a LIB is 15°C to 45°C. [40,41] At temperatures below 15°C, a lithium-ion battery may exhibit suboptimal performance, as electrochemical reactions are hindered. At low temperatures, lithium dendrites can compromise the film during the charging phase, leading to a short circuit in the lithium-ion battery. Furthermore, the voltage of the LIB rises until it attains the cut-off voltage, at which point it encounters capacitive loss. [42] Excessive temperatures can induce battery thermal runaway, resulting in catastrophic failures, including combustion and explosion. The non-uniform temperature distribution within a battery pack leads to localised degradation and accelerated ageing resulting in the release of carbon monoxide (CO), acetylene (C_2H_2), and hydrogen sulphide (H_2S) from internal chemical reactions. [43]

Thermal management systems in EVs typically employ either active or passive cooling methods, or a combination of both to maintain safe working temperature levels for LIBs. Figure 3.6 illustrates various thermal management methods.

- Liquid cooling: The most popular and efficient technique is liquid cooling. A coolant runs through the battery pack, absorbing heat and transferring it to a heat exchanger for dissipation. This approach delivers effective cooling but introduces complexity and expense to the system. Nanoparticles suspended in coolants enhance thermal conductivity and heat transfer, making cooling systems more effective.

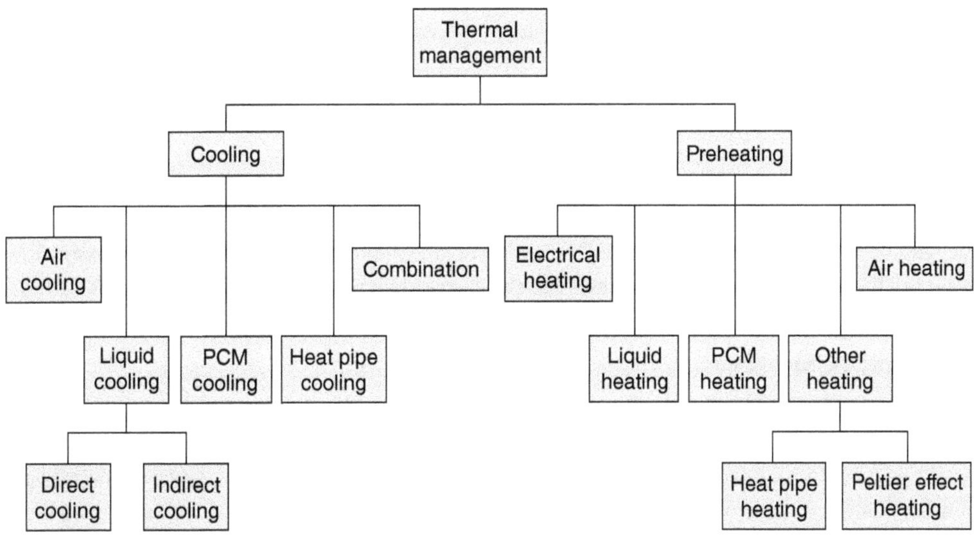

FIGURE 3.6 Thermal management methods for cooling/heating of LIBs.

- Air cooling: Certain EVs employ forced air for battery pack cooling. This approach is more straightforward and cost-effective; but it is not as effective as liquid cooling, particularly for larger battery packs or high-performance automobiles.
- Phase change materials (PCMs) can absorb substantial quantities of heat during phase transitions (e.g., from solid to liquid). This provides a passive cooling mechanism that may manage temperature during high-load operations, although it may necessitate supplementary active cooling for prolonged heat dissipation. A recent development involved the integration of a liquid cooling plate with PCM to provide an economical and modular solution for battery thermal management systems (BTMSs) [44]. In conjunction with the PCM and liquid cooling, an additional BTMS was developed to improve cooling efficiency through the optimisation of coolant flows. [40] Graphene's excellent thermal conductivity can be used to create more efficient heat spreaders and conductors within the battery pack, helping dissipate heat more uniformly.
- Heat pipes and vapour chambers: These systems transfer heat away from the battery cells quickly and efficiently, offering a compact solution for thermal management in high-performance applications.

3.4.5 Fault Diagnosis

The battery management system for EVs is engineered to identify and analyse faults, offering users notifications regarding possible concerns. This function plays a vital role in ensuring safety, as the BMS is capable of isolating or mitigating risks to guarantee dependable vehicle operation. By conducting ongoing diagnostic assessments, the BMS detects malfunctioning cells and delivers insights to inform maintenance and replacement strategies, thereby promoting proactive battery management and enhancing the longevity of EV batteries. There are various types of faults, including sensor faults, internal and external short-circuit faults, BTMS faults, overcharge or over-discharge faults, and actuator malfunctions. [45] Voltage fluctuation represents a common fault response observed in practical scenarios. Consequently, monitoring voltage inconsistencies is essential for the safe and reliable operation of LIB in EVs. To ensure the safe and reliable operation of LIBs in EVs, it is essential to monitor voltage consistency. The entropy approach is not reliant on a specific analytical model or specialised expertise.

3.4.6 Communication and Networking

EV systems integrate the subsystems and networks of a vehicle. The performance of EVs necessitates the electronic monitoring, configuration, and adjustment of the battery management system. It may additionally identify EV charging stations and predict driving range. The BMS records data from the EV data storage system, including voltage and SOC from LIBs, as well as monitoring temperature, charge, discharge, and program control. The BMS transmits and processes stored data related to cell equations, fault diagnostics, and heat management that are being monitored using the microcontroller. The central controller of the EV is linked to multiple control units that compute the battery management system's operations and deliver uniform decision outputs. The data acquisition systems encompass the integration of ESS features and the assessment of conditions via the BMS. [46]

3.5 CONCLUSIONS

Continuous improvements in battery materials and thermal management technologies are essential to the future of EVs. Although lithium-ion technology continues to dominate, novel materials like solid-state electrolytes, lithium-sulfur cathodes, and silicon anodes possess the capacity to transform EV batteries by enhancing energy density, safety, and sustainability. When combined with advanced thermal management techniques, these materials can greatly improve the performance and durability of EVs, facilitating their integration into the mainstream of global transportation.

The way forward entails addressing numerous critical difficulties, including the expansion of production for new materials and the enhancement of thermal management systems for economical and dependable performance. Cooperation among material scientists, engineers, and manufacturers will be crucial in producing next-generation batteries that satisfy the stringent requirements of electric mobility. As these advances progress, EV are anticipated to become increasingly efficient, cost-effective, and secure, facilitating a cleaner and more sustainable future.

REFERENCES

[1] Perera, F. Pollution from fossil-fuel combustion is the leading environmental threat to global pediatric health and equity: Solutions exist. *Int. J. Environ. Res. Public Health* 15, 16 (2017).
[2] Carbonell, T. EPA's Proposed clean power plan: Protecting climate and public health by reducing carbon pollution from the US power sector. *Yale L. & Pol'y Rev.*, 33, 403 (2014).
[3] Where greenhouse gases come from – U.S. Energy Information Administration (EIA). *U.S. Energy Information Administration, Monthly Energy Review, Energy consumption by sector and Environment, April 2024*, pp. 1–5 www.eia.gov/energyexplained/energy-and-the-environment/where-greenhouse-gases-come-from.php.
[4] Koech, A. K., Mwandila, G., Mulolani, F. & Mwaanga, P. Lithium-ion battery fundamentals and exploration of cathode materials: A review. *South African J. Chem. Eng.* 50, 321–339 (2024).
[5] Bhat, P. & Varpe, M. K. Experimental assessment of different air-based battery thermal management system for lithium-ion battery pack. *SAE Int. J. Electrified Veh.* 13, 307–320 (2024).
[6] Nitta, N., Wu, F., Lee, J. T. & Yushin, G. Li-ion battery materials: Present and future. *Mater. Today* 18, 252–264 (2015).
[7] Liu, W., Placke, T. & Chau, K. T. Overview of batteries and battery management for electric vehicles. *Energy Reports* 8, 4058–4084 (2022).
[8] Hossain, M. H., Islam, M. A., Chowdhury, M. A. & Hossain, N. Prospects and challenges of anode materials for lithium-ion batteries–A review. *Clean. Energy Syst.* 9, 100145 (2024).
[9] El Omari, G. El Kindoussy, K., Aqil, M., Dahbi, M., Alami, J., & Makha, M. Advances in physical vapor deposited silicon/carbon based anode materials for Li-ion batteries. *Heliyon* 10, e30431 (2024).
[10] Schmerling, M., Schwenzel, J. & Busse, M. Investigation of the degradation mechanisms of silicon thin film anodes for lithium-ion batteries. *Thin Solid Films* 655, 77–82 (2018).
[11] Batsa, T. E. Understanding battery types, components and the role of battery material testing in development and manufacture. *Technol. Networks – Applied Sciences White paper*, (Online) 30th May 2024 (2023).
[12] Kurc, B., Pigłowska, M. & Rymaniak, Ł. The electrochemical stability of starch carbon as an important property in the construction of a lithium-ion cell. *Entropy* 23, 861 (2021).
[13] Heidari, A. A. & Mahdavi, H. Recent development of polyolefin-based microporous separators for li–ion batteries: A review. *Chem. Rec.* 20, 570–595 (2020).
[14] Zhang, J., Sun, B., Huang, X., Chen, S. & Wang, G. Honeycomb-like porous gel polymer electrolyte membrane for lithium ion batteries with enhanced safety. *Sci. Rep.* 4, 6007 (2014).
[15] Saaid, F. I. et al. Ni-rich lithium nickel manganese cobalt oxide cathode materials: A review on the synthesis methods and their electrochemical performances. *Heliyon* 10, 1–29 (2023).
[16] Pillai, A. M. et al. Synthesis and characterization of Li1.25Ni0.25Mn0.5O2: A high-capacity cathode material with improved thermal stability and rate capability for lithium-ion cells. *J. Alloys Compd.* 938, 168363 (2023).
[17] Ni, E., Uematsu, S. & Sonoyama, N. Lithium intercalation into the polyoxovanadate K7MnV13O38 as cathode material of lithium ion batteries. *Solid State Ionics* 268, 222–225 (2014).
[18] Liu, D. et al. Novel conductive binder for high-performance silicon anodes in lithium ion batteries. *Nano Energy* 36, 206–212 (2017).
[19] Boorboor Ajdari, F. et al. Innovative self-repairing binders tackling degradation and de-lithiation challenges: Structure, mechanism, high energy and durability. *Mater. Sci. Eng. R Reports* 160, 100830 (2024).
[20] Tareq, F. K. & Rudra, S. Enhancing the performance of silicon-based anode materials for alkali metal (Li, Na, K) ion battery: A review on advanced strategies. *Mater. Today Commun.* 39, 108653 (2024).

[21] Yoon, J., Lee, J., Kim, H., Kim, J. & Jin, H. J. Polymeric binder design for sustainable lithium-ion battery chemistry. *Polymer* 16, 254 (2024).
[22] Whitehead, A. H. & Schreiber, M. Current collectors for positive electrodes of lithium-based batteries. *J. Electrochem. Soc.* 152, A2105 (2005).
[23] Park, S. et al. Replacing conventional battery electrolyte additives with dioxolone derivatives for high-energy-density lithium-ion batteries. *Nat. Commun.* 12, 1–12 (2021).
[24] Pan, Y., Chou, S., Liu, H. K. & Dou, S. X. Functional membrane separators for next-generation high-energy rechargeable batteries. *Natl. Sci. Rev.* 4, 917–933 (2017).
[25] Gu, Q. Q., Xue, H. J., Li, Z. W., Song, J. C. & Sun, Z. Y. High-performance polyethylene separators for lithium-ion batteries modified by phenolic resin. *J. Power Sources* 483, 229155 (2021).
[26] Sarfraz, N. et al., Materials advancements in solid-state inorganic electrolytes for highly anticipated all solid Li-ion batteries. *Energy Storage Mater.* 71, 103619 (2024).
[27] Han, S. Y. et al. Stress evolution during cycling of alloy-anode solid-state batteries. *Joule* 5, 2450–2465 (2021).
[28] Lewis, J. A., Cavallaro, K. A., Liu, Y. & McDowell, M. T. The promise of alloy anodes for solid-state batteries. *Joule* 6, 1418–1430 (2022).
[29] Zhang, Q., Liu, K., Ding, F. & Liu, X. Recent advances in solid polymer electrolytes for lithium batteries. *Nano Res.* 10, 4139–4174 (2017).
[30] Lu, M. & Zhang, H. Controllable synthesis of spherical silicon and its performance as an anode for lithium-ion batteries. *Ionics (Kiel).* 19, 1695–1698 (2013).
[31] Zhang, Y. et al. Porous Amorphous silicon hollow nano-boxes coated with reduced graphene oxide as stable anodes for sodium-ion batteries. *ACS Omega* 7, 30208–30214 (2022).
[32] Sun, Y. et al. Reversible potassium-ion alloying storage in crystalline silicene. *Chem. Eng. J.* 435, 134961 (2022).
[33] Dong, H. et al. High-rate performance boron-doped silicon flakes anode using a molten salts method for lithium-ion batteries. *J. Alloys Compd.* 911, 164965 (2022).
[34] McDowell, M. T., Woo Lee, S., Wang, C. & Cui, Y. The effect of metallic coatings and crystallinity on the volume expansion of silicon during electrochemical lithiation/delithiation. *Nano Energy* 1, 401–410 (2012).
[35] Tareq, F. K. & Rudra, S. Enhancing the performance of silicon-based anode materials for alkali metal (Li, Na, K) ion battery: A review on advanced strategies. *Mater. Today Commun.* 39, 108653 (2024).
[36] Okay, K., Eray, S. & Eray, A. Development of prototype battery management system for PV system. *Renew. Energy* 181, 1294–1304 (2022).
[37] Habib, A. K. M. A. et al. Lithium-ion battery management system for electric vehicles: Constraints, challenges, and recommendations. *Batteries* 9, 1–29 (2023). https://doi.org/10.3390/batteries9030152
[38] Samanta, A. & Chowdhuri, S. Active cell balancing of lithium-ion battery pack using dual DC-DC converter and auxiliary lead-acid battery. *J. Energy Storage* 33, 102109 (2021).
[39] Thiruvonasundari, D. & Deepa, K. Optimized passive cell balancing for fast charging in electric vehicle. *IETE J. Res.* 69, 2089–2097 (2023).
[40] Yi, F. et al. Effects analysis on heat dissipation characteristics of lithium-ion battery thermal management system under the synergism of phase change material and liquid cooling method. *Renew. Energy* 181, 472–489 (2022).
[41] Jilte, R., Afzal, A. & Panchal, S. A novel battery thermal management system using nano-enhanced phase change materials. *Energy* 219, 119564 (2021).
[42] Zhang, X., Li, Z., Luo, L., Fan, Y. & Du, Z. A review on thermal management of lithium-ion batteries for electric vehicles. *Energy* 238, 121652 (2022).
[43] Rao, Z. & Wang, S. A review of power battery thermal energy management. *Renew. Sustain. Energy Rev.* 15, 4554–4571 (2011).
[44] Akbarzadeh, M. et al. A novel liquid cooling plate concept for thermal management of lithium-ion batteries in electric vehicles. *Energy Convers. Manag.* 231, 113862 (2021).
[45] Xiong, R., Sun, W., Yu, Q. & Sun, F. Research progress, challenges and prospects of fault diagnosis on battery system of electric vehicles. *Appl. Energy* 279, 115855 (2020).
[46] Zhang, L., Gao, T., Cai, G. & Hai, K. L. Research on electric vehicle charging safety warning model based on back propagation neural network optimized by improved gray wolf algorithm. *J. Energy Storage* 49, 1–22 (2022).

4 Advanced Nanomaterials for Battery Application

Chandrakumar Elumalai, Gopalakrishnan Radhakrishnan, and Satheesh Kumar Radhakrishnan

4.1 INNOVATIVE APPROACHES FOR NEXT-GENERATION ENERGY STORAGE

The fast growth of renewable power, electric vehicles (EVs), and mobile devices has driven up the demand for sophisticated energy storage systems. Although traditional batteries, particularly lithium-ion systems, have provided energy for many contemporary electronics, their drawbacks are beginning to manifest. Researchers are investigating innovative methods that could lead to next-generation batteries in response to the growing need for more efficient and environmentally friendly energy storage [1]. The utilization of nanomaterials and other breakthroughs in materials science, together with investigations into alternative battery chemistries to traditional lithium-ion technologies, are propelling these innovations.

As a result of factors like the exponential growth of the global economy, the extremely rapid increase in the human population, and the rising dependence of people on energy-based equipment, the need to discover and implement new forms of energy has become increasingly apparent [2]. Global energy production will reportedly have to double by the year 2050. In light of this, there is a global effort to develop and test state-of-the-art energy conversion and storage systems. To rise to this challenge, nanotechnology has ushered in new eras of materials research and engineering. Specifically, carbon nanomaterials and nanotechnology have proven to be a game-changer in the development of efficient energy converters and storage devices [3].

For decades, lithium-ion batteries have been the standard for energy storage, but they aren't without their drawbacks. As a measure of energy storage capacity per unit mass or volume, the energy density of modern lithium-ion batteries is almost at its theoretical maximum [4]. As a result, they will be unable to store as much energy from renewable sources like wind and solar or power electric cars for greater distances.

There are two big problems that humans are now facing: the energy crisis and environmental degradation. In order to meet an energy demand that is consistent with the aim of sustainable human growth, it is essential to continuously develop renewable and clean energy sources [5]. To achieve this goal, we need energy storage and conversion devices that are top-notch in order to make use of power in a way that is efficient, flexible, and kind to the environment.

The issue of safety is another important one. There is a danger of overheating, combustion, and leakage with lithium-ion batteries since they use liquid electrolytes, which are volatile. Environmental and ethical difficulties arise from the resource-intensive and ecosystem-harming nature of mining, which is necessary for the production of lithium and other vital metals, such as cobalt [6]. These difficulties have driven the pursuit of safer, more efficient, and environmentally friendly battery solutions.

Nanotechnology is spearheading new developments in energy storage. Because of their extraordinary atomic or molecular-scale characteristics, nanomaterials have game-changing potential for improving battery performance. They are perfect for increasing battery life, charging speed, and

DOI: 10.1201/9781003596097-4

energy density because of their adjustable chemical properties, high surface area-to-volume ratio, and outstanding electrical conductivity.

Nanomaterials have a lot of potential in the creation of electrodes for batteries. Researchers are now investigating potential substitutes for graphite in lithium-ion batteries, such as silicon nanowires, which have the potential to hold a far higher concentration of lithium ions. Silicon anodes suffer from mechanical deterioration due to expansion and contraction during charge and discharge cycles [7]. By improving the material's stability and making it more capable of handling strain produced by variations in volume, nanostructuring silicon at the nanoscale helps alleviate this problem.

In a similar vein, cathodes are using nanomaterials to enhance charge retention and ion diffusion rates. In the past, cathodes made of transition metal oxides were used. Adding nanostructures like graphene or other two-dimensional materials can make the cathode work better by shortening the distance that ions have to travel and increasing the number of places where they can react and exchange ions [8]. This results in reduced charge times and increased energy production.

Anode and cathode ions move through an electrolyte, which nanotechnology has the ability to enhance. Adding nanomaterials to solid-state electrolytes makes them much better and safer than regular liquid electrolytes [9]. Batteries made with nanostructured solid electrolytes are safer and last longer because they have greater thermal stability and ionic conductivity.

Even though lithium-ion batteries are now the standard for energy storage, researchers are working on new battery chemicals that may be able to circumvent lithium's drawbacks. One potential substitute is the sodium-ion battery, which is gaining popularity [10]. Sodium is a more environmentally friendly choice than lithium since it is both more plentiful and cheaper. But traditional materials have a tough time accommodating sodium ions since they are bigger than lithium ions. The development of electrodes that are capable of handling greater ion sizes without sacrificing performance is one way that advances in nanomaterials are contributing to the solution of this problem [11].

The field of solid-state batteries is also experiencing significant activity. The elimination of liquid electrolytes in favor of solid ones makes these batteries safer and less prone to fires. Additionally, they may be able to store more energy in a smaller area, which makes them perfect for uses like electric cars. Nanomaterials are very important for making solid-state batteries because they stabilize the electrode–electrolyte interface and make it easier for ions to flow [12]. Although it is still in its early stages, this technology has the potential to revolutionize energy storage.

Using air-derived oxygen as a reactant instead of a heavy cathode material gives these batteries, especially lithium-air and zinc-air systems, the ability to attain extraordinarily high energy densities. However, they have encountered obstacles that have hindered their advancement, including poor reaction rates and the instability of metal anodes [13]. Nanotechnology is making these problems easier to solve by making the anode materials more stable and the oxygen reactions more effective. This is making metal-air batteries more useful in real life.

This trend is driving innovation in energy storage as power networks integrate more and more renewable energy sources. Solar and wind power, for example, do not provide electricity continually because of their intrinsic intermittency. Installing massive energy storage devices could solve this problem by leveraging periods of high production and releasing the excess energy when needed [14]. If we want to keep the power grid's supply and demand in equilibrium, we need next-generation batteries that can store more energy in less time and have longer lifespans.

Battery technology that improves safety, reduces charging times, and increases driving ranges is in high demand in the automotive industry due to the growing popularity of EVs. Although lithium-ion batteries are now the most popular type of EV battery, their drawbacks are encouraging researchers to look at other, more modern technologies that may provide better performance and longer battery life, such as solid-state batteries or sodium-ion batteries [15].

The commercialization of next-generation energy storage technologies faces substantial obstacles, despite the rapid advancements in research and development (R&D). Scalability is a major challenge. It is a complicated and expensive procedure to produce nanomaterials and incorporate them into batteries on a large enough scale to fulfill the energy demands of the world [16]. Modern batteries still require the extraction and processing of cobalt and nickel in resource-intensive and environmentally harmful ways.

The expense is another obstacle. Although next-generation batteries outperform their predecessors, the production costs of these batteries are now greater than those of regular lithium-ion batteries [17]. In order to make these technologies more accessible and financially feasible, it is crucial to lower the prices of materials, manufacturing processes, and recycling. Figure 4.1 shows innovative approaches for next-generation energy storage.

The promise of next-generation batteries to transform energy storage is enormous, notwithstanding these obstacles. To overcome these obstacles and bring sophisticated energy storage systems to market, there must be ongoing investment in research and advancements in nanomaterials, battery chemistry, and manufacturing procedures.

Changes in energy technology are coming from novel ways of storing energy for the next generation [18]. Advancements in energy storage are well-positioned to address the increasing demands of renewable energy, EVs, and portable electronics. These advancements include the integration of nanomaterials into battery electrodes and electrolytes as well as the development of alternative battery chemistries like sodium-ion and metal-air systems. Ongoing R&D hold the promise of a future where energy storage is safer, more efficient, and more sustainable, despite the fact that scalability, affordability, and sustainability remain issues [19]. These breakthroughs will play a significant role in propelling the shift toward a greener, more electrified world.

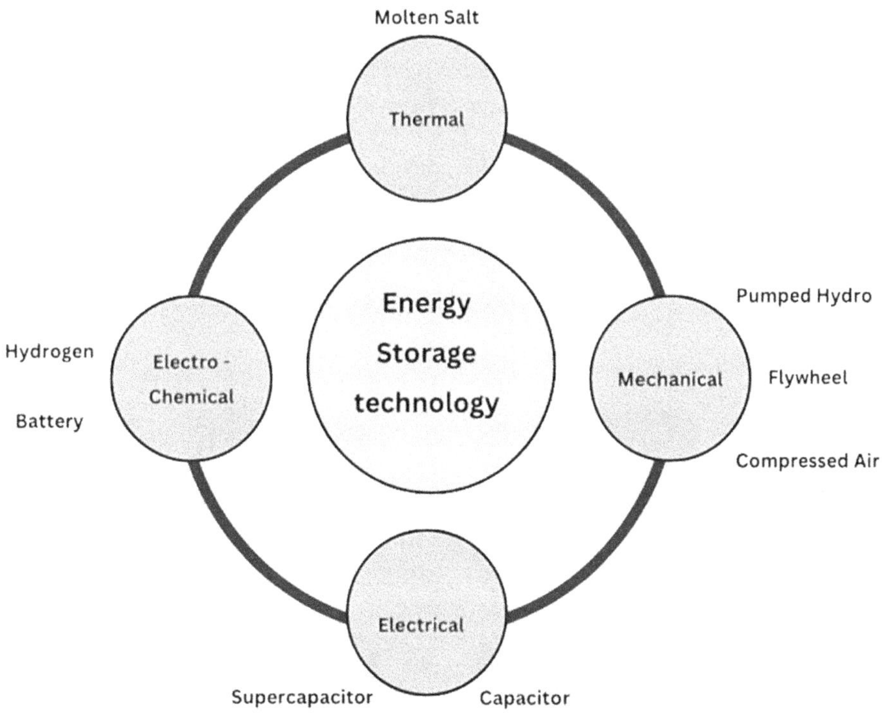

FIGURE 4.1 Innovative approaches for next-generation energy storage.

4.2 REVOLUTIONIZING ENERGY DENSITY AND PERFORMANCE IN MODERN BATTERIES

The widespread use of EVs, the incorporation of renewable energy sources, and innovations in consumer electronics have all contributed to a meteoric rise in the need for improved, high-performance energy storage systems. A revolution in energy density, or the amount of energy a battery can store per unit volume or mass, and an improvement in battery performance across the board in terms of safety, charging speed, and longevity are essential to this progression [20]. Upgrading older battery technology, especially lithium-ion systems, isn't going to cut it for today's energy needs. This is why creative thinkers and doers are delving into novel materials, ideas, and technologies that can improve energy density without sacrificing other important performance indicators.

The capacity of a battery is defined by its energy density. A greater energy density, for example, allows electric cars to have greater ranges without significantly increasing the size or weight of their batteries [21]. One aspect that contributes to customer acceptability and convenience is the reduced need for regular recharging. In the same vein, improvements in energy density have made it possible to create smaller, lighter, and more durable portable gadgets. However, there are always drawbacks to increased energy density, such as shorter lifespans, slower charging times, or safety issues like overheating. Figure 4.2 shows the revolutionizing energy density and performance in modern batteries.

Creating more sophisticated materials is an exciting new frontier in our quest to find solutions to these problems. Specifically, nanomaterials may significantly enhance performance and energy density [22]. To improve battery efficiency, one can make use of the specific physical and chemical characteristics exhibited by nanomaterials, which function on the atomic or molecular scale. For instance, the anode of lithium-ion batteries typically consists of graphite, a material that has served its purpose well but is now exhibiting signs of aging. Silicon has replaced graphite as the material of choice due to its superior lithium ion storage capacity (up to 10 times more). Nevertheless, structural breakdown can occur because of the substantial expansion and contraction of silicon that occurs during charge and discharge cycles [23]. Scientists have developed techniques to reduce this expansion by using nanoparticles or silicon nanowires, which allow for more energy storage with no loss of durability.

To further increase energy density, cathode materials are likewise undergoing revolutionary changes. Materials used for traditional lithium-ion battery cathodes, such as oxides based on cobalt,

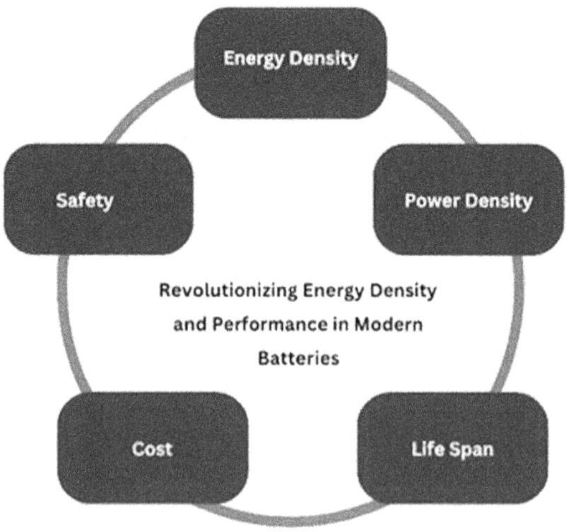

FIGURE 4.2 Revolutionizing energy density and performance in modern batteries.

have a low energy storage capacity and are costly. The use of nanostructured designs in conjunction with materials that are either nickel-rich or cobalt-free has allowed for gains in energy density while simultaneously lowering dependency on precious resources [24]. Furthermore, we can achieve improved ion transport, stability, and storage capacity by incorporating advanced materials into novel structural forms like lithium iron phosphate (LFP) or lithium nickel manganese cobalt oxides (NMC).

Another huge improvement in performance and energy density has come from solid-state batteries. To move ions from the anode to the cathode, solid-state batteries require solid electrolytes, as opposed to liquid electrolytes as used in traditional batteries. This concept has several benefits: it may reduce weight, a crucial factor in EVs, improve safety by doing away with the combustible liquid component, boost energy density by enabling the use of high-voltage cathodes, and so on [25]. Long-term battery performance maintenance relies on solid-state electrolytes, which, when mixed with nanomaterials, provide improved ionic conductivity and mechanical stability.

New battery chemicals are poised to revolutionize the energy storage industry alongside solid-state batteries. Recently, there has been interest in sodium-ion batteries as a potential substitute for lithium-ion systems. A more environmentally friendly alternative would be sodium-ion batteries, as sodium is both more plentiful and cheaper than lithium [26]. While lithium-ion batteries have always had higher energy densities, developments in materials science, such as harder carbon anodes and improved electrolyte formulations, are bringing sodium-ion batteries up to speed. The end product is a battery technology that is less harmful to the environment, cheaper, and just as effective.

The use of lithium-sulfur batteries is another novel strategy; these batteries may one day have a far higher energy density than lithium-ion systems. Theoretically, sulfur-lithium batteries might provide five times the energy density of conventional lithium-ion batteries because sulfur is a plentiful substance. But problems like lithium polysulfide formation which can weaken the battery over time have limited its usefulness in the real world [27]. Researchers have made significant efforts in addressing these constraints by adding nanostructured sulfur cathodes and protective coatings. This makes lithium-sulfur batteries a feasible choice for future energy storage demands.

Efforts to increase energy density also heavily emphasize the development of lithium-metal batteries. For energy storage and discharge, lithium-metal batteries employ pure lithium metal rather than the lithium-ion intercalation method used by traditional lithium-ion batteries. Since lithium metal is a more efficient energy storage material than graphite, this method provides a substantially greater energy density. Dendrites, which are small, needle-like structures that can grow on the surface of lithium while charging, pose a major safety and durability concern for lithium-metal batteries [28]. Dendrites can puncture the battery's separator, leading to short circuits and other hazardous issues. This is why scientists are working on better electrolyte compositions and protective coatings to extend the life of lithium-metal batteries and prevent dendrite development.

Developments in battery design and manufacture, along with materials and chemistry, greatly influence improvements in energy density and performance. As an example, three-dimensional (3D)-structured batteries have the potential to significantly improve power output and energy density by increasing the surface area accessible for ion exchange through a network of interlaced nanomaterials. Micro- and nanoscale battery design allows engineers to build systems with more energy storage capacity, longer charging times, and reduced weight.

Sustainability and recycling are becoming more important factors in the design of new batteries. Reducing the environmental effect of battery manufacture and disposal is becoming more important due to the increasing demand for energy storage. The extraction of essential elements such as nickel, cobalt, and lithium is a significant concern due to its high cost and ecological harm. Researchers are investigating two approaches as potential solutions to these problems: more effective recycling of current battery materials and the development of batteries using more plentiful and less hazardous components. To ensure the long-term viability of energy storage technology, we must shift toward a circular economy in battery production.

To sum up, the increasing power needs of electric cars, integrated renewable energy sources, and portable devices can only be satisfied by radically improving the performance and energy density of contemporary batteries. Novel approaches to improving battery safety, durability, and storage capacity have emerged as a result of developments in materials science, especially with regard to nanomaterials. Future innovations in energy storage hold great potential, including solid-state batteries, lithium-sulfur systems, and lithium metal designs. Improvements in battery design, manufacture, and recycling are laying the foundation for a future of sustainable, high-performance energy.

4.3 ENHANCING BATTERY EFFICIENCY THROUGH NANOTECHNOLOGY

Since renewable energy, electric cars, and portable devices are becoming more and more important, there has been a heightened need for energy storage systems that are more efficient, long-lasting, and powerful. An integral part of this energy shift is batteries, but current technologies, such as lithium-ion batteries, are starting to show their age [29]. Improved battery efficiency in several areas, such as energy density, charging speed, longevity, and safety, may be possible with the use of nanotechnology, which can alter materials on an atomic or molecular scale. By leveraging the distinctive characteristics of nanomaterials, engineers and scientists are paving the way for improved battery performance to meet the demands of modern energy storage.

Designing and manipulating materials on a nanoscale scale, usually between 1 and 100 nanometers, is what nanotechnology is all about. In comparison to their bulk equivalents, materials at this size have distinct mechanical, electrical, and chemical characteristics [30]. We can tune these attributes to boost battery performance by improving ion transport, increasing the response surface area, and stabilizing structures throughout charging and discharging cycles.

The chemical interactions that take place in batteries between the electrolyte, cathode, and anode store and release energy. The reaction efficiency is determined by the mobility of ions, such as lithium ions in lithium-ion batteries. Improved reaction kinetics, quicker ion movement, and more efficient energy storage and release are all possible thanks to nanotechnology in battery material design. Figure 4.3 illustrates the enhancement of battery efficiency through nanotechnology.

Important performance indicators for batteries include their energy density, or the quantity of energy they can store relative to their volume or mass. With a higher energy density, electric cars

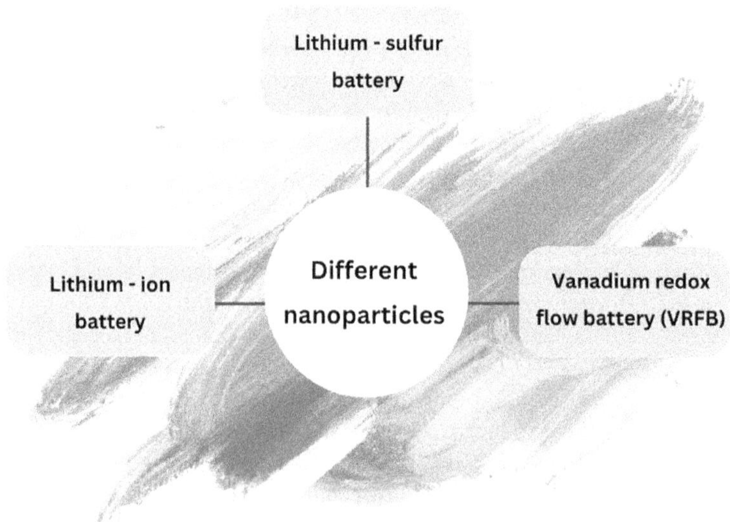

FIGURE 4.3 Enhancing battery efficiency through nanotechnology.

can go further and electronics like laptops and smartphones can last longer on a single charge. Nanotechnology enhances energy density by improving the effectiveness of the electrodes in batteries, which store energy.

The anode of most lithium-ion batteries is graphite, yet this material can only hold so much lithium ion. Silicon anodes are one potential substitute that scientists are looking at since they can store 10 times as much lithium as graphite. Silicon, on the other hand, degrades mechanically because of its high expansion and contraction during charge and discharge cycles. Nanowires and nanoparticles made of silicon allow engineers to increase energy storage capacity without compromising structural integrity by lowering expansion stress [31]. An additional benefit of silicon's nanostructuring is improved ion diffusion, which speeds up the process of energy storage and release.

Cathode materials are also seeing improvements through the application of nanotechnology. The energy storage capacity and durability of traditional cathodes, constructed from materials like lithium cobalt oxide, are both severely restricted and susceptible to degradation. Nanomaterials, such as graphene or transition metal oxides, have allowed scientists to enhance ion transport efficiency and expand the surface area accessible for reactions. The increased capacity to exchange ions between the anode and cathode during each charge cycle results in shorter charging periods and higher energy density.

Another area where nanotechnology has the potential to make a big difference is in battery longevity, which is defined as the number of charge and discharge cycles a battery can endure before its capacity starts to decline considerably [32]. The gradual deterioration of electrode materials is a major contributor to the short life of batteries. Materials can undergo cracking, corrosion, or the formation of undesirable byproducts that reduce performance when subjected to repeated cycles.

Nanotechnology provides a number of answers to these problems. Nanocoatings, for instance, can protect electrode materials against corrosion and mechanical wear. Despite being impermeable to ions, these coatings enhance the durability of the material by acting as a barrier. Also, nanostructured materials may better accommodate the mechanical stress from cycling's frequent expansion and contraction, making them more resistant to breaking and extending their lifespan.

Nanomaterials can improve the performance of the electrolyte medium, which carries ions from the anode to the cathode, and also increase electrode stability. When used for solid-state electrolytes, nanotechnology can improve their stability and safety compared to conventional liquid electrolytes. Researchers have found a way to improve the ionic conductivity of solid electrolytes by adding nanostructured materials. This will make it easier for ions to travel freely and effectively between the different parts of a battery. This improves the battery's charging pace and enhances its overall stability and safety.

Cutting down on charging times is a major obstacle for current battery technologies. In order to maintain a steady supply and demand, businesses like renewable energy storage need quick energy storage and discharge, while consumers want quicker charging for their electronics and electric cars. By enhancing ion transport and expanding the surface area accessible for reactions, nanotechnology presents a promising opportunity for advancements in this field.

A faster charging process is possible with nanostructured materials due to their increased number of active sites for ion exchange and high surface area-to-volume ratio. The use of nanostructured electrodes or nanoporous materials in lithium-ion batteries, for instance, expedites the transfer of lithium ions from the anode to the cathode. This leads to reduced charging durations without sacrificing power output.

Furthermore, we can engineer nanoparticles to enhance battery power delivery. For electric automobiles and other applications that need rapid acceleration, power density—the rate at which a battery can release energy—is crucial. Improved energy release efficiency under high-demand conditions is possible because engineers have optimized the nanostructure of battery components to increase the rate of ion diffusion and decrease internal resistance.

Since lithium-ion systems are prone to overheating, thermal runaway, and even explosion in the worst-case scenario, safety is of the utmost importance when designing batteries. Electric cars and other high-energy applications are especially vulnerable to the fire risks posed by conventional batteries' flammable liquid electrolytes.

Nanotechnology has the potential to enhance battery safety by facilitating the creation of non-flammable, very stable solid-state electrolytes [33]. By enhancing ionic conductivity, nanostructured solid electrolytes can make high-performance batteries safer than those with liquid electrolytes. Dendrites are needle-like structures that can grow on the anode and pierce the electrolyte, resulting in short circuits and possible failure. To prevent this, we could use nanomaterials to build protective barriers within the battery.

Nanotechnology has the potential to enhance heat control in batteries and solve the problem of electrolyte safety. Nanomaterials with high thermal conductivity, such as graphene or carbon nanotubes, can enhance battery components for better heat dissipation. This keeps the battery from becoming too hot or overheating, even while charging quickly or using a lot of power.

Nanotechnology is paving the way for next-generation battery technologies, which promise even more efficiency and performance than existing lithium-ion batteries. For example, problems like sulfur dissolving into the electrolyte have hindered the use of lithium-sulfur batteries, despite their theoretically significantly better energy density compared to lithium-ion systems. Possible solutions, made possible by nanotechnology, include protective nanocoatings and nanostructured sulfur cathodes, both of which aid in battery stabilization and efficiency improvement.

Other battery systems, like sodium-ion and metal-air batteries, which are considered more environmentally friendly and sustainable than lithium-ion systems, are also benefiting from nanotechnology [34]. Although sodium ions are bigger than lithium ions, nanomaterials can enhance the anode and cathode performance of sodium-ion batteries, allowing them to overcome this disadvantage. Nanostructured catalysts can improve the efficiency of oxygen reactions in metal-air batteries, which use air as a reactant, leading to higher power output and energy density.

New materials and designs enabled by nanotechnology are radically changing the efficiency of batteries by improving energy density, longevity, charging speed, and safety. Engineers are addressing many of the shortcomings of conventional battery systems by applying solid electrolytes, protective coatings, nanostructured electrodes, and other novel materials. These developments provide the groundwork for next-generation technologies including lithium-sulfur, sodium-ion, and metal-air batteries, which in turn enhance the performance of lithium-ion batteries. Smaller, lighter, and more powerful batteries will define the future of energy storage as nanotechnology research advances. These batteries will be able to handle the energy demands of a world that is continuously expanding.

4.4 NANOMATERIALS SHAPING THE FUTURE OF BATTERIES

By addressing critical issues that restrict the performance, lifetime, and safety of traditional energy storage devices, nanomaterials are reshaping the area of battery technology. Next-generation batteries, in high demand due to the proliferation of EVs, renewable energy sources, and portable gadgets, are benefiting greatly from nanotechnology advancements. New battery chemistries that are more sustainable and perform better are on the horizon, thanks to nanotechnology's capacity to manipulate materials at the molecular and atomic levels, which greatly improves energy density, charge rates, and overall battery efficiency.

Improving the energy density of batteries, the quantity of energy a battery can store per unit of volume or weight is a primary objective in battery R&D. A higher energy density may enhance the range of electric cars, the battery life of portable electronics, and the efficiency of renewable energy storage systems [35]. Nanomaterials offer a unique opportunity to enhance energy density due to their increased surface areas for electrochemical processes, improved ion transport, and the ability

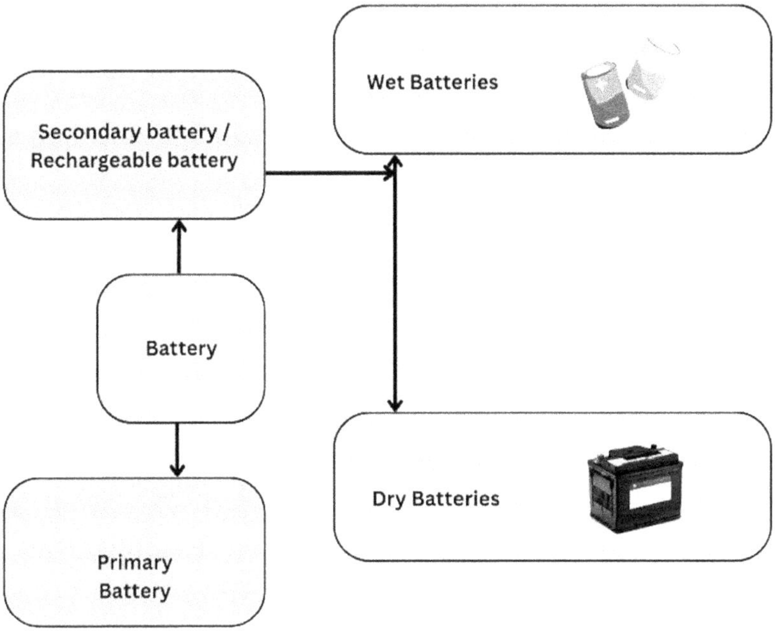

FIGURE 4.4 Nanomaterials shaping the future of batteries.

to employ alternate electrode materials. Figure 4.4 shows the nanomaterials that are shaping the future of batteries.

Traditional lithium-ion batteries typically use graphite, a substance with low energy storage capability, as the anode. One potential substitute is nanomaterials such as silicon, which have a far higher lithium ion storage capacity than graphite. Nevertheless, structural deterioration can occur over time due to silicon's expansion and contraction during charge and discharge cycles. Researchers have created anodes that can withstand this expansion without breaking, thanks to silicon nanowires or nanoparticles [36]. This allows for an increase in energy density without compromising structural integrity. Nanomaterials also have a higher surface area, which speeds up the diffusion of lithium ions and improves charging rates and energy capacity.

Nanotechnology is also facilitating advancements in cathode materials. Traditional lithium-ion cathodes, which use chemicals like lithium cobalt oxide, have limited capacity and availability. LFP and nickel-rich materials are nanostructured alternatives that provide better cycle stability and greater energy densities. These materials can improve ion transport, decrease resistance, and increase the number of active sites for energy storage by adding nanoscale structures; this leads to batteries that are more efficient and last longer.

A battery's performance is heavily dependent on its lifespan, or the amount of charge and discharge cycles it can endure before significantly losing capacity. Since cycling causes mechanical stress and chemical reactions, many traditional batteries degrade over time. By making battery components more structurally stable and chemically resilient, nanomaterials provide answers to these issues.

Nanomaterials play a significant role in the creation of protective coverings for battery electrodes. These nanocoatings can protect electrodes from corrosion and prevent the growth of dendrites, small, needle-like structures that can form on the anode and cause short circuits in lithium-ion batteries. As an example, electrodes coated with nanostructured ceramic materials offer a robust barrier that improves longevity and safety.

Nanotechnology plays a crucial role in the development of a new type of battery that uses solid electrolytes instead of liquid ones. By reducing the possibility of thermal runaway which may cause fires or explosions and eliminating the chance of leaks, solid electrolytes offer enhanced stability and safety, among other benefits [37]. By improving the battery's mechanical characteristics and ion conductivity, nanostructured solid electrolytes pave the way for longer-lasting and more efficient solid-state systems.

The increasing need for fast-charging batteries, widely used in electric cars and portable gadgets, has driven advancements in nanomaterials that enhance the rate of energy storage and release and expedite ion movement. Scientists are developing faster charging batteries without sacrificing capacity or safety by optimizing the nanoscale structure of battery materials and increasing their surface area.

One method takes advantage of nanoporous materials, which may diffuse ions more quickly due to their huge interior surface area. Lithium-ion batteries, for instance, can recharge much faster thanks to nanoporous silicon anodes that facilitate the rapid passage of lithium ions during charging. Similarly, nanostructured cathodes, such as those composed of transition metal oxides, improve charging speed and power supply by facilitating quicker ion and electron transport.

Enhancing the power density of batteries—the rate of energy delivery—is another benefit of nanomaterials. Applications requiring strong bursts of energy, such as electric cars, find this particularly crucial. Engineers boost ion mobility and decrease internal resistance by improving the nanoscale design of battery components. This leads to batteries that can tolerate large loads without overheating or deteriorating and can provide power more effectively.

Particularly for high-energy systems, such as electric cars and large-scale energy storage, safety is a key issue with battery technology. The use of flammable liquid electrolytes in traditional lithium-ion batteries makes them potentially dangerous due to their propensity to overheat and short circuit under specific circumstances. By facilitating the creation of non-combustible solid-state systems and enhancing the thermal management of battery components, nanotechnology is helping to advance the creation of safer batteries.

The use of solid materials in place of liquid electrolytes in solid-state batteries is a huge step forward in terms of battery safety. Ceramic and polymer nanostructured solid electrolytes have outstanding thermal stability and excellent ionic conductivity. The use of solid-state batteries is safer for high-performance applications due to the nanomaterials' ability to resist greater temperatures and avoid the creation of harmful dendrites.

Nanomaterials like graphene and carbon nanotubes, along with solid electrolytes, can enhance the thermal conductivity of batteries. To improve heat dissipation and avoid overheating during high-power operation or rapid charging, experts included these nanoparticles into the battery's structure. By keeping the battery at the ideal operating temperature, this not only makes it safer, but it also keeps it running for longer.

New battery chemistries made possible by nanomaterials provide much more energy efficiency and sustainability than what is now possible with lithium-ion technology. When compared to lithium-ion systems, the energy density of lithium-sulfur batteries is much higher; nonetheless, problems like sulfur dissolving into the electrolyte have limited their use. Potentially addressing these issues, nanotechnology has led to the development of nanostructured sulfur cathodes and protective coatings that extend the life of the cycle and stop deterioration.

Since sodium is abundant in nature, sodium-ion batteries provide a viable alternative to lithium-ion technology in terms of both cost and environmental sustainability. However, the larger size of sodium ions compared to lithium ions may result in slower ion transport and a decrease in energy density. To get around these constraints, scientists are using nanomaterials like transition metal oxide cathodes and nanoporous carbon anodes to make sodium-ion batteries more structurally stable and improve ion transport.

Nanotechnology is developing metal-air batteries using air-derived oxygen as a reactant. These batteries might have extremely high energy densities. To increase energy production and operating lifespans of metal-air batteries, catalysts with nanostructures, such as graphene or platinum, can enhance the efficiency of oxygen processes.

Nanomaterials are revolutionizing the battery industry by providing new approaches to age-old problems with safety, longevity, charging speed, and energy density. Scientists are harnessing the unique properties of nanoscale materials to propel energy storage technology to unprecedented heights [38]. Improving upon current lithium-ion systems and introducing groundbreaking new battery chemistries, nanotechnology is propelling the creation of future-critical batteries that are more efficient, powerful, and environmentally friendly. The possibility that next-generation batteries may transform sectors as diverse as transportation and renewable energy is growing as nanomaterial development quickens.

4.5 THE ROLE OF NANOSTRUCTURES IN HIGH-PERFORMANCE ENERGY STORAGE SYSTEMS

With their revolutionary advances in power output, energy density, charge–discharge rates, and total battery longevity, nanostructures have become an essential component of high-performance energy storage systems. As renewable energy, electric cars, and portable gadgets continue to expand, there is a growing demand for energy storage solutions that are efficient, durable, and sustainable [39]. This underscores the critical importance of these technologies. Engineers and scientists are finding ways to circumvent many of the shortcomings of conventional materials utilized in energy storage systems by modifying them on the microscopic level. This includes creating porous structures, nanowires, nanotubes, or nanoparticles. In this chapter, we look at how nanostructures could affect energy in the future and how they improve the efficiency of current energy storage methods.

Enhancement vital to the efficiency of any energy storage system is the energy density, or the quantity of energy that a battery can retain in relation to its volume or mass. Portable electronics with more battery life, electric cars with greater range, and renewable energy storage systems with greater efficiency are all benefits of high energy density. Graphite and other conventional electrode materials in lithium-ion batteries have a finite amount of ion storage capacity, which lowers the system's energy density [40]. By increasing the surface area available for electrochemical processes, nanostructured materials overcome this barrier and allow for more efficient energy storage.

When compared to graphite, silicon's ability to hold lithium ions makes it a prime material for high-capacity anodes. The rapid degradation of bulk silicon occurs during charge–discharge cycles because of its expansion and contraction properties, which cause cracking and capacity loss. Researchers have found a way to address these concerns by modifying silicon into nanowires or nanoparticles. By preventing the material from cracking under the pressure of silicon's changing volume, these nanostructures prolong the material's useful life and greatly increase the energy density of the battery. Figure 4.5 shows some energy storage systems.

Nanostructuring has also been useful for cathode materials. LFP or nickel-rich compounds that are nanostructured are slowly replacing their more common cobalt oxide counterparts in lithium-ion batteries. The nanoscale structure of these materials allows for increased capacity, less internal resistance, and quicker ion movement, all of which lead to better energy density. To improve the cycle stability and energy storage capacity of cathodes, scientists are experimenting with different particle sizes and arrangements of these materials.

Among the many critical performance criteria for energy storage systems is power density, or the rate at which a battery can release energy. This is particularly true for uses such as electric automobiles and grid storage, which necessitate significant power output for brief durations. By facilitating quicker charge and discharge rates through the provision of shorter paths for ion and

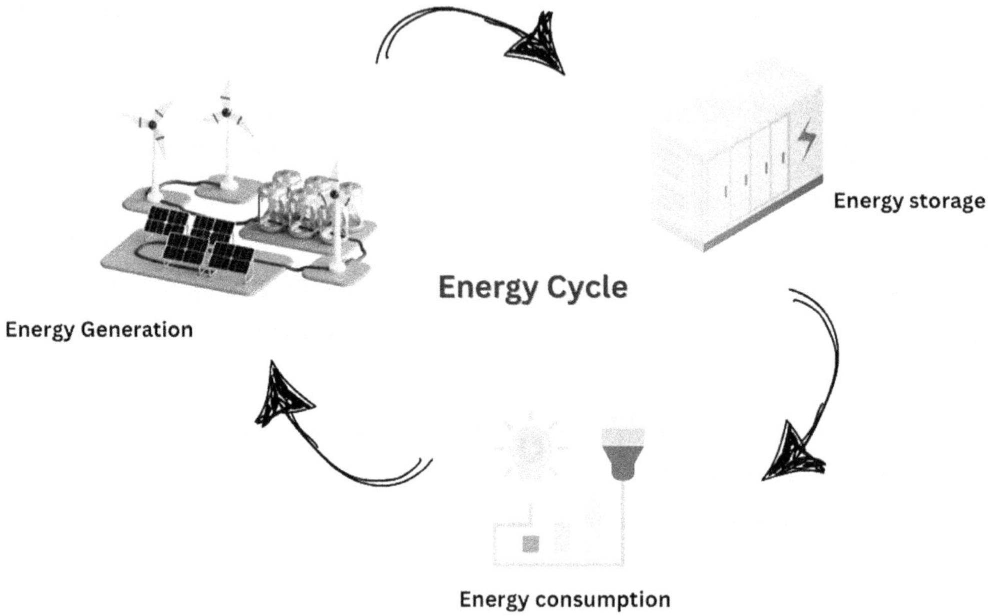

FIGURE 4.5 Energy storage systems.

electron transport, nanostructured materials have the potential to greatly increase the power density of batteries.

Increasing power production is a specialty of nano-porous materials, which are characterized by their large surface area and network of linked pores. These materials shorten the charging period of the battery by facilitating faster ion diffusion across the electrode. Graphene and carbon nanotubes are two examples of nanostructured carbon materials that have attracted a lot of attention due to their potential to enhance electrical conductivity and speed up electron transport [41]. Additional active materials can combine these carbon nanostructures to create electrodes that have a large capacity and excellent conductivity, enabling efficient and rapid energy delivery.

Similarly, transition metal oxides, commonly used as cathodes, benefit from nanostructuring. Nanoparticles and nanowires made of these oxides allow scientists to improve the material's ion diffusion characteristics, which in turn allows for more efficient and rapid energy storage. Applications requiring quick energy supply, such as power tools or hybrid electric cars, place a premium on this.

The number of charge–discharge cycles that a battery can withstand before losing some of its function is a common metric for evaluating the longevity of energy storage devices. Mechanical stress, the creation of unwanted byproducts, and other degradation factors cause the capacity of many traditional batteries to decline with time. Nanostructures provide solutions to these issues by making battery materials more chemically resilient and structurally stable.

Nanostructures improve the stability of electrode materials, which in turn leads to longer battery life spans. Nano-coatings made from titanium dioxide or aluminum oxide can help protect the electrode surface from rusting, dendrite growth, and the formation of solid–electrolyte interphase layers. A short circuit can occur if dendrites, which are needle-like structures that develop on the anode as a result of frequent cycling, pierce the separator between the electrodes. Nanostructured coatings, which inhibit dendrite development, preserve the safety and performance of lithium-ion batteries after lengthy usage.

The use of solid electrolytes with nanostructures in solid-state batteries is another strategy for extending their life. Although solid-state batteries are generally less effective in conducting ions, they are more stable than their liquid-electrolyte equivalents. Improved ion mobility and preserved mechanical strength are two goals of nanostructured solid electrolytes made of materials like polymers or ceramics. This eliminates the risk of leakage and flammability associated with liquid electrolytes, thereby enhancing the battery's cycle stability.

Nanostructures are helping to develop new battery chemistries that not only improve the performance of current lithium-ion batteries but also make them more sustainable and energy efficient. As an alternative to lithium-ion technology, lithium-sulfur batteries are very promising due to their ability to store far more energy than traditional systems. On the other hand, these batteries' sulfur cathodes have a tendency to dissolve into the electrolyte when cycling, which causes a quick decrease in capacity. To improve the efficiency and longevity of the battery, researchers have found ways to structure sulfur at the nanoscale. Researchers can achieve this by forming nano-spheres or coated particles that contain sulfur and prevent its dissolution.

Nanostructures are also having an effect on sodium-ion batteries, which substitute sodium for lithium. Compared to lithium-ion batteries, sodium-ion batteries are more environmentally friendly since sodium is both more plentiful and cheaper. However, because sodium ions are bigger than lithium ions, ion transport may be slower and energy density lower [42]. To get around these problems, sodium-ion batteries use nanostructuring to make their anode and cathode materials more porous and better able to transport ions.

Researchers are now investigating the potential application of metal-air batteries with nanostructured catalysts. These batteries provide extremely high energy densities by reacting with airborne oxygen. The sluggish response kinetics and poor cycle stability of these batteries have prevented their widespread use, despite the revolutionary promise they possess for energy storage in general and electric cars in particular. Nanocatalysts like graphene or platinum can enhance the efficiency and practicality of metal-air batteries by speeding up the oxygen processes.

Improved high-performance energy storage technologies owe a significant deal to nanostructures. Nanomaterials are revolutionizing energy storage technology by boosting power output, extending the lifespan of batteries, improving energy density, and allowing novel battery chemistries. The development and integration of nanostructured materials will be necessary to address future energy issues, given the growing demand for sustainable, high-capacity, and efficient energy storage. These advancements could be the catalyst for future developments in energy storage technology and have the potential to radically improve battery performance in many uses.

4.6 PIONEERING NANOTECHNOLOGY FOR SUSTAINABLE AND SCALABLE BATTERY SOLUTIONS

Nanotechnology is driving innovations in energy storage, holding great promise for the development of long-term, scalable battery systems. An increase in the use of renewable energy sources, portable devices, and EVs has led to a surge in the need for batteries that are both durable and ecologically benign [43]. There are several problems with traditional battery technologies like lithium-ion, including low energy density, high costs, limited availability of resources, and large environmental impacts. Nanotechnology provides novel answers to these problems by facilitating the design of atomic- and molecular-scale battery materials and systems, which in turn improves their performance, longevity, and sustainability. In this chapter, we'll look at how nanotechnology is paving the way for future scalable and environmentally friendly battery solutions.

Battery technology aims to improve the energy density of batteries to store more energy in less bulky and lightweight batteries. Due to the direct relationship between energy density and EV range, EVs and portable devices, which have limited space, are two areas where improving the energy density of batteries is crucial. Nanomaterials enhance the efficiency of ion storage and transport by augmenting the surface area for electrochemical processes, thereby increasing energy density [44]. Figure 4.6 shows some sustainable and scalable battery solutions.

Anodes made of nanostructured silicon, for instance, have a far greater energy capacity than their more conventional graphite counterparts [45]. A big hurdle has been silicon's propensity to expand and shrink during charge–discharge cycles, despite the fact that it can store 10 times more lithium ions. Nanotechnology solves this problem by forming nanowires or nanoparticles out of silicon, which can withstand these variations in volume without degrading, leading to greater energy density and more stable batteries.

Researchers are creating new nanostructured options like LFP and nickel-rich materials to replace the traditional lithium cobalt oxide in cathode materials. These nanomaterials improve battery performance by enabling quicker ion diffusion, more energy storage, and higher cycle stability. Scientists are improving the anodes and cathodes of batteries using nanotechnology, which increases their energy density and makes them live longer.

There will likely be a dramatic increase in the demand for battery materials like lithium and cobalt as the world moves toward EVs and renewable energy. This brings up questions regarding the sustainability of these resources, the harm they cause to the environment, and the geopolitical dangers that come with their supply networks [46]. By making it possible to employ more plentiful and less ecologically harmful materials in battery manufacture, nanotechnology offers a chance to alleviate these problems.

One such innovation is the sodium-ion battery, which uses sodium as the charge carrier instead of the more expensive and scarce lithium. Nanotechnology provides answers to the problems that have plagued sodium-ion batteries in the past, such as poorer energy density and slower charge rates when contrasted with lithium-ion systems. We can engineer sodium ion anodes and cathodes at the

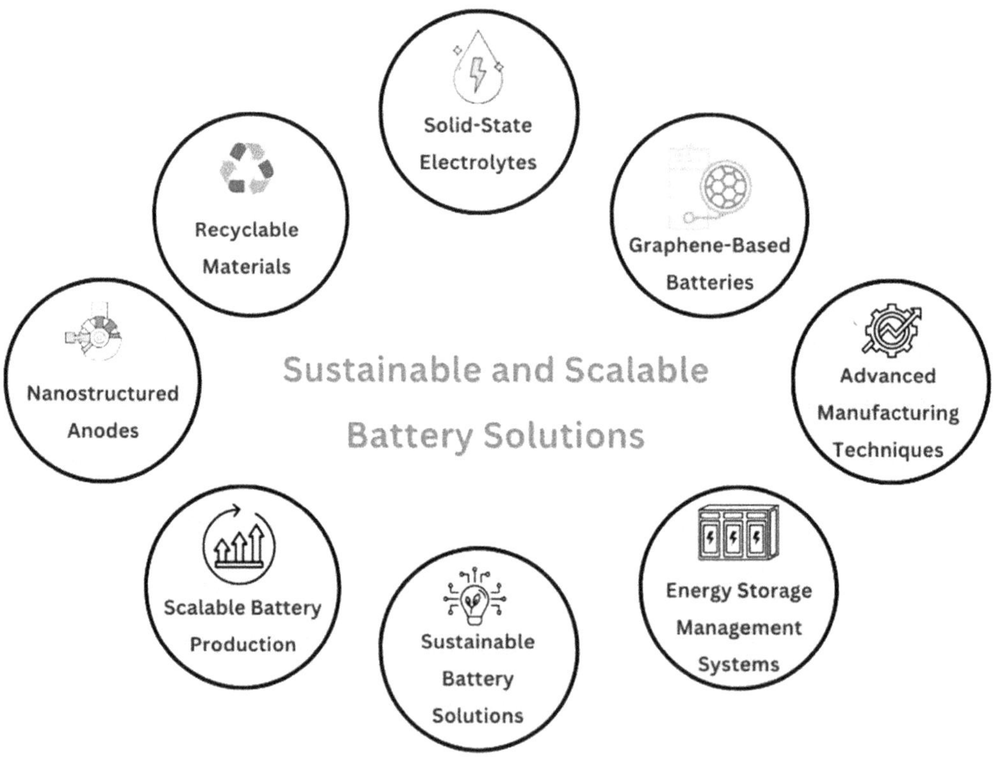

FIGURE 4.6 Sustainable and scalable battery solutions.

nanoscale to enhance energy capacity, structural stability, and ion transport. This makes sodium-ion batteries a suitable candidate for large-scale energy storage.

Additionally, nanotechnology is facilitating the creation of lithium-sulfur batteries, which have the potential to provide energy densities that are substantially greater than those of present-day lithium-ion technologies [47]. Sulfur is a cheap and plentiful material, which makes lithium-sulfur batteries a better long-term investment. Sulfur, on the other hand, dissolves into the electrolyte when cycling, which causes the capacity to deplete quickly. We can improve the battery's cycle life and stop this dissolution by adding nanoscale coatings or confinement mechanisms to the sulfur cathode. This makes lithium-sulfur technology more commercially viable.

The extraction and processing of raw materials as well as the disposal of spent batteries pose significant environmental concerns in the battery production process. Nanotechnology has opportunities to enhance the recyclability of battery components while reducing the environmental impact of battery production.

Nanotechnology improves material efficiency, which in turn decreases the environmental effect of batteries. Nanostructured materials like graphene or carbon nanotubes, which increase mechanical strength and conductivity, can enhance battery electrodes to improve performance while using less material [48]. By minimizing the demand for mining and processing raw materials, this decrease in material utilization lowers the cost of batteries and lessens the environmental load.

Batteries are becoming more recyclable thanks in part to nanotechnology. Academics are developing nanostructured recycling procedures to enhance the effectiveness and ecological friendliness of battery recycling. One example is selective extraction methods for recovering important metals like cobalt and lithium. By reducing energy usage and waste, these techniques make battery recycling a more practical and ecological option.

There is a new, safer, and more eco-friendly option to the old, dangerous lithium-ion batteries: solid-state batteries. These employ solid electrolytes instead of the flammable liquid ones. Better, longer-lasting, and more efficient solid-state batteries are now within reach, thanks to nanostructured solid electrolytes that enhance ion conductivity and mechanical characteristics [49]. As an alternative to traditional, potentially dangerous chemical additions, these batteries provide a safer, more environmentally friendly way to store high-performance energy without sacrificing performance.

Before using nanotechnology for energy storage, we must address the issue of scalability, or the ability to generate these sophisticated materials and processes on a wide scale while keeping costs down. Nevertheless, new developments in nano-manufacturing are allowing for the scalability of nanomaterial production without sacrificing their performance advantages.

For instance, researchers are refining techniques for synthesizing carbon nanotubes and graphene, such as solution-based procedures and chemical vapor deposition (CVD), to facilitate their large-scale utilization in batteries. These nanomaterials are highly sought-after for their use in high-performance batteries due to their enhanced conductivity, mechanical strength, and flexibility. The possibility of commercializing graphene-based batteries is growing as manufacturing methods become more efficient and affordable.

Developments in additive manufacturing and 3D printing have made it possible to precisely manufacture nanostructured battery components. These strategies enable the design of custom high-performance batteries with optimized topologies that enhance power output, durability, and energy storage. Battery manufacturers may make use of nanotechnology-enabled manufacturing to increase output without sacrificing the efficiency and performance that come from using nanoscale materials.

To keep up with today's ever-increasing energy needs, nanotechnology is essential in developing more long-lasting and scalable battery solutions. Nanomaterials are propelling the advancement of next-generation batteries by increasing energy density, boosting material efficiency, decreasing environmental effects, and allowing the utilization of more plentiful resources [50]. Along with solving important problems with scalability and sustainability, these breakthroughs have the potential to radically alter sectors, including electric cars, renewable energy storage, and portable gadgets.

An exciting new chapter in the evolution of nanotechnology might usher in a period of environmentally benign, highly efficient, and broadly available energy storage technologies, which would have a profound impact on the future of our world's power grid.

4.7 CONCLUSION

There is an immediate need to build next-generation batteries due to the fast progress in energy storage technologies. Better battery performance is possible by making use of nanomaterials' one-of-a-kind features, such as their large surface area, better electron transport, and adjustable electrochemical properties. Anodes, cathodes, and electrolytes that use nanomaterials are solving important problems such as energy density limits, cycle stability, and safety issues. The performance of batteries has already been significantly enhanced using nanomaterials. As an example, nanostructured silicon anodes have demonstrated a tenfold enhancement in lithium-ion storage capacity as compared to traditional graphite anodes, with theoretical capacities of about 4200 mAh/g as opposed to 372 mAh/g for graphite. With more than three times the energy density of conventional lithium-ion batteries (~250 Wh/kg), nanostructured lithium-sulfur batteries have the potential to increase the range of EVs from 400 km to more than 1000 km per charge. The environmentally friendly alternative to lithium-ion batteries, sodium-ion batteries can cut material prices by as much as 30%. Making large-scale energy storage more feasible, sodium-ion technology seeks to bring the production costs of lithium-ion batteries below $80/kWh, compared to the current estimates of $100–150/kWh. The long-term replacement costs of solid-state batteries using nanostructured electrolytes are substantially lower than those of traditional lithium-ion batteries, which have cycle lifetimes of 1500–3000 cycles. Nanotechnology is also working to make things safer and more sustainable. The yearly failure rate of conventional liquid electrolyte lithium-ion batteries is 1–2% because of leakage and overheating concerns; however, the safety of EVs and grid storage applications is greatly improved by using solid-state batteries with nanostructured ceramic electrolytes, which virtually eliminate thermal runaway incidents. Furthermore, using cathodes improved with nanomaterials might reduce dependence on cobalt—an element that is both costly and ecologically problematic—by 50–70%, which would lower prices and reduce hazards in the supply chain. The problem of commercialization persists notwithstanding these encouraging developments. While conventional cathode materials cost $10–50/kg, sophisticated nanomaterials can cost $500–2000/kg, making large-scale nanomaterial manufacturing prohibitively costly. Costs are predicted to drop by 40–60% over the next decade due to economies of scale and continued research. Energy storage is poised for a safer, more efficient, and more sustainable future because of ongoing investment in R&D, advancements in nanotechnology, and improvements in battery production. A cleaner and more energy-secure future, with reduced carbon footprints and more accessible high-performance energy storage systems, will be shaped by next-generation battery technologies incorporating nanomaterials, which are in high demand due to the increasing demand for electrification and renewable energy.

REFERENCES

[1] Li, Wei, Ruyi Fang, Yang Xia, Wenkui Zhang, Xiuli Wang, Xinhui Xia, and Jiangping Tu. "Multiscale porous carbon nanomaterials for applications in advanced rechargeable batteries." *Batteries & Supercaps* 2, no. 1 (2019): 9–36.

[2] Mei, Jun, Ting Liao, Liangzhi Kou, and Ziqi Sun. "Two-dimensional metal oxide nanomaterials for next-generation rechargeable batteries." *Advanced Materials* 29, no. 48 (2017): 1700176.

[3] Dai, Liming, Dong Wook Chang, Jong-Beom Baek, and Wen Lu. "Carbon nanomaterials for advanced energy conversion and storage." *Small* 8, no. 8 (2012): 1130–1166.

[4] Cong, Lina, Haiming Xie, and Jinghong Li. "Hierarchical structures based on two-dimensional nanomaterials for rechargeable lithium batteries." *Advanced Energy Materials* 7, no. 12 (2017): 1601906.

[5] Guo, Yu-Guo. "Nanostructures and nanomaterials for batteries." In You, Y. and Y. G. Guo (Eds.) *Nanostructures and Nanomaterials for Batteries*. Springer, 2019. https://doi.org/10.1007/978-981-13-6233-0.
[6] Cui, Qiuhong, Yeteng Zhong, Lu Pan, Hongyun Zhang, Yijun Yang, Dequan Liu, Feng Teng, Yoshio Bando, Jiannian Yao, and Xi Wang. "Recent advances in designing high-capacity anode nanomaterials for Li-ion batteries and their atomic-scale storage mechanism studies." *Advanced Science* 5, no. 7 (2018): 1700902.
[7] Xu, Zheng-Long, Jang-Kyo Kim, and Kisuk Kang. "Carbon nanomaterials for advanced lithium sulfur batteries." *Nano Today* 19 (2018): 84–107.
[8] Ghiyasiyan-Arani, Maryam. "Active nanomaterials for Li-ion batteries and advanced nanomaterials for supercapacitors." In Sahar Zinatloo-Ajabshir and Ardashir Mohammadzadeh (Eds.) *Renewable and Clean Energy Systems Based on Advanced Nanomaterials*, pp. 125–153. Elsevier, 2024.
[9] Liu, Yayuan, Guangmin Zhou, Kai Liu, and Yi Cui. "Design of complex nanomaterials for energy storage: Past success and future opportunity." *Accounts of Chemical Research* 50, no. 12 (2017): 2895–2905.
[10] Korgel, Brian A. "Nanomaterials developments for higher-performance lithium ion batteries." *The Journal of Physical Chemistry Letters* 5, no. 4 (2014): 749–750.
[11] Ahmad, Awais, Shumila Shaheen, Saadat Majeed, Muhammad Pervaiz, Zohaib Saeed, Umer Younas, Muhammad Sufyan Javed, Rafael Luque, and Lalitha Gnanasekaran. "Recent developments in metal/metalloid nanomaterials for battery applications; A comparative review." *Fuel* 340 (2023): 127399.
[12] Geckeler, Kurt E., and Hiroyuki Nishide, (Eds.). *Advanced Nanomaterials*. John Wiley, 2009.
[13] Peng, Lele, Zhiwei Fang, Yue Zhu, Chunshuang Yan, and Guihua Yu. "Holey 2D nanomaterials for electrochemical energy storage." *Advanced Energy Materials* 8, no. 9 (2018): 1702179.
[14] Peng, Lele, Zhiwei Fang, Yue Zhu, Chunshuang Yan, and Guihua Yu. "Holey 2D nanomaterials for electrochemical energy storage." *Advanced Energy Materials* 8, no. 9 (2018): 1702179.
[15] Hur, Jaehyun. "Nanomaterials for ion battery applications." *Nanomaterials* 12, no. 13 (2022): 2293.
[16] Su, Xin, Qingliu Wu, Juchuan Li, Xingcheng Xiao, Amber Lott, Wenquan Lu, Brian W. Sheldon, and Ji Wu. "Silicon-based nanomaterials for lithium-ion batteries: A review." *Advanced Energy Materials* 4, no. 1 (2014): 1300882.
[17] Muñoz-Batista, Mario J., Daily Rodriguez-Padron, Alain R. Puente-Santiago, and Rafael Luque. "Mechanochemistry: Toward sustainable design of advanced nanomaterials for electrochemical energy storage and catalytic applications." *ACS Sustainable Chemistry & Engineering* 8, (2018): 9530–9544.
[18] Long, Yong, Mei Ding, and ChuankunJia. "Application of nanomaterials in aqueous redox flow batteries." *ChemNanoMat* 7, no. 7 (2021): 699–712.
[19] Wu, Songping, Rui Xu, Mingjia Lu, Rongyun Ge, James Iocozzia, Cuiping Han, Beibei Jiang, and Zhiqun Lin. "Graphene-containing nanomaterials for lithium-ion batteries." *Advanced Energy Materials* 5, no. 21 (2015): 1500400.
[20] Li, Shuang, Lang Ma, Yi Xia, Hongrong Luo, Li Qiu, and Chong Cheng. "Advanced nanomaterials for efficient oxygen electrodes in metal–air batteries." In Fen Ran and Shaowei Chen (Eds.) *Advanced Nanomaterials for Electrochemical-Based Energy Conversion and Storage*, pp. 191–222. Elsevier, 2020.
[21] Vijayakumar, Vidyanand, Meena Ghosh, Kiran Asokan, Santhosh Babu Sukumaran, Sreekumar Kurungot, Jonas Mindemark, Daniel Brandell, Martin Winter, and Jijeesh Ravi Nair. "2D layered nanomaterials as fillers in polymer composite electrolytes for lithium batteries." *Advanced Energy Materials* 13, no. 15 (2023): 2203326.
[22] Liu, Jian. *Development of Advanced Nanomaterials for Potential Lithium-Ion Battery Application*. The University of Western Ontario, 2013.
[23] Sharma, Ved Prakash, Utkarsh Sharma, Mahadev Chattopadhyay, and V. N. Shukla. "Advance applications of nanomaterials: A review." *Materials Today: Proceedings* 5, no. 2 (2018): 6376–6380.
[24] Yu, LePing, XiaoHong Zhou, Lu Lu, XiaoLi Wu, and FengJun Wang. "Recent developments of nanomaterials and nanostructures for high-rate lithium ion batteries." *ChemSusChem* 13, no. 20 (2020): 5361–5407.
[25] Lu, Yong, Yanying Lu, Zhiqiang Niu, and Jun Chen. "Graphene-based nanomaterials for sodium-ion batteries." *Advanced Energy Materials* 8, no. 17 (2018): 1702469.

[26] Knoop, James E., and SeongkiAhn. "Recent advances in nanomaterials for high-performance Li–S batteries." *Journal of Energy Chemistry* 47 (2020): 86–106.
[27] Zoller, Florian, Daniel Böhm, Thomas Bein, and Dina Fattakhova-Rohlfing. "Tin oxide based nanomaterials and their application as anodes in lithium-ion batteries and beyond." *ChemSusChem* 12, no. 18 (2019): 4140–4159.
[28] Poorshakoor, Ehsan, and Mahdi Darab. "Advancements in the development of nanomaterials for lithium-ion batteries: A scientometric review." *Journal of Energy Storage* 75 (2024): 109638.
[29] Jin, Ting, Qingqing Han, Yijing Wang, and Lifang Jiao. "1D nanomaterials: Design, synthesis, and applications in sodium–ion batteries." *Small* 14, no. 2 (2018): 1703086.
[30] Zhu, Yue, Lele Peng, Zhiwei Fang, Chunshuang Yan, Xiao Zhang, and Guihua Yu. "Structural engineering of 2D nanomaterials for energy storage and catalysis." *Advanced Materials* 30, no. 15 (2018): 1706347.
[31] Chen, Jun Song, and Xiong Wen Lou. "SnO2-based nanomaterials: Synthesis and application in lithium-ion batteries." *Small* 9, no. 11 (2013): 1877–1893.
[32] Yang, Zhuoran. "Nanomaterials as well their applications and effects in batteries." *Highlights in Science, Engineering and Technology* 32 (2023): 73–82.
[33] Mateti, Srikanth, Irin Sultana, Ying Chen, Manikantan Kota, and MdMokhlesur Rahman. "Boron nitride-based nanomaterials: Synthesis and application in rechargeable batteries." *Batteries* 9, no. 7 (2023): 344.
[34] Wei, Qiulong, Fangyu Xiong, Shuangshuang Tan, Lei Huang, Esther H. Lan, Bruce Dunn, and Liqiang Mai. "Porous one-dimensional nanomaterials: Design, fabrication and applications in electrochemical energy storage." *Advanced Materials* 29, no. 20 (2017): 1602300.
[35] Ni, Jiangfeng, and Yan Li. "Carbon nanomaterials in different dimensions for electrochemical energy storage." *Advanced Energy Materials* 6, no. 17 (2016): 1600278.
[36] Geng, Hongya, Yan Peng, Liangti Qu, Haijiao Zhang, and Minghong Wu. "Structure design and composition engineering of carbon-based nanomaterials for lithium energy storage." *Advanced Energy Materials* 10, no. 10 (2020): 1903030.
[37] Yu, Shijin, Vincent Ming Hong Ng, Fajun Wang, Zhuohao Xiao, Cuiyun Li, Ling Bing Kong, Wenxiu Que, and Kun Zhou. "Synthesis and application of iron-based nanomaterials as anodes of lithium-ion batteries and supercapacitors." *Journal of Materials Chemistry A* 6, no. 20 (2018): 9332–9367.
[38] Liu, Tao, Liuyang Zhang, Bei Cheng, and Jiaguo Yu. "Hollow carbon spheres and their hybrid nanomaterials in electrochemical energy storage." *Advanced Energy Materials* 9, no. 17 (2019): 1803900.
[39] Zhao, Qinqin, Lisha Ma, Qiang Zhang, Chenggang Wang, and Xijin Xu. "SnO2-based nanomaterials: Synthesis and application in lithium-ion batteries and supercapacitors." *Journal of Nanomaterials* 2015, no. 1 (2015): 850147.
[40] Liu, Pengcheng, Kongjun Zhu, YanfengGao, Hongjie Luo, and Li Lu. "Recent progress in the applications of vanadium-based oxides on energy storage: From low-dimensional nanomaterials synthesis to 3D micro/nano-structures and free-standing electrodes fabrication." *Advanced Energy Materials* 7, no. 23 (2017): 1700547.
[41] Xia, Chenfeng, Yansong Zhou, Chaohui He, Abdoulkader IbroDouka, Wei Guo, Kai Qi, and Bao Yu Xia. "Recent advances on electrospun nanomaterials for zinc–air batteries." *Small Science* 1, no. 9 (2021): 2100010.
[42] Devaraju, Murukanahally Kempaiah, and Itaru Honma. "Hydrothermal and solvothermal process towards development of LiMPO4 (M= Fe, Mn) nanomaterials for lithium-ion batteries." *Advanced Energy Materials* 2, no. 3 (2012): 284–297.
[43] Ma, Lin, Kenville E. Hendrickson, Shuya Wei, and Lynden A. Archer. "Nanomaterials: Science and applications in the lithium–sulfur battery." *Nano Today* 10, no. 3 (2015): 315–338.
[44] Yun, Qinbai, Liuxiao Li, Zhaoning Hu, Qipeng Lu, Bo Chen, and Hua Zhang. "Layered transition metal dichalcogenide-based nanomaterials for electrochemical energy storage." *Advanced Materials* 32, no. 1 (2020): 1903826.
[45] Wang, Yiran, Huige Wei, Yang Lu, Suying Wei, Evan K. Wujcik, and ZhanhuGuo. "Multifunctional carbon nanostructures for advanced energy storage applications." *Nanomaterials* 5, no. 2 (2015): 755–777.

[46] Alcántara, Ricardo, Pedro Lavela, Gregorio Ortiz, and José L. Tirado. "Carbon nanomaterials for advanced lithium and sodium-ion batteries." In Rajib Paul, Vinodkumar Etacheri, Yan Wang and Cheng-Te Lin (Eds.) *Carbon Based Nanomaterials for Advanced Thermal and Electrochemical Energy Storage and Conversion*, pp. 335–355. Elsevier, 2019.

[47] Aziam, Hasna. "Recovery of nanomaterials for battery applications." *Waste Recycling Technologies for Nanomaterials Manufacturing. Topics in Mining, Metallurgy and Materials Engineering*, pp. 125–145. Springer, Cham. 2021. https://doi.org/10.1007/978-3-030-68031-2_5

[48] Baldwin, Richard S., J. Wu, and W. Bennett. *Nanomaterials for Advanced Lithium-Ion Battery Anodes*. Jenny Stanford Publishing, 2014.

[49] Wu, Songping, Cuiping Han, James Iocozzia, Mingjia Lu, Rongyun Ge, Rui Xu, and Zhiqun Lin. "Germanium-based nanomaterials for rechargeable batteries." *Angewandte Chemie International Edition* 55, no. 28 (2016): 7898–7922.

[50] Bruce, Peter G., Bruno Scrosati, and Jean-Marie Tarascon. "Nanomaterials for rechargeable lithium batteries." *Angewandte Chemie International Edition* 47, no. 16 (2008): 2930–2946.

5 Lightweight Materials Contribution for the Development of Battery Management Systems in Electric Vehicles

Usha Raghunathagouda Patil, Rudresha S.J., Sekar R., Kiran Kumar G.R., and Shanmuka M.C.

5.1 INTRODUCTION

5.1.1 Overview of Electric Vehicles (EVs) and Battery Management Systems (BMSs)

The BMS assures safe and best possible operations of EVs and energy storage systems actively functioning in the background to carry out mostly nonequilibrium functions such as state monitoring, charge control, cell balancing, and thermal management [1–3]. Various kinds, including electric, thermal, and electrothermal, enhance BMS performance through precise estimation of state of charge (SoC), state of health (SoH), and internal temperature for efficient battery monitoring and predictive maintenance [2,4]. Model-based and data-driven technologies, including machine learning (ML), also enhance SoC and SoH estimation, and optimize the battery's charging cycle and lifespan [3]. Real-time monitoring and safety are bolstered by the Internet of Things (IoT) and cloud computing, particularly for EVs [5]. A functional block diagram of a BMS is shown in Figure 5.1.

These advancements facilitate remote diagnostics, predictive analysis, and autonomous control systems, improving overall battery reliability. Furthermore, the increase in renewable energy-based charging stations and fast-charging stations facilitates the swift growth of the EV market, reducing charging time and improving energy efficiency. The development of bidirectional charging and vehicle-to-grid (V2G) technology is also transforming EVs as energy storage assets, facilitating grid stability and utilizing sustainable energy solutions.

The present research problems are mainly related to battery modeling and address various challenges in managing thermal conditions and BMS design for the enhancement of the performance, lifespan, and safety of future energy storage systems [3,4]. These technologies will find immense use in the automobile and energy future as EVs are being accepted in larger numbers.

5.1.2 Need for Lightweight Materials in EV Design

Lightweight materials support increasing drive and reducing the cost of the battery in an EV [6]. Suggested materials include aluminum alloy, high-strength steel alloys, and composites with carbon fibers [7]. When there is an advancement in the new version of batteries and motors then

 DOI: 10.1201/9781003596097-5

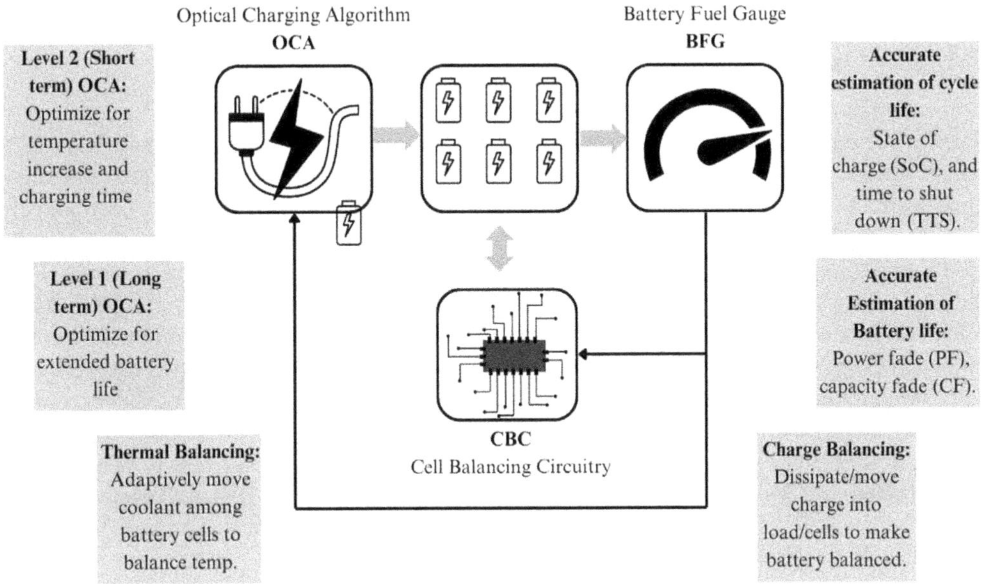

FIGURE 5.1 Functional block diagram of a battery management system with three important components of a BMS which are the battery fuel gauge, optimal charging algorithm, and cell-balancing circuitry [5].

the cost of expensive materials will be reduced [6]. Life cycle assessments (LCAs) indicate that the replacement of steel with aluminum or glass fiber-reinforced plastics reduces the emissions of greenhouse gases, and carbon fiber demonstrates outstanding performance in structural parts. EV-specific characteristics will be considered for future research and full vehicle lifecycles will be assessed.

With the shift in the car industry toward EVs for better performance and longer driving range, 10% savings in weight would result in 6–8% savings in energy [8,9]. Sustainable materials can also reduce safety risks since fiber-reinforced polymers have a high strength-to-weight ratio [8]. Even though there is a lot of promise in lightweight technology and enhancements on the powertrain, eventually these are where the bottlenecks for zero-carbon electricity will be with such an accelerating rate of adoption for EVs. According to Zhou et al. [10], other options include battery reusability, interoperable charging, shared mobility, and new frame designs. The main objective of this chapter is to introduce innovative lightweight materials by examining advanced lightweight materials such as advanced polymer composites, metal–organic frameworks (MOFs), and nanostructured components that can improve the design and functionality of the BMS for EVs. The enhanced BMS design and performance emphasizes the capability of lightweight materials to enhance structural integrity, thermal performance, and electrical efficiency of the BMS, thus maximizing energy efficiency and safety in EVs. Environmental impacts are assesse by discussing how sustainable and recyclable products can minimize the environmental impact of EV production and use, according to global sustainability goals. Challenges and opportunities are highlighted by determining the manufacturing challenges, which include machinability and cost-effectiveness of materials; and also highlighting advanced techniques, including additive manufacturing (AM) and artificial intelligence (AI)-driven material design, for addressing the challenges. The future of EVs are discussed including how lightweight materials are going to change the course of EV technology and their future in enhancing energy density, improving range, and developing the overall sustainability of EVs.

5.2 TECHNOLOGICAL DEMANDS ON BATTERY MANAGEMENT SYSTEMS

5.2.1 Key Challenges in Next-Generation BMS

Next-generation BMSs for EVs will face challenges of increasing energy density, range, thermal management, efficiency, and safety. These are all compounded by the limited knowledge of the internal states of the battery, lack of adaptability to extreme conditions, and ineffective predictive maintenance. Advanced BMS solutions are researching combined battery modeling, heat and electric modeling, and ML to enhance state estimation and fault diagnosis. Temperature management and cell balancing are also critical for the longevity and better performance of batteries. Future BMS technologies could encompass new computer solutions, data models, and smart tools, such as the IoT and blockchain, to further upgrade the functions. After lithium, post-lithium batteries and wireless charging systems will also contribute to overcoming the present problems associated with energy storage and EV charging systems. Managing heat, especially when power changes, is essential to prolonging the life of lithium-ion batteries and keeping them safe [11,12]. Good BMS designs should monitor battery voltage, current, charge/discharge levels, and heat levels to better improve the performance of EVs and solve the problems being faced currently by EVs and renewable energy storage systems [13]. More research must be done to solve such problems and create better, more sustainable BMS solutions.

5.2.2 Role of Lightweight Materials in Overcoming Challenges

Lightweight materials are vital for enabling the use of less fuel during car manufacture, while reducing toxic emissions, and enhancing safety and sustainability [14,15]. New materials, from strong steels to alloys of aluminum and magnesium combined with carbon fiber composites, are used in the body of vehicles to enable them to be lightweight and also to enhance their efficiency in braking capability and safety in collisions [15,16]. Today, while vehicle use is energy-intensive, future vehicle electrification could shift the focus of attention to material production [17]. Research continues to evolve versatile materials and also optimize manufacturing processes to achieve a safer, more environment-friendly vehicle [15,16].

The car industry is experiencing pressure to become efficient due to fuel economy rules and regulations on emissions. There is an increased usage of lightweight materials [14,17]. Aluminum, magnesium alloy, and composites are used as substitutes for regular metals. This is attributed to safety requirements, environmental concerns, and also due to the demand of consumers for better designs and less noise [14,17]. Hybrid and electrical cars also need these materials so that the overall efficiency can be improved [14, 17].

5.3 ADVANCED LIGHTWEIGHT MATERIALS FOR BATTERY SYSTEMS

5.3.1 Polymer Composites

The polymer composite strength suggests their growing application in BMS structural as well as functional components. Composites made of polymers have high strength-to-weight ratios, corrosion resistance, and material design flexibility. Therefore, their uses include in enclosures, separators, and other thermal management systems [18].

The mechanical characteristics have also been enhanced by carbon or glass fiber reinforcements in composites. Functional composites may include phase change materials (PCMs) for heat storage and release to control temperature, thus improving the thermal stability of battery packs [19]. These properties have increased the performance and reliability of BMS in energy storage systems and EVs.

Polymer composites have various significant advantages. They are also lighter in weight, reducing the overall weight of battery packs, thus making energy consumption better and enhancing the capacity to travel greater distances. In addition to rust prevention, they preserve cabling within their indigenous insulated state for extensive periods, thereby prolonging component life even with harsh working environments. The combined behavior of these stiff materials and flexible configuration gives us the potential to fabricate complicated structures meeting specific performance and safety needs. It includes hard cases to pass through battery cells in order to make the system safe. The polymer composites, particularly the boron nitride nanosheet-based composites, exhibit superior thermal management characteristics. They allow heat to flow but stop electricity from flowing.

The materials offer adequate pliability and heat conductivity to use in high-power devices, for instance, EV battery packs. The thin films had a maximum thermal conductivity of 16.3 W/(m·K), making them highly efficient in the regulation of heat from batteries, thereby avoiding overheating and enhancing safety. The main evolution that can be seen to enhance performance, safety, and longevity in BMS is the addition of these composites to their thermal and structural components.

5.3.2 Metal–Organic Frameworks (MOFs)

New materials, specifically MOFs, improve BMSs. They offer superior energy storage and safety. Large surface area and special holey structures of MOFs can be easily tailored, which can further entrap more ions; therefore, they are ideal for boosting energy density in batteries. This ability is especially useful for EVs. Higher energy density allows the use of smaller and more efficient batteries, which helps performance and driving range. Also, their design improves how ions move, increasing battery efficiency and electrochemical performance.

Using MOFs in BMSs makes battery packs lighter, thus improving the efficient use of energy by vehicles. They are lighter and their energy storage is better, which enables the use of smaller yet high-performance batteries. Using MOFs also makes the batteries safer since they stabilize the system and prevent overheating, which is critical for EVs, particularly in hot or stressful situations. MOFs also enhance electrolyte properties that facilitate the easy movement of ions. They lead to fast charging and efficient energy. The chemical stability ensures that the battery lasts for longer periods, thus there is less frequent maintenance required. In a nutshell, MOFs are a good solution to the improvement of energy density, safety, and efficiency in the battery systems of EVs and other advanced applications.

5.3.3 Nanomaterials (Graphene, Carbon Nanotubes)

Nanomaterials are critical; however, most critical among them are the uses of graphene and carbon nanotubes (CNTs) to improve the features of lithium-ion batteries. They exhibit excellent electrical and thermal properties, which help determine how the battery will dissipate its heat. The incorporation of CNTs in PCMs greatly increases the heat transfer rate. This enables maintaining appropriate temperatures within the cells of the battery and ensures that the cells do not get overheated [20]. Nanomaterials are thus effectively used in air, liquid, and PCM-based cooling techniques to maintain excellent heat control and keep the batteries running [21].

Besides thermal control, nanomaterials are also useful in preventing thermal runaway, which is a significant safety issue found in battery systems. Such encapsulated inorganic PCMs with CNTs have dual functionality in which the increase in temperature is decelerated, also, temperature uniformity improves on account of battery operation. Such types of materials also serve to prevent thermal runaway as, through them, the cell triggering time is delayed, which rapidly propagates heat forward, which greatly improves battery safety [22].

TABLE 5.1
Materials Used in Battery Management Systems (BMSs) for Electric Vehicles

Material	Properties	Advantages in BMS	References
Polymer composites	High strength-to-weight ratio, corrosion resistance, phase change materials (PCMs), fiber reinforcement (e.g., carbon, glass fibers)	- Enhance thermal management and safety - Lightweight - Improve performance and reliability of BMS	[17,21,23]
Metal–organic frameworks (MOFs)	Large surface area, holey structures, ion trapping capabilities	- Improve energy density and battery efficiency - Lightweight - Increase safety and stability	[19,23]
Nanomaterials (graphene, CNTs)	High heat and electrical conductivity, enhance thermal and electrical properties	- Improve heat transfer - Prevent thermal runaway - Enhance battery safety and efficiency	[20,22]

Recent advancements in lithium-ion battery technology have focused on blending nanomaterials into electrodes to make performance better. Nanostructured electrodes having carbon nanotube fillers improve heat transfer; this is crucial for high-power batteries. Materials used in BMSs for EVs are tabulated in Table 5.1. Moreover, core–shell nanostructures have been prepared to circumvent issues of low mass density and poor electrical conductivity while retaining the merits of nanostructures. Coatings with conformal graphene on nanostructured cathodes increase both the packing density and cycling stability, and hence address the main constraints in battery design [23]. As research advances, cost–performance–stability trade-offs of nanomaterials will be crucial for their sustainable commercialization in battery technology.

5.4 THERMAL MANAGEMENT AND SAFETY: THE ROLE OF LIGHTWEIGHT MATERIALS

5.4.1 Heat Dissipation in EV Batteries

Thermal management is very important for optimal battery performance and lifespan for lithium-ion batteries used in EVs. They typically operate between 15–35 °C, and different thermal management systems such as conductive cooling, liquid cooling, and other cooling methods using phase-change technology have been proposed by adding technology available already for temperature control and efficiency [25]. PCM systems are particularly favored because of their lightweight properties and ability to maintain uniform temperatures during charging and discharging cycles [9]. Incorporation of thermally conductive particles and metal fins may further enhance the heat transfer in these systems.

Heat pipes are another promising battery thermal management system (BTMS) technology, which offers effective heat transfer and uniform temperature distribution [26]. Further enhancements of cooling efficiency can be obtained by incorporating heat pipes with PCMs and nanofluids, which eliminates the ambient temperature effects and scales up [27]. These technologies complement the prevailing air and liquid cooling that commercial EVs use today. Lightweight materials enhance the heat conduction and dissipation in BTMS, which reduces the overall weight with improved heat transfer. Distributed forced convection cooling, along with these materials, is an economical method for uniform temperature and voltage distributions [28]. Digital solutions for thermal

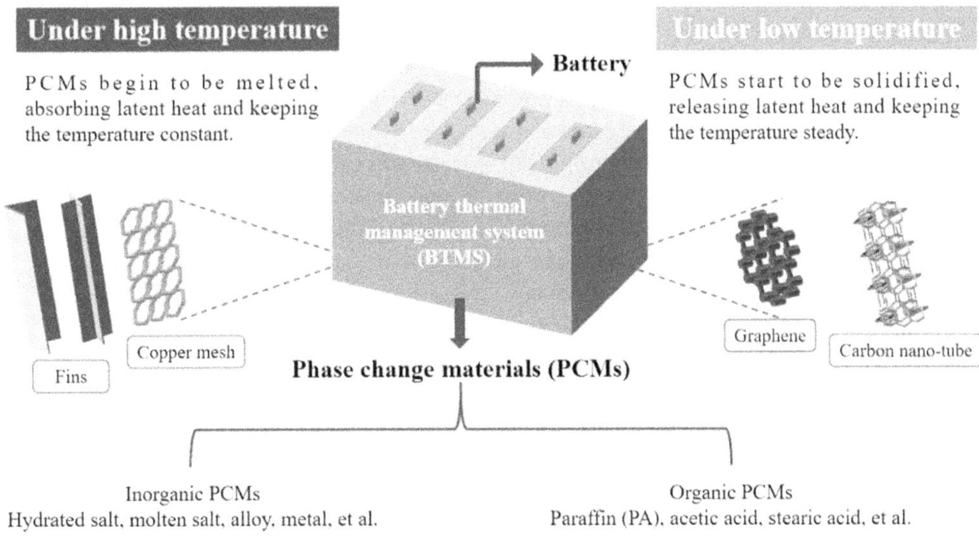

FIGURE 5.2 Operating principles of PCMs in battery thermal management systems. (Adapted from Ref. [30].)

management are also being pursued to improve temperature regulation under varying conditions, further improving battery safety and performance [29]. The operating principles of PCMs in BTMSs [30] are shown in Figure 5.2.

5.4.2 Reducing Risks of Thermal Runaway

Reducing the risk of thermal runaway (TR) in lithium-ion batteries is very significant for safety reasons in EVs and has massive applications. TR, which is attributed to mechanical, electrical, or thermal abuse, can lead to fire and heat emission [31]. Advanced materials used include ceramic-hydrogel nanocomposites for cooling and insulation against fires as well as responsive polymer membranes that suppress premature thermal release [16].

These enhanced safety, delaying the onset of TR and reducing its impact using passive and active control methods and predictive models [32]. Graphene-enhanced composite PCMs (CPCMs) enhance thermal conductivity and extend the TR triggers by more than 113 minutes in battery modules [33]. Structure-enhanced CPCM modules with flame retardants show better prevention of TR propagation, and tubular designs outperform traditional modules. With an increase in the energy density of lithium-ion batteries, safety risks also increase. Balancing performance and safety is always a crucial goal; there is a research focus on the incorporation of advanced materials and innovative designs that not only enhance safety but will also maintain structural integrity.

5.5 SUSTAINABILITY AND RECYCLING: BIO-BASED POLYMERS AND COMPOSITES

5.5.1 Environmental Impact of Lightweight Materials

Sustainable composites incorporating natural and recycled carbon fibers sourced from bio-based and recyclable materials are gaining increasing significance in environmental applications, particularly within the automotive and aviation industries. Agricultural residues are frequently utilized as raw materials, aligning with circular economy principles by conserving energy and minimizing emissions, especially in EVs and aircraft [8,34]. The evaluations of environmental impacts are conducted through LCA so that proper selection of materials and product design is required for greater sustainability. Cropped biocomposites, made of crop-based plastic and plant fibers, provide an opportunity for the reduction of carbon emissions and energy use. Additionally, ultralight cellulose-based materials, including foams and aerogels, exhibit great potential as packaging and cleaning materials in green manufacturing [35]. With respect to process efficiency and mechanical strength, AM is an efficient method of utilizing materials and energy [36]. There are further examples of improvements in material properties, addition of multifunctionality, and enhancement to sustainability that are valid areas for future advances in different industries for green manufacturing and lightweight engineering applications.

5.5.2 Challenges and Solutions for Bio-Based Materials in EVs

Bio-based materials are picking up pace toward the sustainability of EVs, however durability, machinability, and cost-effectiveness remain a big challenge. Polymers from lignocellulosic biomass help minimize the environmental footprint, but their mechanical properties and economic feasibility need to be enhanced for them to be used extensively [37]. Among the most environmentally viable alternatives, these plant fiber polymer composites remain the appropriate substitutional candidates for automotive components, while raw material supply constrictions and high production costs imply limited commercial application [38]. Multi-disciplinary research fused with innovative strategies will take a prime role in addressing this failure of the increased use of bio-based materials for noise and vibration control in EVs and hybrid electric vehicles [8].

Such novel breakthroughs in bio-based materials provide new potential for EVs. Researchers developing high-performance bio-based polymers from cellulose, lignin, and vegetable oils are attempting to match or exceed the properties of conventional synthetic polymers for appropriate usage. The mechanical strength, thermal stability, and environmental endurance are further improved by the application of nanostructured additives and polymer blending, making these materials even more competitive. In the bio-composite engineering synthesis, weight-to-strength ratio optimization can further enhance their suitability as structural components for EVs. As an outcome, the development of bio-composite materials fabricated by means of additive (three-dimensional, 3D) printing forms that are easy to process also contribute significantly to the reduction in waste and energy consumption. Various bio-based plastic materials are being evaluated to find their applications in lithium-ion batteries as binders, separators, and electrolytes to serve as safer and more sustainable alternatives for such batteries. Such progress is in line with the prospective demand for cleaner energy storage systems. The future of bio-based materials for advanced manufacturing and engineering points toward a new era of a sustainable and high-performance EV industry.

5.6 NEXT-GENERATION BATTERY TECHNOLOGIES

5.6.1 Solid-State Batteries

Solid-state batteries are a future-generation energy storage device. They illustrate high energy density, enhanced safety, and exceptional performance. With excellent thermal stability and wide electrochemical windows, they are well-suited for applications in EVs and large-scale energy storage systems [39]. Significant progress in the design of solid-state electrolytes, including composites comprising a combination of ceramic scaffolds and polymer layers, have enabled remarkably high energy densities, for example, up to 506 Wh kg^{-1} and 1514 Wh L^{-1}, in pursuit of light and powerful applications. Some key challenges remain, mainly at the electrode–electrolyte interface, such as chemical and mechanical instabilities, ionic transport limitations, and lithium dendrite growth, which reduce the performance. High-capacity lithium-rich oxide cathodes and advanced interface engineering are targeted for enhanced cycling stability and energy storage capabilities [40].

These problems are being tackled by innovative interface designs, strain-tolerant architectures, and advanced fabrication methods. Advanced characterization techniques also help to solve interfacial instabilities and optimize ionic pathways. Such developments are important for obtaining reliable, scalable solid-state battery technologies for EVs and renewable energy systems [41,42].

5.6.2 Lithium-Sulfur (Li-S) Batteries

Due to their advantage of energy density-promoting sustainability, Li-S batteries are being explored by researchers around the world due to the high theoretical energy density of 2600 Wh kg^{-1} and the specific capacity of 1675 mAh g^{-1}, which makes them suitable for next-generation energy storage systems [43]. They serve as a sustainable departure from conventional lithium-ion-based batteries.

Role of Lightweight Materials in Battery Longevity and Safety: Despite all these possibilities, their commercialization faces some challenges such as polysulfide dissolution, low sulfur conductivity, and lithium dendrite formation [44]. Lightweight materials and nanostructured innovations are paving the way forward in overcoming these issues.

The key progress includes:

- Nanostructured Carbon Hosts: The sulfur hosts include porous carbon-based materials and 3D nanocomposites, which help improve conductivity, immobilize polysulfides, and increase efficiency in general [45].
- Exploring New Structures: In efforts to obtain both high sulfur loading and reduced volumetric shrinkage, sulfur-carbon composites and sulfur-polymer composites offer better cycling stabilities [46].
- Next-Generation Materials: Nitrogen/sulfur-codoped graphene sponge electrodes reported high S utilizations and stable cycling over 500 cycles with more than 1200 mAh g^{-1} at 0.2 °C [45].
- Separator and Anode Modifications: Separator coatings and lithium anode protection are techniques that address polysulfide shuttling and improve safety [47].

However, waiting for such optimism is a bittersweet thing that really makes one frown since a long, arduous path must still be taken before a fully-fledge working platform-stable, long-term and scalable-would unfold to exploit all the potentially magnificent features of Li-S batteries into EVs and portable gadgets.

5.7 MANUFACTURING TECHNIQUES FOR LIGHTWEIGHT MATERIALS

5.7.1 Innovative Production Routes

AM or 3D printing revolutionized the world of production and brought into existence extremely complex and light structures from metals, polymers, and ceramics that could never be created using conventional processes. These processes consist of the material jetting, extrusion, and powder bed fusion of substances like metals, thermoplastics, and ceramics, which also minimizes the formation of waste as well as availing the scope of creating unique designs [21].

AM finds particular application in industries such as the aerospace and automotive sectors, where energy-efficient and lighter weight high-strength materials such as aluminum alloys and carbon-fiber composites contribute to decreased emissions and improved efficiency [14]. Recent technologies, such as composite structures and volumetric printing, can print with improved scalability and accuracy even using advanced materials such as live cells and spacecraft-grade metals [49]. However, accuracy, surface finish, and material uniformity remain issues. Advances in computational design and improved feedstocks need to be worked on to mitigate such challenges [50]. Polymer-derived ceramics (PDCs) are receiving interest due to their potential to be tailored to specific properties with high performance, but they need to be optimized for uniformity and scalability [51]. Such challenges notwithstanding, the strength of AM lies in being able to create lightweight high-performance parts.

5.7.2 Cost-Efficiency and Sustainability in Production

The application of lightweight materials in vehicle production is good for sustainability in a number of ways, including mass reduction and improvement of strength, among others, and also enhancing production efficiency [52]. Issues regarding the environmental effects of such materials at every phase of the product's life cycle exist. At times, these effects can be negative [53]. Material choice should be guided by cost, ecological impact, and the processes used to make them [53]. As more EVs are adopted, the stage of producing materials will become more important in terms of lifecycle effects, while fuel efficiency is paramount in reducing total energy use and emissions [17].

The innovative design of any automobile component often involves reinforcing automotive parts using high-strength steel-to-aluminum alloys or carbon-fiber-reinforced polymers for increased efficiency [14]. AM of components enables enhancing the designs while minimizing emissions [54]. However, advancements in battery and motor technologies may reduce the need to replace traditional materials such as steel [6]. Yet, since the development of the batteries and motors is an ongoing area, the environmental benefit of weight reduction materials could be lost through the ongoing consideration of traditional materials such as steel [6]. Furthermore, although latticework designs reduce weight, they may increase the production time and contribute to the environmental production cost [55]. An even more radical strategy would be to create long-lived, shared EVs with significantly reduced mass and aerodynamic drag, which can drastically minimize material consumption and environmental expenses. This discussion shows the intricate trade-offs required in attaining sustainability in car design and production.

5.8 COMPUTATIONAL MODELING AND AI IN MATERIAL OPTIMIZATION

5.8.1 AI-Driven Material Discovery and Optimization

ML and AI are transforming the discipline of materials science, speeding up material discovery and optimization for lightweight solutions. These technologies substantially obviate the requirement to conduct time-consuming, as well as costly, physical experimentation, by efficiently surveying large material search spaces that would otherwise be required [56]. ML techniques are critical in material discovery, such as the prediction of material properties, inverse design, and characterization of new materials. Evolutionary algorithms, such as genetic algorithms, are very effective in optimizing materials for specific performance requirements, and discovering new materials with tailored properties [56].

ML also facilitates reducing material dimensionality in material design, accelerates the evaluations of property, and will even suggest previously completely unheard-of material architectures [57]. Introducing artificial intelligence coupled with computational models such as density functional theory (DFT) into material research further assists with intensifying discovery efforts for further heightening both efficacy and precision. Surrogate models created by ML algorithms can simulate the material's behavior rapidly, but in the process do not sacrifice accuracy.

It also assists researchers with experimental design, data analysis, and extraction of crucial information from large datasets; the result is often a breakthrough that wouldn't have occurred even based on human intuition [58]. This push in AI and ML toward the development of materials will revolutionize industries by potentially making material discoveries faster, more efficient, and even more cost-effective. Challenges do still exist, but these new technologies are making materials research turn into a process of teaming human expertise with that of machine-driven innovation.

5.8.2 Simulation and Design Tools

Computational modeling and AI significantly cut the time and cost of prototyping; therefore, the design of materials is getting revolutionized. Digital materials design (DMD) platforms use fundamental scientific principles combined with advanced computing capabilities to discover material behaviors never known before, thereby allowing designs for experiments [59]. As stated by Kivi et al. [60], tools in multi-scale modeling such as DFT and atomistic simulations accelerate the research on complex alloys without costly experimental work. AI techniques such as ML and deep learning have a greater advantage over large datasets for the prediction of mechanical properties and optimal designs in material applications in most fields, as stated by Huang et al. [61].

Advanced simulations with nonlinear behavior physics enable the study of how materials and structures respond to various conditions for a more profound understanding of their dynamic changes [62]. This integrated approach leads to faster, less expensive material development processes. A new development has been working toward predicting physical properties and optimizing compositions with the help of ML models, which has greatly reduced the design cycle time [63]. It has been transforming material informatics, molecular dynamics, and finite element analysis and has been essential to optimize the manufacturing process. Smart manufacturing will also include the development of their respective digital twins [64] to expedite the development process with potential novel high-performance materials within the design-to-manufacture process so that it incurs no friction.

5.9 CASE STUDIES AND REAL-WORLD APPLICATIONS

5.9.1 Commercial EVs Using Advanced Lightweight Materials

Advanced lightweight materials are necessary for the efficiency and performance of EVs. Some of the most critical materials include high-strength steels, aluminum, magnesium alloys, and carbon-fiber-reinforced polymers. The weight is reduced, and subsequently, energy consumption decreases and smaller battery sizes can be used [14]. However, with further advancements in battery energy density and electric motor technologies, the demand for expensive lightweight materials may also decline with time [6]. In addition, the use of smart materials with mechanical systems and electronics improves the functionality of vehicles, thereby making the advanced systems more efficient and adaptive [14]. BMSs are becoming smart by including AI and ML, which enables higher accuracy in predictions for the battery, such as SOH, SOC, and state of power (SOP) [65]. All such technological advancements in materials and BMS lead to overcoming challenges in terms of energy efficiency, safety, emissions, and cost-effectiveness in the automotive sector.

The phenomenal progress in BMS is also reflected in recent work, especially the innovative usage of a single DC-DC converter for quick transfer and balancing between the battery cells [66,67]. The DEMOBASE project is important, wherein the development with seamless integration has established digitalization as an essential element to enhance the design, safety, and management of EV batteries. Engineers are coming up with lightweight, fire-resistant polymer composites that enable safety and the circular economy in EVs. These will not only help extend driving ranges and raise the safety levels of EVs but will also ensure a key position in fostering the global transition toward zero-emission mobility [8].

5.9.2 Key Insights and Future Prospects

BMSs are the backbone for the safe and efficient operation of EV batteries. However, the present systems have several limitations associated with sensing accuracy, state estimation, and fault diagnosis. This limitation puts a restriction on maximizing the battery performance and safety. It is expected that improvements in computing power, vehicle communication technologies, and multiphysics modeling will change the game in the functionality of BMSs. Improved systems could facilitate closer monitoring of battery health, real-time performance adaptation, and predictive maintenance, among other things, thereby dealing with extreme environmental and operational conditions [31].

Major drivers while developing EVs include their lightweight materials. Advanced high-strength steels along with aluminum- and magnesium-based alloys help in reducing the weight to optimum; carbon-fiber-reinforced polymers are essential for structure integrity and safety [14,48]. Such materials also make for improved energy efficiency and longer driving ranges. Expanding the EV market necessitates such improvements; the price and scalability of such materials, however, is currently a major challenge. Developments in manufacturing processes—including AM and hybrid material solutions—promise significant strides toward overcoming these obstacles to their wider industrial adoption.

New battery technologies such as lithium-metal and post-lithium seem to portend a sea change in the realization of higher energy density and therefore charging efficiency. Wireless charging systems are being accepted increasingly quickly as yet another complementary innovation, promoting user convenience and integration with the system [12]. Combining these advanced battery technologies with lightweight materials could optimize design for performance, energy efficiency, and sustainability. This synergy can address key issues like degradation of the battery and inefficiency in energy usage so that the ecosystem for the EV is more reliable and environmentally friendly.

Smart technologies such as the IoT, blockchain, and cloud computing transform BMSs into intelligent and multifunctional systems, enabling real-time data collection, advanced fault detection, and decision-making capabilities. Digital twins and AI-driven modeling are also essential tools for simulating and testing new materials and new technologies, cutting prototyping costs and time by substantial amounts [12,68]. As these innovations advance further, they will accelerate sustainable, high-performance EV use and promote a global switch to zero-emission modes of transportation.

5.10 CONCLUSION

Advanced BMSs and the integration of lightweight materials are two key drivers for the development of EVs. With such integration, the critical issues of energy efficiency, optimized performance, and a healthy environment can be addressed. Materials such as advanced high-strength steels, aluminum and magnesium alloys, and carbon-fiber-reinforced polymers can bring significant weight savings to a vehicle. At the same time, these materials provide higher structural strength and safety so that improved performance will not compromise reliability or occupant protection.

At the same time, advancements in BMS technology have totally reengineered the handling of EV batteries with highly sophisticated features, such as very accurate SoC and SoH estimation, very efficient energy balancing, and prognostic fault detection. It also involves predictive maintenance and thermal management, and all of these are totally integrated within IoT platforms. The union of light materials and advanced BMS technologies is propelling the rapid pace with which EVs are becoming increasingly common and transforming transport globally.

The future of EVs will be shaped by advancements in light materials and emerging battery technologies. The main areas are structural battery composites, bio-based polymers, and recyclable flame-retardant materials, which are all set to enhance performance and sustainability. Battery technology breakthroughs like lithium-metal batteries and wireless charging technology will provide higher energy density and greater user convenience.

Academia, industry, and policymakers have to work together continuously to surmount such scalability and cost barriers. Standardization of procedures, and optimizing manufacturing processes and expenses will be feasible, allowing such transformational technologies to gain wider acceptance. Advanced BMS and lightweight materials continue to be at the heart of the realization of a greener and more sustainable transportation future.

REFERENCES

[1] Xing, Y.; Ma, E. W. M.; Tsui, K. L.; Pecht, M. Battery Management Systems in Electric and Hybrid Vehicles. *Energies* **2011**, *4* (11), 1840–1857. https://doi.org/10.3390/en4111840

[2] Liu, K.; Li, K.; Peng, Q.; Zhang, C. A Brief Review on Key Technologies in the Battery Management System of Electric Vehicles. *Frontiers of Mechanical Engineering* **2018,** *14* (1), 47–64. https://doi.org/10.1007/s11465-018-0516-8

[3] Bhushan, N.; Mekhilef, S.; Tey, K. S.; Shaaban, M.; Seyedmahmoudian, M.; Stojcevski, A. Overview of Model- and Non-Model-Based Online Battery Management Systems for Electric Vehicle Applications: A Comprehensive Review of Experimental and Simulation Studies. *Sustainability* **2022**, *14* (23), 15912. https://doi.org/10.3390/su142315912

[4] Shaik, N .; Dhanamjayulu, C. A Review of Battery Energy Storage Systems and Advanced Battery Management System for Different Applications: Challenges and Recommendations. *Journal of Energy Storage* **2024**, *86*, 111179–111179. https://doi.org/10.1016/j.est.2024.111179

[5] Balasingam, B.; Ahmed, M.; Pattipati, K. Battery Management Systems—Challenges and Some Solutions. *Energies* **2020**, *13* (11), 2825. https://doi.org/10.3390/en13112825

[6] Burd, J. T. J.; Moore, E. A.; Ezzat, H.; Kirchain, R.; Roth, R. Improvements in Electric Vehicle Battery Technology Influence Vehicle Lightweighting and Material Substitution Decisions. *Applied Energy* **2020**, *283*, 116269. https://doi.org/10.1016/j.apenergy.2020.116269

[7] Lipman, T. E.; Maier, P. Advanced Materials Supply Considerations for Electric Vehicle Applications. *MRS Bulletin* **2021**, *46* (12), 1164–1175. https://doi.org/10.1557/s43577-022-00263-z

[8] Mohanty, A. K.; Vivekanandhan, S.; Tripathi, N.; Roy, P.; Snowdon, M. R.; Drzal, L. T.; Misra, M. Sustainable Composites for Lightweight and Flame Retardant Parts for Electric Vehicles to Boost Climate Benefits: A Perspective. *Composites Part C: Open Access* **2023**, *12*, 100380. https://doi.org/10.1016/j.jcomc.2023.100380

[9] Wazeer, A.; Das, A.; Abeykoon, C.; Sinha, A.; Karmakar, A. Phase Change Materials for Battery Thermal Management of Electric and Hybrid Vehicles: A Review. *Energy Nexus* **2022**, *7*, 100131. https://doi.org/10.1016/j.nexus.2022.100131

[10] Zhou, X.; Jiang, J.; Hu, Z.; Hua, L. Lightweight Materials in Electric Vehicles. *International Journal of Automotive Manufacturing and Materials* **2022**, *3*. https://doi.org/10.53941/ijamm0101003

[11] Yue, Q. L.; He, C. X.; Wu, M. C.; Zhao, T. S. Advances in Thermal Management Systems for Next-Generation Power Batteries. *International Journal of Heat and Mass Transfer* **2021**, *181*, 121853. https://doi.org/10.1016/j.ijheatmasstransfer.2021.121853

[12] Waseem, M.; Ahmad, M.; Parveen, A.; Suhaib, M. Battery Technologies and Functionality of Battery Management System for EVs: Current Status, Key Challenges, and Future Prospectives. *Journal of Power Sources* **2023**, *580*, 233349. https://doi.org/10.1016/j.jpowsour.2023.233349

[13] Habib, A. K. M. A.; Hasan, M. K.; Issa, G. F.; Singh, D.; Islam, S.; Ghazal, T. M. Lithium-Ion Battery Management System for Electric Vehicles: Constraints, Challenges, and Recommendations. *Batteries* **2023**, *9* (3), 152. https://doi.org/10.3390/batteries9030152

[14] Taub, A. I.; Luo, A. A. Advanced Lightweight Materials and Manufacturing Processes for Automotive Applications. *MRS Bulletin* **2015**, *40* (12), 1045–1054. https://doi.org/10.1557/mrs.2015.268

[15] Guo, H.; Zhou, X.; Liu, Z. Advanced Lightweight Structural Materials for Automobiles: Properties, Manipulation, and Perspective. *Science of Advanced Materials* **2024**, *16* (5), 563–580. https://doi.org/10.1166/sam.2024.4686

[16] Zhang, W.; Xu, J. Advanced Lightweight Materials for Automobiles: A Review. *Materials & Design* **2022**, *221* (110994), 110994. https://doi.org/10.1016/j.matdes.2022.110994

[17] Keoleian, G. A.; Sullivan, J. L. Materials Challenges and Opportunities for Enhancing the Sustainability of Automobiles. *MRS Bulletin* **2012**, *37* (4), 365–373. https://doi.org/10.1557/mrs.2012.52

[18] Arankalle, A. Advances in Light Weight Materials for Body-In-White (BIW). Edited by: Bajpai, R.P., Chandrasekhar, U., *Innovative Design and Development Practices in Aerospace and Automotive Engineering*. Lecture Notes in Mechanical Engineering, pp. 517–525, **2016**. Springer, Singapore. https://doi.org/10.1007/978-981-10-1771-1_55

[19] Pegoretti, A.; Fredi, G. Thermal Management with Polymer Composites. *Express Polymer Letters* **2019**, *13* (10), 844. https://doi.org/10.3144/expresspolymlett.2019.73

[20] Shivram, S.; Harish, H. Impact of Dual Nano-Enhanced Phase Change Materials on Mitigating Thermal Runaway in Lithium-Ion Battery Cell. *Case Studies in Thermal Engineering* **2024**, *60*, 104667. https://doi.org/10.1016/j.csite.2024.104667

[21] Kumar, P.; Chaudhary, D.; Varshney, P.; Varshney, U.; Yahya, S. M.; Rafat, Y. Critical Review on Battery Thermal Management and Role of Nanomaterial in Heat Transfer Enhancement for Electrical Vehicle Application. *Journal of Energy Storage* **2020**, *32*, 102003. https://doi.org/10.1016/j.est.2020.102003

[22] Dai, X.; Ping, P.; Kong, D.; Gao, X.; Zhang, Y.; Wang, G.; Peng, R. Heat Transfer Enhanced Inorganic Phase Change Material Compositing Carbon Nanotubes for Battery Thermal Management and Thermal Runaway Propagation Mitigation. *Journal of Energy Chemistry* **2023**, *89*, 226–238. https://doi.org/10.1016/j.jechem.2023.10.001

[23] Chen, J.; Huang, X.; Sun, B.; Jiang, P. Highly Thermally Conductive yet Electrically Insulating Polymer/Boron Nitride Nanosheets Nanocomposite Films for Improved Thermal Management Capability. *ACS Nano* **2018**, *13* (1), 337–345. https://doi.org/10.1021/acsnano.8b06290

[24] Zhang, Y.; Yu, L.; Zhang, X.-D.; Wang, Y.-H.; Yang, C.; Liu, X.; Wang, W.-P.; Zhang, Y.; Li, X.-T.; Li, G.; Xin, S.; Guo, Y.-G.; Bai, C. A Smart Risk-Responding Polymer Membrane for Safer Batteries. *Science Advances* **2023**, *9* (5), 1–9. https://doi.org/10.1126/sciadv.ade5802

[25] Hwang, F. S.; Confrey, T.; Reidy, C.; Dorel Picovici; Callaghan, D.; Culliton, D.; Nolan, C. Review of Battery Thermal Management Systems in Electric Vehicles. *Renewable & Sustainable Energy Reviews* **2024**, *192*, 114171. https://doi.org/10.1016/j.rser.2023.114171

[26] Abdelkareem, M. A.; Maghrabie, H. M.; Abo-Khalil, A. G.; Adhari, O. H. K.; Sayed, E. T.; Radwan, A.; Rezk, H.; Jouhara, H.; Olabi, A. G. Thermal Management Systems Based on Heat Pipes for Batteries in EVs/HEVs. *Journal of Energy Storage* **2022**, *51*, 104384. https://doi.org/10.1016/j.est.2022.104384

[27] Bernagozzi, M.; Georgoulas, A.; Miché, N.; Marengo, M. Heat Pipes in Battery Thermal Management Systems for Electric Vehicles: A Critical Review. *Applied Thermal Engineering* **2023**, *219*, 119495. https://doi.org/10.1016/j.applthermaleng.2022.119495

[28] Karimi, G.; Li, X. Thermal Management of Lithium-Ion Batteries for Electric Vehicles. *International Journal of Energy Research* **2012**, *37* (1), 13–24. https://doi.org/10.1002/er.1956

[29] Lin, J.; Liu, X.; Li, S.; Zhang, C.; Yang, S. A Review on Recent Progress, Challenges and Perspective of Battery Thermal Management System. *International Journal of Heat and Mass Transfer* **2021**, *167*, 120834. https://doi.org/10.1016/j.ijheatmasstransfer.2020.120834

[30] Liu, C.; Xu, D.; Weng, J.; Zhou, S.; Li, W.; Wan, Y.; Jiang, S.; Zhou, D.; Wang, J.; Huang, Q. Phase Change Materials Application in Battery Thermal Management System: A Review. *Materials* **2020**, *13* (20), 1–39. https://doi.org/10.3390/ma13204622

[31] Hu, X.; Gao, F.; Xiao, Y.; Wang, D.; Gao, Z.; Huang, Z.; Ren, S.; Jiang, N.; Wu, S. Advancements in the Safety of Lithium-Ion Battery: The Trigger, Consequence and Mitigation Method of Thermal Runaway. *Chemical Engineering Journal* **2024**, *481*, 148450–148450. https://doi.org/10.1016/j.cej.2023.148450

[32] Shahid, S.; Agelin-Chaab, M. A Review of Thermal Runaway Prevention and Mitigation Strategies for Lithium-Ion Batteries. *Energy Conversion and Management: X* **2022**, *16*, 100310. https://doi.org/10.1016/j.ecmx.2022.100310

[33] Talele, V.; Moralı, U.; Khaboshan, H. N.; Patil, M. S.; Panchal, S.; Fraser, R.; Fowler, M. Improving Battery Safety by Utilizing Composite Phase Change Material to Delay the Occurrence of Thermal Runaway Event. *International Communications in Heat and Mass Transfer* **2024**, *155*, 107527. https://doi.org/10.1016/j.icheatmasstransfer.2024.107527

[34] Phiri, R.; Sanjay, M. R.; Siengchin, S.; Oladijo, O. P.; Dhakal, H. N. Development of Sustainable Biopolymer-Based Composites for Lightweight Applications from Agricultural Waste Biomass: A Review. *Advanced Industrial and Engineering Polymer Research* **2023**, *6* (4), 436–450. https://doi.org/10.1016/j.aiepr.2023.04.004
[35] Weiss, M.; Haufe, J.; Carus, M.; Brandão, M.; Bringezu, S.; Hermann, B.; Patel, M. K. A Review of the Environmental Impacts of Biobased Materials. *Journal of Industrial Ecology* **2012**, *16*, S169–S181. https://doi.org/10.1111/j.1530-9290.2012.00468.x
[36] Colorado, H. A.; Velásquez, E. I. G.; Monteiro, S. N. Sustainability of Additive Manufacturing: The Circular Economy of Materials and Environmental Perspectives. *Journal of Materials Research and Technology* **2020**, *9* (4), 8221–8234. https://doi.org/10.1016/j.jmrt.2020.04.062
[37] O'Dea, R. E.; Willie, J. A.; Epps, T. H. 100th Anniversary of Macromolecular Science Viewpoint: Polymers from Lignocellulosic Biomass. Current Challenges and Future Opportunities. *ACS Macro Letters* **2020**, *9* (4), 476–493. https://doi.org/10.1021/acsmacrolett.0c00024
[38] Mehta, J.; Gupta, K.; Shilpi Lavania; Kumar, P.; Chaudhary, V.; Gupta, P. Inherent Roadmap in Synthesis and Applications of Sustainable Materials Using Oil Based and Microbial Polymers. *Materials Today Sustainability* **2024**, *25*, 100615. https://doi.org/10.1016/j.mtsust.2023.100615
[39] Wang, Y.; Feng, X.; Huang, W.; He, X.; Wang, L.; Ouyang, M. Challenges and Opportunities to Mitigate the Catastrophic Thermal Runaway of High-Energy Batteries. *Advanced Energy Materials* **2023**, *13* (15), 1–29. https://doi.org/10.1002/aenm.202370061
[40] Xu, L.; Chen, S.; Su, Y.; He, J.; Wang, L.; Shen, X.; Chen, L.; Cao, D.; Lu, Y.; Wang, M.; Bao, L.; Li, N.; Wu, F. Building Better Batteries: Solid-State Batteries with Li-Rich Oxide Cathodes. *Energy Material Advances* **2023**, *4*, 1–24. https://doi.org/10.34133/energymatadv.0045
[41] Chen, R.; Li, Q.; Yu, X.; Chen, L.; Li, H. Approaching Practically Accessible Solid-State Batteries: Stability Issues Related to Solid Electrolytes and Interfaces. *Chemical Reviews* **2019**, *120* (14), 6820–6877. https://doi.org/10.1021/acs.chemrev.9b00268
[42] Sergiy Kalnaus; Dudney, N. J.; Westover, A. S.; Herbert, E. G.; Hackney, S. Solid-State Batteries: The Critical Role of Mechanics. *Science* **2023**, *381* (6664), 1–8. https://doi.org/10.1126/science.abg5998
[43] Kang, W.; Deng, N.; Ju, J.; Li, Q.; Wu, D.; Ma, X.; Li, L.; Minoo Naebe; Cheng, B. A Review of Recent Developments in Rechargeable Lithium-Sulfur Batteries. *Nanoscale* **2016**, *8* (37), 16541–16588. https://doi.org/10.1039/c6nr04923k
[44] Manthiram, A.; Fu, Y.; Su, Y.-S. Challenges and Prospects of Lithium-Sulfur Batteries. *Accounts of Chemical Research* **2012**, *46* (5), 1125–1134. https://doi.org/10.1021/ar300179v
[45] Zhou, G.; Paek, E.; Hwang, G. S.; Manthiram, A. Long-Life Li/Polysulphide Batteries with High Sulphur Loading Enabled by Lightweight Three-Dimensional Nitrogen/Sulphur-Codoped Graphene Sponge. *Nature Communications* **2015**, *6* (1), 1–11. https://doi.org/10.1038/ncomms8760
[46] Manthiram, A.; Chung, S.-H.; Zu, C. Lithium-Sulfur Batteries: Progress and Prospects. *Advanced Materials* **2015**, *27* (12), 1980–2006. https://doi.org/10.1002/adma.201405115
[47] Seh, Z. W.; Sun, Y.; Zhang, Q.; Cui, Y. Designing High-Energy Lithium-Sulfur Batteries. *Chemical Society Reviews* **2016**, *45* (20), 5605–5634. https://doi.org/10.1039/c5cs00410a
[48] Taub, A. I. Automotive Materials: Technology Trends and Challenges in the 21st Century. *MRS Bulletin* **2006**, *31* (4), 336–343. https://doi.org/10.1557/mrs2006.74
[49] Karakurt, I.; Lin, L. 3D Printing Technologies: Techniques, Materials, and Post-Processing. *Current Opinion in Chemical Engineering* **2020**, *28*, 134–143. https://doi.org/10.1016/j.coche.2020.04.001
[50] Babu, S. S.; Love, L.; Dehoff, R.; Peter, W.; Watkins, T. R.; Pannala, S. Additive Manufacturing of Materials: Opportunities and Challenges. *MRS Bulletin* **2015**, *40* (12), 1154–1161. https://doi.org/10.1557/mrs.2015.234
[51] Lacelle, T.; Sampson, K. L.; Sarvestani, H. Y.; Rahimizadeh, A.; Barroeta Robles, J.; Mirkhalaf, M.; Rafiee, M.; Jakubinek, M. B.; Paquet, C.; Ashrafi, B. Additive Manufacturing of Polymer Derived Ceramics: Materials, Methods, and Applications. *APL Materials* **2023**, *11* (7). https://doi.org/10.1063/5.0151661
[52] Del Pero, F.; Delogu, M.; Kerschbaum, M. Design of a Lightweight Rear Crash Management System in a Sustainable Perspective. *Sustainability* **2020**, *12* (13), 5243. https://doi.org/10.3390/su12135243

[53] Hashemi Sohi, F. S.; Mansour, S.; Dehghanian, A. Multi-Objective Optimization for Selecting Sustainable Materials with Simultaneous Consideration of Several Components in a Product. *International Journal of Sustainable Engineering* **2022**, *15* (1), 109–124. https://doi.org/10.1080/19397038.2022.2080888
[54] Priarone, P. C.; Catalano, A. R.; Settineri, L. Additive Manufacturing for the Automotive Industry: On the Life-Cycle Environmental Implications of Material Substitution and Lightweighting through Re-Design. *Progress in Additive Manufacturing* **2023**, *8* (6), 1229–1240. https://doi.org/10.1007/s40964-023-00395-x
[55] Colombini, G.; Rosa, R.; Ferrari, A. M.; Defanti, S.; Bassoli, E. Life Cycle Assessment of Lattice Structures: Balancing Mass Saving and Productivity. *Journal of Cleaner Production* **2024**, *445*, 141390. https://doi.org/10.1016/j.jclepro.2024.141390
[56] Le, T. C.; Winkler, D. A. Discovery and Optimization of Materials Using Evolutionary Approaches. *Chemical Reviews* **2016**, *116* (10), 6107–6132. https://doi.org/10.1021/acs.chemrev.5b00691
[57] Kadulkar, S.; Sherman, Z. M.; Ganesan, V.; Truskett, T. M. Machine Learning–Assisted Design of Material Properties. *Annual Review of Chemical and Biomolecular Engineering* **2022**, *13* (1), 235–254. https://doi.org/10.1146/annurev-chembioeng-092220-024340
[58] Ball, P. Using Artificial Intelligence to Accelerate Materials Development. *MRS Bulletin* **2019**, *44* (5), 335–344. https://doi.org/10.1557/mrs.2019.113
[59] Wu, P. Digital Materials Design: Computational Methodologies as a Discovery Tool. *MRS Bulletin* **2006**, *31* (12), 995–998. https://doi.org/10.1557/mrs2006.227
[60] Beyramali Kivi, M.; Hong, Y.; Asle Zaeem, M. A Review of Multi-Scale Computational Modeling Tools for Predicting Structures and Properties of Multi-Principal Element Alloys. *Metals* **2019**, *9* (2), 254. https://doi.org/10.3390/met9020254
[61] Huang, J. S.; Liew, J. X.; Ademiloye, A. S.; Liew, K. M. Artificial Intelligence in Materials Modeling and Design. *Archives of Computational Methods in Engineering* **2020**. https://doi.org/10.1007/s11831-020-09506-1
[62] Weinstock, M.; Stathopoulos, N. Advanced Simulation in Design. *Architectural Design* **2006**, *76* (2), 54–59. https://doi.org/10.1002/ad.240
[63] Suzuki, A.; Chen, S.; Kumar, N. Application of Computational Tools in Alloy Design. *Mrs Bulletin* **2019**, *44* (4), 247–251. https://doi.org/10.1557/mrs.2019.70
[64] Papadimitriou, I.; Gialampoukidis, I.; Vrochidis, S.; Kompatsiaris, I. AI Methods in Materials Design, Discovery and Manufacturing: A Review. *Computational Materials Science* **2024**, *235*, 112793–112793. https://doi.org/10.1016/j.commatsci.2024.112793.
[65] Ghalkhani, M.; Habibi, S. Review of the Li-Ion Battery, Thermal Management, and AI-Based Battery Management System for EV Application. *Energies* **2022**, *16* (1), 185. https://doi.org/10.3390/en16010185
[66] Krishna, T. N. V.; Kumar, S. V. S. V. P. D.; Srinivasa Rao, S.; Chang, L. Powering the Future: Advanced Battery Management Systems (BMS) for Electric Vehicles. *Energies* **2024**, *17* (14), 3360. https://doi.org/10.3390/en17143360
[67] Wang, L.; Zhou, Z.; Yan, X.; Hou, F.; Wen, L.; Luo, W.-B.; Liang, J.; Dou, S. X. Engineering of Lithium-Metal Anodes towards a Safe and Stable Battery. *Energy Storage Materials*, **2018**, *14*, 22–48. https://doi.org/10.1016/j.ensm.2018.02.014

6 Role of Additive Manufacturing and Thermal Spray Processed Materials in Electric Vehicle (EV) and Hybrid Electric Vehicle (HEV) Applications

C. Durga Prasad, Pradeep Kumar G.S., Shashi Kumar M.E., Krupa R., Rajesh M.N., and Veeresha G.

6.1 INTRODUCTION

Hybrid electric vehicles (HEVs) and electric vehicles (EVs) have emerged as strategic measures of cutting fossil fuel and environmental depletion. EVs only use electrical power, while HEVs have both systems. Increased customer concern for environmental conservation through modest use of automobiles and appropriate governmental actions have also boosted the utilization of electric and hybrid vehicles all over the globe. Technological innovation has therefore triggered the use of lithium-ion batteries, electric motors, and lighter materials in vehicles, which has shifted the use of gasoline-based transport to electrified transport. In EVs, lithium-ion batteries play important roles in the performance, while at the same time HEVs have a more efficient way to use fuel, through hybrid systems. Nonetheless, the efficiency and durability of EVs/HEVs depend significantly on material choices for regional packages, electric paths, structural parts, and tribological counterparts. Choosing heavy-duty and lightweight high strength-to-weight ratio, thermal and electrical conductivity along with high wear resistance material is possible only with proper selection of materials for these vehicles [1–6].

However, the emerging technology of additive manufacturing (AM), more commonly referred to as 3D printing, has modified the automotive industry, especially concerning the creation of EVs and HEVs. AM technology can build practical, elegant, lightweight, and tailored parts that would be otherwise challenging and costly in terms of production using more conventional manufacturing processes. With an increased focus on sustainability and high performance in automotive applications, AM is a key enabler for forthcoming EVs and HEVs. The prominent benefits of employing AM include the reduction of the vehicles weight through light and thin structures and specific geometry, efficiency increases, and opportunities for producing individual parts. It also minimizes material usage due to the layer-by-layer construction, hence using only the required material for construction, and it also reduces the costs and amount of pollution caused. AM also provides the capability to rapidly prototype, further helping to reduce the time it takes to develop new ideas and products, which also combines to positively impact research and development. The automotive industry is a highly saturated and competitive market, thus the ability to innovate as well

DOI: 10.1201/9781003596097-6

as the time to market are crucial. AM is also applied to producing battery parts, electrical contacts, cooling systems, light heat exchangers, and intricate cooling channels. Furthermore, AM methodologies can be employed to create new materials with improved performance characteristics or different organic–inorganic combinations of stronger polymers or metals [7–13].

Thermal spray coatings are critical in automotive applications, especially in EVs and HEVs. They provide a shield as well as better performance to parts prone to depressed conditions, such as heat, wear, corrosion, as well as mechanical stress [14]. The automotive industry is gradually transitioning toward designs that promote low energy consumption and lightweighting, making thermal spray coatings more valuable for this sector. Thermal spray coatings also provide wear resistance and extended durability to the surfaces because these coatings help alleviate friction and wear on components such as bearings, shafts, pistons, and gears [15]. They also deliver anticorrosive features, especially within conditions that are less ideal, such as in regions surrounding large water bodies or with treated roads. Thermal efficiency and electrical efficiency are also another advantage of thermal spray coatings where ceramic coatings like alumina and zirconia are used; they act as an insulating material in thermal applications. Another benefit of thermal spray coatings is embodied in the lightweighting and efficiency that caters for higher fuel economy in HEVs and longer battery back-up in EVs. In addition, thermal spray coatings enhance the sustainability of the automotive business since they do not need part substitution, therefore cutting the utilization of raw materials. New-generation coatings are developed to incorporate lower risk factors and are less toxic in comparison with previous generation coatings [16–20].

In this chapter, the focus will be given to the analysis of the materials and technologies related to EVs and HEVs with regard to thermal spray coatings and AM technologies. It becomes useful in its approach to underscore these technologies in improving the performance, looks, dependability, and durability of automotive parts. AM capabilities of creating lightweight parts with intricate and complex geometry and accurate tolerance mean a better fit with customized design. In addition to wear properties, the application of thermal spray coatings provide anticorrosion properties, thermal barrier features, and electrical barrier properties. Most of the challenges which automotives have to face are as follow: scalability, cost, and materials compatibility. The industry outlook and emerging technologies are also discussed herein. Finally, the chapter delineates real-life case studies and discusses feasible applications of thermal spray coatings and AM in EV and HEV engineering.

6.2 OVERVIEW OF ADDITIVE MANUFACTURING TECHNOLOGIES

6.2.1 Additive Manufacturing Principle

AM is a family of advanced manufacturing technologies for creating parts and objects through the layer-by-layer addition of materials according to a computer model or digital design. AM builds a part layer by layer, fundamentally changing the way parts are made by adding materials in a layer-wise manner, which makes it possible to make world-class shapes and structures that would otherwise take more than one regular process to make.

Figure 6.1 illustrates the principle of AM which is predicated on layer-by-layer build-up of material to form a three-dimensional structure [21]. It adopts a computer-aided design (CAD) model that is then sliced into layers to provide the aviation hardware of the AM machine. Depending on what is programmed or designed on the computer, the machine layers the material and packs it until the product shape is complete. This digital to physical transformation feature enables engineers to modify and enhance designs using virtual means with no physical limitations in engineering manufacturing [22]. AM is compatible with a vast array of materials including metals, polymers, ceramics, and composite materials. It provides a way to make geometries which cannot be made or are very difficult to be made through traditional methods, and is therefore very useful in industries such

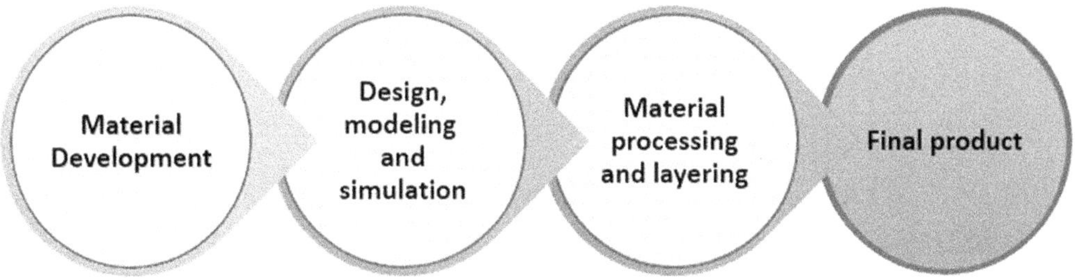

FIGURE 6.1 Principles of additive manufacturing.

as the automobile, aerospace, and medical fields. AM adds only the required material, and therefore its efficiency in the use of material is high, and thus contributes to addressing concerns about sustainability. AM is most suitable in industries that call for modifications including the manufacturing of tailormade medical implants, automotive parts, and consumer goods. It also makes it possible to perform rapid prototyping and manufacturing, which makes it easier to close the gap between the two processes [23–26].

AM is an innovative technology that has changed the practice of designing and manufacturing of parts across different sectors such as the automotive, aerospace, and medical areas. Analyzing its applications in-depth is important to comprehend the developments implemented in manufacturing procedures. AM technologies are revolutionizing industries owing to the abilities they present for refining part design, for creating prototypes, and for manufacturing the final product; enhanced manufacturing solutions for efficiency, sustainability, and customization are made possible through the use of AM technologies. Currently, the automotive industry is the most unique AM beneficiary since it increases the efficiency and sustainability of manufacturing solutions while making them more customizable [27–29]. It is this transition from theory to practice that we need to understand, when assessing the value of AM in the modern manufacturing environment.

6.2.2 AM Technologies: Classification by Material Types

The classification of AM techniques used in the automobile sector is based on materials. Metal-based AM results in high strength and high durability of parts suitable for use in engine parts and structures. Polymer-based AM has low density and low cost for interior and non-load-bearing applications [30]. Ceramic-based AM builds high-temperature wear-resistant components for extreme application environments. Polymer matrix AM hinges on the strengthening of polymer with fibers to produce lightweight, durable parts [31]. Figure 6.2 illustrates the classification of diverse materials and specific applications enabled by AM in automotive manufacturing.

6.2.2.1 *Direct Metal Laser Sintering (DMLS)*

This technique employs a high-power laser to melt and bond metal powder particles in a layer-by-layer fashion to produce a solid dense metal component. DMLS does not fuse powder to a molten state, but sinter the particles, which allows it to produce intricate shapes and strong components mostly for use in the aerospace and automobile sectors [32].

6.2.2.2 *Electron Beam Melting (EBM)*

In EBM, an electron beam is utilized instead of a laser, enabling the full melting of the metal powder. The process is performed under vacuum to minimize oxidation and to be most effective on high-temperature components like titanium. The EBM process is best known for delivering components that offer clients excellent mechanical properties; it is applied in making medical implants and components for aerospace projects [33].

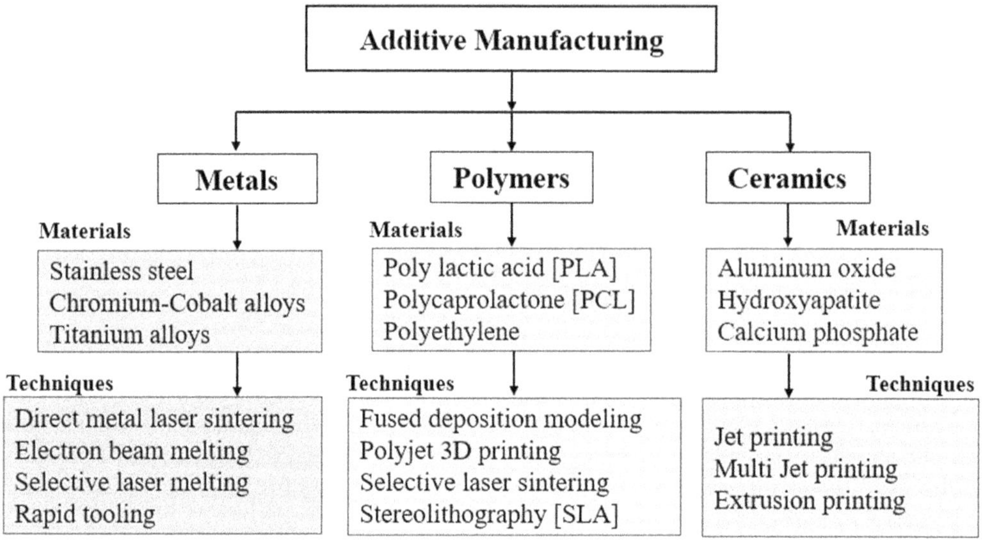

FIGURE 6.2 Classification of AM techniques based on materials used for the automobile industry.

6.2.2.3 *Selective Laser Melting (SLM)*

SLM, like DMLS, uses a laser to selectively dissolve a metal powder and therefore gives a better density to the final product. SLM is able to produce components with the properties of wrought metals, fine details, and with good mechanical properties. Automotive and aerospace applications and industries where the performance of the part may be crucial are where it is most widely used for tooling [34].

6.2.2.4 *Rapid Tooling*

Rapid tooling applies AM to quickly and cheaply make molds, dies, or other tooling items as required. This approach can include both direct and indirect methods. For instance, tooling integrated by AM entails fabrication of the tooling itself, while tooling produced under AM entails creating a mold that forms the basis on which to design the true tool. Rapid tooling shortens the lead time to within a minimal number of hours and is used in automotive and manufacturing industries for prototyping and low-volume production [35].

6.2.2.5 *Fused Deposition Modeling (FDM)*

FDM is one of the most popular types of AM techniques and is preferred in multiple industries including for prototyping and functional applications. It accomplishes this by melting a thermoplastic and extruding it through a nozzle, depositing each layer of material one on top of the other. FDM is appreciated for its usability, affordability, and for manufacturing robust products from thermal plastic materials. This means that it is fairly common in the automotive industry, consumer products industry, and specifically for educational models that need a prototype or concept explants [36].

6.2.2.6 *PolyJet 3D Printing*

PolyJet printing is an AM technique that uses a jetting process in order to deposit photopolymer resin layers which are cured by UV light. It can also create thin walls, smooth surfaces, and parts with multiple materials or colors in one build. PolyJet is particularly suitable for models with numerous delicate features and complicated surfaces, which may include medical-related models, dental models, and clear models of consumer products [37].

6.2.2.7 *Selective Laser Sintering (SLS)*

SLS requires a high-energy laser to soften the powder, which usually comprises nylon or other varieties of thermoplastics to create solid three-dimensional forms. However, SLS does not need support structures since the surrounding powder offers support to the part being built during the build time in contrast to the FDM and PolyJet methods. This makes SLS perfect for curved and functional parts as it will cut costs greatly compared with other methods. The method is widely employed in various industries, particularly in manufacturing of parts that require high strength and low density, as well as models of functional parts [38].

6.2.2.8 *Stereolithography (SLA)*

SLA is one of the oldest AM technologies and works by projecting an ultraviolet laser, which cures the liquid resin which solidifies to form a layer for a high-precision part. SLA has among the best surface finish accuracies, which is why it is suitable for applications such as part prototypes and molds, among others. It is predominantly applied in areas such as dentistry, jewelry, and the creation of prototype automobile parts, since the process results in a fine surface finish [39].

6.2.2.9 *Jet Printing*

Jet printing in AM can be generally defined as a process for which a print head selectively dispenses droplets of material, such as liquid photopolymer or binder, on to a build platform. The material is hardened or solidified on the build platform to fabricate the complete three-dimensional structure. The jet printing method provides accuracy, fine detail, and an exacting surface finish which means that it is a favorite for detail reproduction, particularly in color, form, fit, and function prototypes, for example [40].

6.2.2.10 *Multi Jet Printing (MJP)*

Multi jet printing is a new kind of jet printing technique where three nozzles dispense minute droplets of a liquid photopolymer resin. UV light subsequently hardens each layer when it is applied. MJP offers high resolution, and multiple materials and color capability, making it suitable for the creation of models with fine features. It is often applied where very fine features and surface finish are required, for example, models for prosthetic limbs, dentures, and accurate prototypes during product formulation [41].

6.2.2.11 *Extrusion Printing*

This technique usually means technologies such as FDM in which the material, usually thermoplastic, is extruded through a nozzle to create a part layer by layer. Extrusion printing is quite common, affordable, and provides great results, which is why it is used for rapid prototyping or creating useful objects. It is primarily used in the automotive, aerospace, and consumer product industries where parts that are required to have great strength, sturdiness, and are not cosmopolitan in nature and do not require surface finishing [42].

6.3 APPLICATIONS OF ADDITIVE MANUFACTURING IN EVS AND HEVS

Sustainable technology, such as AM printing, plays a critical role in the development of EVs or HEVs. It allows for the development of thin-walled parts, like battery enclosures, motor parts, as well as some structural parts of the car body, that can enhance energy intake and driving distance. AM also finds its application in the battery and energy storage systems by offering the opportunity for systems designers to elaborate and create power-dense battery modules and cooling systems that directly determine the battery cell cluster geometry and channel the heat out. In addition, it facilitates the development of components for solid-state rechargeable batteries, which provide greater energy density and increased safety [43–45].

AM is also revolutionizing the manufacture of electric motors and powertrain parts for both EVs and HEVs. Conventional motors have numerous components, which need to be assembled, but with AM many of these can be merged into one printed part, which decreases complexity. Motor components produced through AM are not only lighter but can be designed with inherent cooling paths. Also, the complex internal channel and geometry of compact and efficient cooling systems can be produced employing AM. Optimizing thermal management using AM has positive knock-on effects on safety, electrical and mechanical efficiency, and longevity of EV and HEV components. For EV and HEV manufacturers, several advantages using AM technology are provided, such as integrated design of interior and exterior parts, short time-to-market, environmental attractiveness, and lightweight construction with fewer materials. Additional options, such as customized dashboard, seats, and trims, with regard to shape as well as texture, offer a wider range of appearances and uses than normally produced ones. AM also involves the fabrication of body panels, bumpers, and aerodynamics components that allow car manufacturers to achieve and design the aerodynamics required to improve energy efficiency [46–49].

Intense competition and new requirements for EV and HEV models have dramatically boosted the pace of car creation through the use of prototypes for rapid development that allows engineers to try new designs without waiting for time-consuming mold or tooling production. Such a rapid iterative capacity results in enhanced innovation and speed to market, which are very valuable innovations in highly competitive segments such as EVs and HEVs [50–52]. AM also helps in the low material usage during manufacturing, also using materials that are recycled or even bio-based. It also enables easier repair and maintenance with constructed replaceable components through three-dimensional printing that eliminate the need for large stock holdings and protracted maintenance of components. Several applications of AM in EVs and HEVs are presented in Figure 6.3.

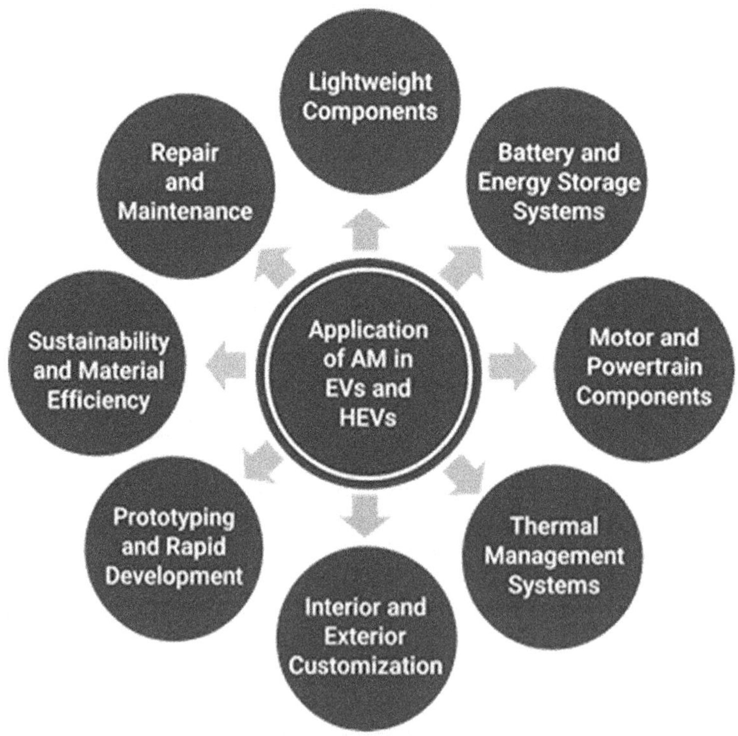

FIGURE 6.3 Applications of AM in EVs and HEVs.

6.4 ROLE OF THERMAL SPRAY COATINGS (TSCS) IN AM COMPONENTS

6.4.1 *Thermal Spray Coatings*

TSC is a technique of applying several coating materials to a surface that has to be coated; this material can be in a solid, semi-solid, or liquid state. Thermal spray uses conductivity powder or wire forms that are fused at a specific temperature point. This is a coating material which is in the form of a solid that is manufactured and applied in abundance on bodies made from aluminum [53]. Here, metal particles or a combination of both metal and/or non-metal particles in solid or semisolid state impinge on the base material (substrate) to build a layer, as shown in Figure 6.4.

Thermal spray coating techniques are applied in a broad spectrum of industries due to the fact that they have different advantages depending on the kind of material deposited and the kind of properties needed. Among surface coating methods, plasma spraying is widely used; it employs a plasma jet to melt and accelerate particles onto a substrate, producing layers with a thermal barrier and wear-resistant properties. The technique holds significant promise for ceramic and metallic thin films and is ideal in applications such as the aerospace and automotive sectors, where heat is a factor [54]. High-velocity oxy-fuel (HVOF) spraying uses fuel and oxygen combinations in which a high-velocity gas stream is aerodynamically propelled, thus offering dense, hard, wear- and corrosion-resistant coatings [55]. HVOF coatings are well suited for components that exist in very harsh operating conditions such as the oil and gas and marine and mining industries. Cold spraying is a method of coating a substrate using powder without high-temperature fusion, with special reference to aluminum and titanium. Flame spraying or flame coating is an inexpensive and efficient process of using an oxygen and fuel flame to heat the coating material and accelerate it toward the substrate [56–58]. Pre-treatment is often used to deposit zinc or aluminum on AM components to provide a protective coating according to their industrial and structural natures.

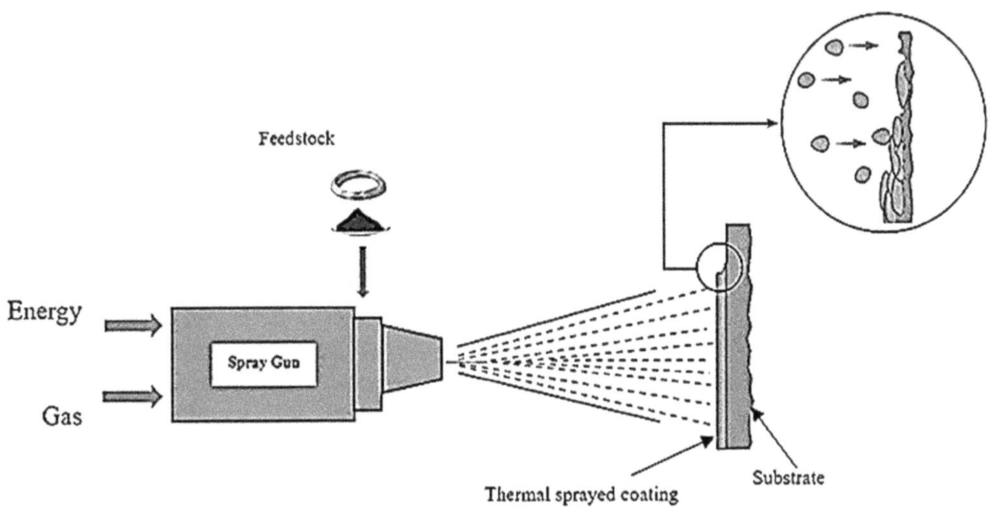

FIGURE 6.4 Concept of thermal spraying.

6.5 INTEGRATION OF THERMAL SPRAY COATINGS IN AM COMPONENTS

The application of thermal spray coatings to AM components has already been proven to enhance part characteristics and holds promise for use in the durability of automotive parts in EV and HEV systems. New studies also encompass conceivable advantages of thermal spray protection such as increased tribological characteristics, better corrosion resistance, as well as thermal insulation. For instance, Wang et al. proved that employing tungsten carbide thermal spray coatings to AM-manufactured parts effectively served to increase wear protection and provided adequate improvements to meet the needs of the automotive industries for those metallic parts. Moreover, some paints, including zirconia, provided reasonable thermal barrier resistance for battery casings, and therefore, the heat dissipation and performance of EVs. Through the thermal spray technique, it is possible to modify the properties of the surface of the AM parts, therefore providing performance characteristics that fit the need of automobile parts' life cycle for reliability [59–62].

However, several issues are associated with surface treatment using this process together with AM parts. Adhesion is the most critical factor, and many researchers have noted that the baseline surface morphology and cleaning steps must be optimized to achieve sufficient adhesion before coating deposition. In addition, it is important that the thermal properties of the substrate and coating materials are compatible to avoid problems such as delamination or thermal stress during operation. Other research in post-processing technologies, such as machining and heat treatments, to improve the coating quality and performance are also in progress. In sum, the ability to incorporate thermal spray coatings into the new AM components is promising for the automotive field that offers a vast area for the future innovation of lightweight high-performance vehicles tailored to the requirements of an eco-friendly future [63–65].

The integration of AM and TSC is a significant advancement in the production of high-performance components across various industries. This synergy addresses several inherent limitations of AM-produced parts, resulting in enhanced functionality and performance. The combination of these two technologies allows manufacturers to optimize component properties, tailoring them to meet specific operational requirements and performance standards. One of the primary challenges faced by AM components is surface roughness, which can adversely affect the operational efficiency of the finished product. Thermal spray coatings can significantly improve this situation by providing a smooth, uniform surface finish. Through processes such as plasma spraying or HVOF, TSC can fill in the surface irregularities and create a dense, well-adhered layer on top of the AM part, enhancing its aesthetic appeal and optimizing its performance by reducing friction and improving contact characteristics [66,67]. Additionally, the microstructural integrity of AM-produced metals can pose challenges, as parts created using AM techniques may have residual stresses or porosity, leading to mechanical weaknesses and increased likelihood of stress-induced failures. By applying thermal spray coatings, manufacturers can provide a protective layer that mitigates the effects of these inherent flaws, enhancing the mechanical strength of the AM part and increasing its resistance to fatigue and failure over time, ultimately contributing to the reliability of the component in demanding applications. Another significant advantage of combining AM with TSC is the capability for customized performance. AM is renowned for its ability to produce complex geometries that would be challenging or impossible to achieve with traditional manufacturing methods. When paired with TSC, manufacturers can apply targeted coatings that enhance specific functional properties of the part, enabling engineers to design components that are not only lighter and more efficient but also specifically tailored to meet the varied demands of their operating conditions [68–72].

6.6 BENEFITS OF TSC FOR AM COMPONENTS

TSC plays a critical role in improving the performance of AM parts by improving their properties. They enhance the surface wear, resistance of the product to various corrosive agents, heat-insulating features, and mechanical strength. TSC provides a hard and wear-resistant surface, for example, WC-Co coatings that may be deposited using HVOF which enhance the durability of tools used under severe conditions, such as mining machinery. In aerospace applications, ceramic-based plasma-sprayed coatings prove to be effective against wear caused by flow velocities and particle strikes [73].

Microstructural defects are prevented or minimized through picking, plating, anodizing, or providing a chromium oxide, aluminum oxide, or nickel-based alloy coating on the component, thus prolonging the service life of the former. For AM components where parts of the structures are exposed to high temperature, TSC functions as thermal insulation so that the base material is protected from thermal degradation and oxidation [74].

TSC is playing an increasingly important role in AM surface engineering, making it easier for components to conform to industry requirements. It can also increase mechanical characteristics such as the strength of cold-sprayed aluminum coatings and increase the capacity by offering a homogeneous and defect-free surface leading to higher fatigue resistance. TSC is very effective in improving the performance and increasing the life expectancy of components produced using AM technology [75].

6.7 TSC ENHANCING AM-PRODUCED AUTOMOTIVE COMPONENTS

The automotive industry is increasingly using AM to create lightweight and complex parts, enhancing fuel efficiency and design flexibility. However, these components face significant challenges due to the harsh automotive environment, such as high wear, thermal cycling, and corrosion. TSCs are employed to mitigate these challenges, ensuring that AM-produced components can perform reliably and efficiently throughout their service life. Exhaust systems, engine components, brake discs, and pistons are all critical for vehicle performance and emissions control [76]. Ceramic thermal barrier coatings and oxidation-resistant materials are used to protect against high temperatures and extend component lifespan. High-temperature engine components, such as turbine blades, valves, and pistons, are also crucial for overall vehicle efficiency and performance. Yttria-stabilized zirconia (YSZ) and aluminum coatings are commonly used as thermal insulation coatings on high-temperature components to reduce heat loss and prevent thermal fatigue [77]. Brake discs and pistons, made via AM, are subjected to severe wear and high temperatures during operation. Tungsten carbide-cobalt-chromium coatings applied via HVOF spraying enhance the wear resistance and thermal stability of brake components, improving their service life by providing a hard, durable surface. The application of TSC not only enhances wear resistance but also ensures that brake components maintain their performance under high temperatures, reducing the risk of failure and enhancing vehicle safety. Overall, TSC provides protective layers that enhance the durability and performance of AM parts in the automotive industry [78–80].

Figure 6.5 provides details of wear loss estimates for various kinds of HVOF and atmospheric plasma spray (APS) ceramic coatings with different compositions enacted on wear. The wear loss is defined by the ability and longevity of each coating under the wear scenario. The wear loss, which is imparted into the HVOF-sprayed coatings, is less as compared to APS coatings, which proves a higher wear-resistant nature of the HVOF coatings [81]. Concentration differences play a huge role in the wear performance of materials, though wear protection can be enhanced by incorporating specific combinations of materials. This comparison suggests that one is able to choose the best coatings for uses which call for better wear capacity.

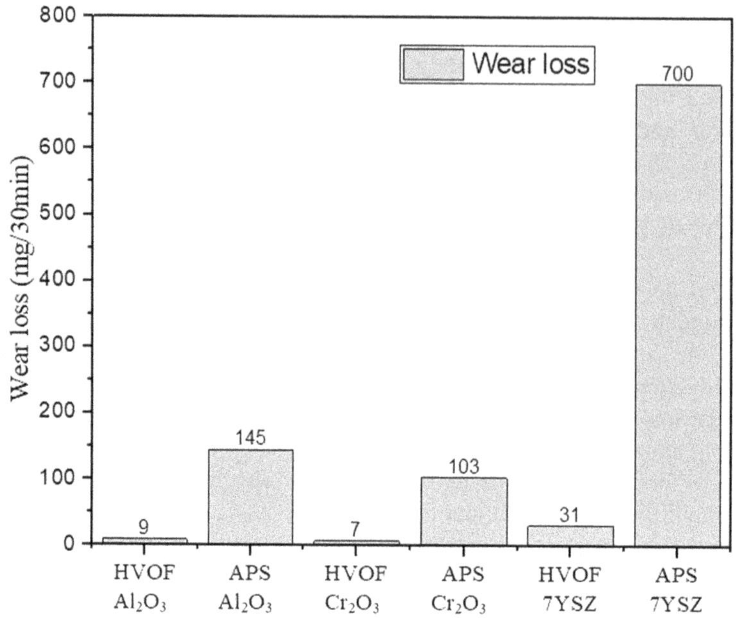

FIGURE 6.5 Wear loss measured for various HVOF and APS sprayed ceramic coatings with various compositions.

6.8 DISCUSSION

AM is also known as 3D printing or direct tooling and is a completely transformative technology within automotive manufacturing, especially for EVs and HEVs. AM makes it possible to design lightweight, highly personalized, and effective parts which impact on managing the power, and density of these vehicles. As AM technologies are expensive and energy-consuming, they are entering EVs and HEVs over time and primarily driven by the need for weight reduction, new battery improvements, and increasing geometric complexity.

Since EVs and HEVs are relatively recent innovations, the adaptability of AM has one of the principal advantages of being able to manufacture lightweight components with complex geometry which can be difficult to achieve using other manufacturing methods. For instance, the use of lattice structures and hollow designs integrated by virtue of AM reduces mass in components such as battery casings, brackets, and structural members, without sacrificing toughness. It also reduces weight, offloads charge on the battery, and increases power consumption efficiency and battery recharging capability. One clear advantage of AM is the fact that part production can occur rapidly and be fashioned to coordinate with the peculiarities of automobile design, which is particularly beneficial when developmental cycles are compressed. Designers and engineers can quickly generate and improve solutions and make sure these integrate, perform, and are fit for use to specifications more swiftly, getting improved solutions to the marketplace more quickly.

Given the broad classification of AM materials straddling metals, polymers, to composites, different manufacturers are able to simplify the process of selecting the right material type for each application in EVs and hybrids. In detail, specific advantages of metal-based AM include high-strength, heat sinks, motor covers, powertrain, and battery contact pins. The polymer-based AM techniques are capable of fabricating interior and aesthetic designs, and reduce total weight as well as cost. Therefore, the integration of AM in EV and HEV production is leading to an innovation of the automotive industry through the use of lightweight products, customized car parts production, and sustainable manufacturing.

6.9 CONCLUSIONS

Therefore, several key components, designs, and production prospects in EVs and HEVs can be revolutionized through AM with lightweight, multi-material, and made-to-order structural parts and components.

Thermal spray coatings are used to upgrade AM part functionality and include additional features such as surface finish, antiwear, and anticorrosion properties. This symbiosis of AM and TSC enhances battery enclosure, thermal systems, and structural parts.

The flexibility of the choices of material in AM also helps make AM suitable for use in EVs and HEVs. From metals, polymers, to the most advanced composites, AM enables engineers to choose materials that meet precisely the requirements of each part.

The combination of AM and TSC technologies creates opportunities for producing more advanced and efficient EVs and HEVs, while ensuring that components and materials align with global efforts toward clean energy and sustainable manufacturing..

ACKNOWLEDGMENTS

Author Dr. Durga Prasad C. would like to thank the Science and Engineering Research Board (SERB) for financial support to carry out this research work. Project File no: CRG/2022/004140, under Core Research Grant (CRG) scheme, Government of India.

REFERENCES

[1] Frazier, William E. "Metal Additive Manufacturing: A Review." *Journal of Materials Engineering and Performance* 23, no. 6 (2014): 1917–28.

[2] Ligon, Sebastian C., et al. "Polymers for 3D Printing and Customized Additive Manufacturing." *Chemical Reviews* 117, no. 15 (2017): 10212–90.

[3] Thompson, Mary K., et al. "Design for Additive Manufacturing: Trends, Opportunities, Considerations, and Constraints." *CIRP Annals* 65, no. 2 (2016): 737–60.

[4] Lu, Jun, et al. "Additive Manufacturing for Next-Generation Electric Vehicles." *Additive Manufacturing* 30 (2019): 100830.

[5] Gebhardt, Andreas, et al. *Additive Manufacturing: 3D Printing for Prototyping and Manufacturing*. Carl Hanser Verlag, 2016.

[6] Shamsaei, Nima, et al. "An Overview of Additive Manufacturing of Titanium Components by Directed Energy Deposition: Microstructure and Mechanical Properties." *Journal of Materials Processing Technology* 252 (2018): 197–231.

[7] Bai, Chao, et al. "Additive Manufacturing of Metal–Ceramic–Metal Sandwich Structure Using Selective Laser Melting." *Journal of Alloys and Compounds* 834 (2020): 155197.

[8] Rashed, Md G., et al. "Additive Manufacturing of High-Performance 3D Printing Materials for the Production of Lightweight EV Components." *Materials & Design* 191 (2020): 108586.

[9] Liu, Peter, et al. "On the Role of Additive Manufacturing for Intelligent Electric Vehicle Design." *Journal of Advanced Transportation* 2020 (2020): 8736845.

[10] Raut, Dipen, et al. "Additive Manufacturing Processes for Electric Vehicles: Current Status and Prospects." *Journal of Cleaner Production* 238 (2019): 117956.

[11] Ma, Yuanzhi, et al. "Revolutionizing Electric Vehicle Design through Additive Manufacturing." *International Journal of Precision Engineering and Manufacturing* 21, no. 4 (2020): 659–76.

[12] Pinho, Thiago M., et al. "Additive Manufacturing Applications for Electric Vehicle Lightweight Structures." *Procedia CIRP* 99 (2021): 349–54.

[13] Du, Zhijiang, et al. "Additive Manufacturing Technologies for Multi-Functional EV and HEV Components." *Additive Manufacturing* 31 (2019): 100936.
[14] Shamsaei, Nima, et al. "Additive Manufacturing of Alloys for Automotive Applications." *Journal of Materials Processing Technology* 274 (2019): 116282.
[15] Bikas, Haralampos, Panagiotis Stavropoulos, and George Chryssolouris. "Additive Manufacturing Methods and Modeling Approaches: A Critical Review." *International Journal of Advanced Manufacturing Technology* 83, no. 1 (2016): 389–405.
[16] Li, Yuankun, et al. "Additive Manufacturing in Energy Storage and Conversion for Electric Vehicles." *Energy Storage Materials* 21 (2019): 58–78.
[17] Prakash, Chandan, et al. "3D Printing and Additive Manufacturing for Automotive Applications: The Industry Perspective." *Progress in Additive Manufacturing* 4, no. 2 (2019): 77–92.
[18] Rengier, Fabian, et al. "3D Printing Based on Additive Manufacturing Technologies: State of the Art and Future Perspectives." *Applied Radiology* 41, no. 5 (2010): 42–50.
[19] Herrmann, Christoph, et al. "Sustainability in Additive Manufacturing of EV and HEV Components." *Sustainable Production and Consumption* 26 (2021): 1014–29.
[20] Jackson, Benjamin, et al. "Additive Manufacturing for the Aerospace and Automotive Industries: Synergies and Challenges." *Additive Manufacturing* 31 (2019): 100837.
[21] Brandt, Milan, ed. *Laser Additive Manufacturing: Materials, Design, Technologies, and Applications*. 2nd ed. Woodhead, 2022.
[22] Bandyopadhyay, Amit, and Susmita Bose, eds. *Additive Manufacturing*. 2nd ed. Cambridge University Press, 2020.
[23] Galati, Manuela, and Luca Iuliano. "A Literature Review of Powder-Based Electron Beam Melting Focusing on Numerical Simulations." *Additive Manufacturing* 35 (2020): 101220.
[24] Ngo, Tuan D., Alireza Kashani, Gabriele Imbalzano, Kate T. Q. Nguyen, and David Hui. "Additive Manufacturing (3D Printing): A Review of Materials, Methods, Applications, and Challenges." *Composites Part B: Engineering* 143 (2019): 172–96.
[25] DebRoy, T., et al. "Additive Manufacturing of Metallic Components – Process, Structure, and Properties." *Progress in Materials Science* 92 (2019): 112–224.
[26] Herzog, Dirk, et al. "Additive Manufacturing of Metals." *Acta Materialia* 117 (2019): 371–92.
[27] Parthasarathy, Mandrel, and Trevor Dean. "Additive Manufacturing for Aerospace: A Review on Opportunities and Constraints." *Materials & Design* 224 (2022): 111335.
[28] Mohan, N., and K. Sivaprakasam. "Recent Advances in Polymer Additive Manufacturing and Challenges in Applications." *Journal of Polymer Research* 28, no. 2 (2021): 50.
[29] Zhang, Minghao, and Sudarsanam S. Babu. "A Review of Steel Alloys for Additive Manufacturing: Microstructure, Properties, and Challenges." *Additive Manufacturing* 49 (2022): 102479.
[30] Ranjan, Rajeev, Deepak Kumar, Manoj Kundu, and Subhash Chandra Moi. "A Critical Review on Classification of Materials Used in 3D Printing Process." *Materials Today: Proceedings* 61 (2022): 43–9.
[31] Awasthi, Ankita, Kuldeep K. Saxena, and Ravi K. Dwivedi. "An Investigation on Classification and Characterization of Bio Materials and Additive Manufacturing Techniques for Bioimplants." *Materials Today: Proceedings* 44 (2021): 2061–8.
[32] Attarzadeh, Faridreza, Behzad Fotovvati, Michael Fitzmire, and Ebrahim Asadi. "Surface Roughness and Densification Correlation for Direct Metal Laser Sintering." *The International Journal of Advanced Manufacturing Technology* 107 (2020): 2833–42.
[33] Murr, Lawrence E., Sara M. Gaytan, Diana A. Ramirez, Edwin Martinez, Jennifer Hernandez, Krista N. Amato, Patrick W. Shindo, Francisco R. Medina, and Ryan B. Wicker. "Metal Fabrication by Additive Manufacturing using Laser and Electron Beam Melting Technologies." *Journal of Materials Science & Technology* 28, no. 1 (2012): 1–14.
[34] Kurzynowski, T., A. Pawlak, and I. Smolina. "The Potential of SLM Technology for Processing Magnesium Alloys in Aerospace Industry." *Archives of Civil and Mechanical Engineering* 20, no. 1 (2020): 23.

[35] Ahn, Dong-Gyu. "Applications of Laser Assisted Metal Rapid Tooling Process to Manufacture of Molding & Forming Tools—State of the Art." *International Journal of Precision Engineering and Manufacturing* 12 (2011): 925–38.
[36] Mazurchevici, Andrei Danut, Dumitru Nedelcu, and Ramona Popa. "Additive Manufacturing of Composite Materials by FDM Technology: A Review." *Indian Journal of Engineering and Materials Sciences (IJEMS)* 27, no. 2 (2021): 179–92.
[37] Mostafaei, Amir, Amy M. Elliott, John E. Barnes, Fangzhou Li, Wenda Tan, Corson L. Cramer, Peeyush Nandwana, and Markus Chmielus. "Binder Jet 3D Printing—Process Parameters, Materials, Properties, Modeling, and Challenges." *Progress in Materials Science* 119 (2021): 100707.
[38] Tagliaferri, Vincenzo, Federica Trovalusci, Stefano Guarino, and Simone Venettacci. "Environmental and Economic Analysis of FDM, SLS and MJF Additive Manufacturing Technologies." *Materials* 12, no. 24 (2019): 4161.
[39] Msallem, Bilal, Neha Sharma, Shuaishuai Cao, Florian S. Halbeisen, Hans-Florian Zeilhofer, and Florian M. Thieringer. "Evaluation of the Dimensional Accuracy of 3D-Printed Anatomical Mandibular Models using FFF, SLA, SLS, MJ, and BJ Printing Technology." *Journal of Clinical Medicine* 9, no. 3 (2020): 817.
[40] Cooper, Casey, and Bruce Hughes. "Aerosol Jet Printing of Electronics: An Enabling Technology for Wearable Devices." In *2020 Pan Pacific Microelectronics Symposium (Pan Pacific)*, pp. 1–11. IEEE, 2020.
[41] Muthuram, N., P. Sriram Madhav, D. Keerthi Vasan, Manav Eshwar Mohan, and G. Prajeeth. "A Review of Recent Literatures in Poly Jet Printing Process." *Materials Today: Proceedings* 68 (2022): 1906–20.
[42] Faes, Matthias, H. Valkenaers, F. Vogeler, J. Vleugels, and E. Ferraris. "Extrusion-Based 3D Printing of Ceramic Components." *Procedia Cirp* 28 (2015): 76–81.
[43] Liu, Peter, et al. "On the Role of Additive Manufacturing for Intelligent Electric Vehicle Design." *Journal of Advanced Transportation* 2020 (2020): 8736845.
[44] Lu, Jun, et al. "Additive Manufacturing for Next-Generation Electric Vehicles." *Additive Manufacturing* 30 (2019): 100830.
[45] Pinho, Thiago M., et al. "Additive Manufacturing Applications for Electric Vehicle Lightweight Structures." *Procedia CIRP* 99 (2021): 349–54.
[46] Du, Zhijiang, et al. "Additive Manufacturing Technologies for Multi-Functional EV and HEV Components." *Additive Manufacturing* 31 (2019): 100936.
[47] Raut, Dipen, et al. "Additive Manufacturing Processes for Electric Vehicles: Current Status and Prospects." *Journal of Cleaner Production* 238 (2019): 117956.
[48] Ma, Yuanzhi, et al. "Revolutionizing Electric Vehicle Design through Additive Manufacturing." *International Journal of Precision Engineering and Manufacturing* 21, no. 4 (2020): 659–76.
[49] Herrmann, Christoph, et al. "Sustainability in Additive Manufacturing of EV and HEV Components." *Sustainable Production and Consumption* 26 (2021): 1014–29.
[50] Rashed, Md G., et al. "Additive Manufacturing of High-Performance 3D Printing Materials for the Production of Lightweight EV Components." *Materials & Design* 191 (2020): 108586.
[51] Jackson, Benjamin, et al. "Additive Manufacturing for the Aerospace and Automotive Industries: Synergies and Challenges." *Additive Manufacturing* 31 (2019): 100837.
[52] Song, Yingjie, et al. "Potential and Challenges of Additive Manufacturing for Lightweight Electric Vehicle Parts." *Materials* 13, no. 19 (2020): 4223.
[53] Kumar, S. Suresh, C. Durga Prasad, and Harish Hanumanthappa. "Role of Thermal Spray Coatings on Erosion, Corrosion, and Oxidation in Various Applications: A Review." *Journal of Bio-and Tribo-Corrosion* 10, no. 2 (2024): 22.
[54] Reddy, G. Madhu Sudana, C. Durga Prasad, Pradeep Patil, Gagan Shetty, M. R. Ramesh, and T. Nageswara Rao. "Investigation of Thermally Sprayed NiCrAlY/TiO2 and NiCrAlY/Cr2O3/YSZ Cermet Composite Coatings on Titanium Alloys." *Engineering Research Express* 4, no. 2 (2022): 025049.
[55] Madhu, G., K. M. Mrityunjaya Swamy, Durga Ajay Kumar, C. Durga Prasad, and U. Harish. "Evaluation of Hot Corrosion Behavior of HVOF Thermally Sprayed Cr3C2-35NiCr Coating on SS 304 Boiler Tube Steel." In *AIP Conference Proceedings*, vol. 2316, no. 1. AIP Publishing, 2021.

[56] Nayak, Haridasa, K. R. Varun, Sonal Shamkuwar, R. Suresh Kumar, T. A. Sudarshan, Abhijeet Malge, M. Vijay, and C. Durga Prasad. "Thermal Cycle Behaviour of Plasma Sprayed Thermal Barrier Coatings on Cast Iron Substrate for the Application of Liner of Internal Combustion Engine." *Results in Surfaces and Interfaces* 17 (2024): 100297.

[57] Prasad, C. Durga, Pradeep Patil, Naresh Kakur, and M. R. Ramesh. "Elevated Temperature Erosion Performance of Plasma Sprayed NiCrAlY/TiO2 Coating on MDN 420 Steel Substrate." *Surface Topography: Metrology and Properties* 10, no. 2 (2022): 025010.

[58] Prasad, C. Durga, Sharnappa Joladarashi, M. R. Ramesh, and M. S. Srinath. "Microstructural and Tribological Resistance of Flame-Sprayed CoMoCrSi/WC-CrC-Ni and CoMoCrSi/WC-12Co Composite Coatings Remelted by Microwave Energy." *Journal of Bio-and Tribo-Corrosion* 6 (2020): 1–15.

[59] Dufresne, Eric, et al. "Thermal Spray Coatings on Additive Manufacturing Components: Challenges and Opportunities." *Journal of Thermal Spray Technology* 30, no. 6 (2021): 941–57.

[60] Wang, T. H., et al. "Integration of Thermal Spray Coatings in Additive Manufacturing Components for High-Temperature Applications." *Surface and Coatings Technology* 408 (2021): 126868.

[61] Bamberger, Marco, et al. "Thermal Spray Coatings for Additive Manufacturing: A Review of Recent Developments." *Materials* 13, no. 5 (2020): 1153.

[62] Valek, David, et al. "Enhancing the Surface Properties of Additively Manufactured Components with Thermal Spray Coatings." *Surface Engineering* 36, no. 12 (2020): 1457–67.

[63] Ferreira, Jorge A., et al. "Additive Manufacturing and Thermal Spray Coatings: A Synergistic Approach to Improve Performance of Metallic Components." *Journal of Manufacturing Science and Engineering* 142, no. 11 (2020): 111004.

[64] Jafari, M., et al. "Development of Hybrid Coating Systems for Additive Manufacturing Components Using Thermal Spray Processes." *Materials and Design* 187 (2020): 108336.

[65] Pinte, Emilie, et al. "Plasma Spray Coating of Additively Manufactured Parts: Influence on Performance and Durability." *Journal of Materials Processing Technology* 273 (2019): 116262.

[66] Blaker, J., et al. "Applications of Thermal Spray Coatings in Additive Manufacturing: Enhancing Wear and Corrosion Resistance." *Additive Manufacturing* 31 (2019): 100925.

[67] Zhang, Yi, et al. "Cold Spray Coating on Additively Manufactured Parts: A Novel Approach for Enhanced Material Performance." *Materials & Design* 179 (2019): 107919.

[68] Yao, J., et al. "The Role of High-Velocity Oxy-Fuel (HVOF) Coatings in Additive Manufacturing for High-Performance Components." *Surface and Coatings Technology* 384 (2020): 125293.

[69] Zhao, Zhenyu, et al. "Integration of Thermal Spray Coatings for Surface Engineering in Additive Manufacturing: Progress and Challenges." *Journal of Thermal Spray Technology* 28, no. 3 (2019): 453–71.

[70] Yadav, Amit, et al. "Enhancing the Performance of AM Components Using Thermal Spray Coatings." *International Journal of Advanced Manufacturing Technology* 109 (2020): 473–84.

[71] Guo, Hong, et al. "Thermal Spray Coatings for Additive Manufacturing: The Effect of Coating Process on Performance and Applications." *Coatings* 10, no. 3 (2020): 266.

[72] Zhou, Yang, et al. "Integration of Additive Manufacturing and Thermal Spray Coatings for Aerospace Components: A Review of Processing Techniques and Performance." *Aerospace Science and Technology* 104 (2020): 105927.

[73] Dwivedi, Praveen Kumar, Girish Kumar, and R. C. Singh. "Justification Framework for Adoption of Additive Manufacturing to Automotive Supply Chain: AHP Approach." *International Journal of Maritime Engineering* 1, no. 1 (2024): 455–62.

[74] Champion, Jesse D., Kayleigh M. Dodd, Hilaire C. Lam, Mohammad AM Alzahrani, Sara Seifan, Ellie Rad, David Oliver Scourfield et al. "Drug Inhibition of Redox Factor-1 Restores Hypoxia-Driven Changes in Tuberous Sclerosis Complex 2 Deficient Cells." *Cancers* 14, no. 24 (2022): 6195.

[75] Franz, D. N., J. J. Bissler, and F. X. McCormack. "Tuberous Sclerosis Complex: Neurological, Renal and Pulmonary Manifestations." *Neuropediatrics* 41, no. 5 (2010): 199–208.

[76] Raut, Dipen, et al. "Enhancement of Wear and Corrosion Resistance of Additive Manufactured Automotive Components Using Thermal Spray Coatings." *Journal of Manufacturing Processes* 61 (2020): 182–93.

[77] Yadav, Amit, et al. "Thermal Spray Coatings for Additively Manufactured Automotive Components: Challenges and Solutions." *Surface and Coatings Technology* 386 (2020): 125458.
[78] Cicala, Giacomo, et al. "Role of Thermal Spray Coatings in the Enhancement of Mechanical Properties for Additive Manufactured Automotive Components." *Materials & Design* 195 (2020): 108964.
[79] Tavares, Sara, et al. "Plasma Spray Coatings on 3D Printed Automotive Components: Performance Improvements for High-Temperature Applications." *Journal of Thermal Spray Technology* 30, no. 7 (2021): 1131–44.
[80] Huang, Shao-Feng, et al. "High-Velocity Oxy-Fuel (HVOF) Coatings for Additive Manufactured Automotive Components: Advances and Challenges." *Journal of Materials Science & Technology* 57 (2021): 124–32.
[81] Nowakowska, Monika, Leszek Łatka, Paweł Sokołowski, Mirosław Szala, Filofteia-Laura Toma, and Mariusz Walczak. "Investigation into Microstructure and Mechanical Properties Effects on Sliding Wear and Cavitation Erosion of Al2O3–TiO2 Coatings Sprayed by APS, SPS and S-HVOF." *Wear* 508 (2022): 204462.

7 Polymer Hollow Fiber Heat Exchanger

Novel Approach to Battery Cooling

Ananth S. Iyengar, Prakash C.H., Abhilash N., Pavan Kumar Reddy, Nimona Hailu, and Jeyaprakash N.

7.1 INTRODUCTION

Heat exchangers are indispensable components in a wide range of industrial and domestic applications, serving as critical devices for transferring thermal energy between fluids. This fundamental functionality underpins systems such as power plants, chemical processing facilities, refrigeration units, and modern electric vehicle (EV) battery cooling systems [1]. Traditionally, heat exchangers have been manufactured using metals like copper, aluminum, and stainless steel, owing to their exceptional thermal conductivity, mechanical strength, and durability. While these attributes have solidified metals as the material of choice for heat exchangers over decades, they are not without significant drawbacks. Metallic heat exchangers are often susceptible to corrosion when exposed to certain environments or chemicals, require complex and energy-intensive manufacturing processes, and are inherently heavy and expensive [2]. These limitations have spurred innovation in alternative materials and designs, leading to the emergence of polymeric heat exchangers (PHEs). PHEs leverage polymers as their primary construction material, presenting an attractive alternative to traditional metallic designs. Polymers offer numerous advantages, such as resistance to corrosion, lightweight properties, and cost-effectiveness. Furthermore, their inherent electrical insulating properties make them particularly suitable for applications where electrical safety is a concern. In the context of EV battery cooling systems, these characteristics offer transformative potential. The design and manufacturing processes for PHEs are generally more flexible compared to metals, enabling the creation of innovative geometries and integrated solutions. This design freedom not only reduces the overall weight of the heat exchanger but also contributes to enhanced performance by optimizing heat transfer and fluid dynamics. These heat exchangers are challenging the conventional heat exchangers not only in the field of heating, ventilation, and air conditioning (HVAC) and automobile radiators, but also in low-temperature-range applications such as EV car battery cooling.

7.1.1 Heat Generation and Thermal Management

Understanding the electrical behavior and thermal characteristics of lithium-ion batteries is crucial for optimizing performance, safety, and longevity. Various lithium battery technologies, based on electrochemical reactions, are commercially available. These include lithium iron phosphate (LFP $LiFePO_4$) having a long cycle life and lower energy density, lithium nickel manganese cobalt oxide (NMC, $LiNiMnCoO_2$) with high energy density but which uses expensive cobalt, lithium cobalt oxide (LCO, $LiCoO_2$) with high energy density but shorter life span, lithium manganese oxide (LMO, $LiMn_2O_4$) with high thermal stability, but shorter life cycle, lithium titanate (LTO, $Li_4Ti_5O_{12}$) with exceptional cycle life but low energy density, and lithium nickel cobalt aluminum oxide (NCA, $LiNiCoAlO_2$) with high energy density but which is expensive. Based on the common

 DOI: 10.1201/9781003596097-7

characteristics of these cells, an equivalent circuit model is created to accommodate the individual differences in the internal chemistry and overall lithium cell module working. Experimental measurement and theoretical assessment are some of the common methods to determine the heat generation in battery packs. In general, experimental methods involve measurement of fluid temperature in contact with the cells [3], while core temperature measurement is also reported in the literature [4]. This method does not give a physical insight into the chemical reaction mechanism of heat generation.

7.1.2 Equivalent Circuit Model (ECM)

A model that mimics the heat generation of lithium cells has been developed to understand the inner workings of the cell. This model is an electrical circuit equivalent in behavior that provides a simplified yet effective representation of the cell's electrical dynamics, while providing heat generation analysis. The model helps in managing temperature rises and preventing thermal runaway. Heat generation in lithium-ion cells occurs due to various electrochemical and resistive effects. The primary sources of heat generation are, Joule heating (ohmic losses), entropic heat, and electrochemical polarization heat generation. The total heat generated can be estimated using Eq. (7.1).

$$Q_{total} = I^2 R_o + I \frac{dU}{dT} \tag{7.1}$$

where Q_{total} is the total heat generated. Joule heating ($I^2 R_o$) arises due to the internal resistance (R_o) of the cell when current (I) flows through it either for charging or discharging. The term ($\frac{dU}{dT}$) is the entropic coefficient (V/K), which varies with the state of charge (SOC). Joule heating dominates at high charge or discharge rates (< 1 C). At high SOCs, some lithium-ion chemistries show exothermic behavior, while at lower SOCs, it can be endothermic. Typically, heat generation calculations result in a low value of heat generation for moderate loads in batteries with low charge/discharge cycles. However, due to dense packing of these batteries, the heat generation density per volume poses engineering challenges to maintain the battery at safe temperatures [5]. Equation (7.1) is a first-order approximation of heat generation and can be represented with Thevenin's equivalent circuit model. This circuit is shown in Figure 7.1, where R_o is further divided into ohmic resistance, the charge transfer resistance R_1, and the polarization resistance occurring due to the charging and discharging R_2. These resistances are also coupled with the electrolyte capacitance C_1 and capacitance due to the charge separation at electrode–electrolyte interfaces C_2.

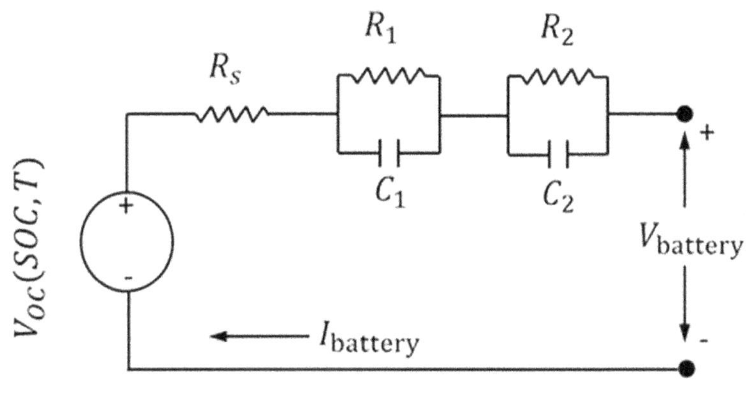

FIGURE 7.1 The Thevenin model showing the equivalent circuit model for heat generation.

7.2 THERMAL MANAGEMENT OF LITHIUM-ION BATTERIES

Liquid cooling of EV batteries is widely utilized and is the most effective technique for maintaining optimal thermal conditions within the battery pack [6,7]. Figure 7.2 shows a common configuration of liquid cooling for 18650 batteries, with glycol-based coolant flowing through ribbed cooling tubes. Other techniques based on air flow, mist, and phase-change materials are all reported in the literature and are used for applications where energy performance is not critical. High-performance applications, such as EV operation, require effective thermal management ensuring the quick discharge, longevity, and safety of EV batteries, particularly under high-power operation or extreme environmental conditions. The liquid cooling method typically involves circulating a non-corrosive, electrically insulating ethylene glycol-based nanofluid through the battery pack to absorb and dissipate heat. This approach leverages the high thermal conductivity and specific heat capacity of the cooling fluid, ensuring efficient temperature regulation even during peak loads. Traditional liquid cooling systems rely on welded metal tubes to direct the coolant flow through the battery modules. While this design is effective in achieving thermal management, it introduces several challenges, particularly with regard to weight. The metal components, combined with additional hardware required for the final assembly and integration into the vehicle chassis, significantly increase the overall weight of the cooling system [7]. This added weight can reduce the energy efficiency and range of the EV, counteracting the fundamental objectives of lightweight vehicle design. The need for a more efficient and lightweight solution has led to increased interest in alternatives to traditional metal-based cooling systems. Lightweight polymers present a promising avenue for achieving comparable thermal management performance while addressing the limitations associated with metal components. Polymers offer several advantages, including lower density, corrosion resistance, and design flexibility, making them ideal candidates for modern EV applications [8]. By replacing metallic tubes with polymer-based cooling channels or components, significant weight reductions can be achieved without compromising the efficiency of the thermal management system.

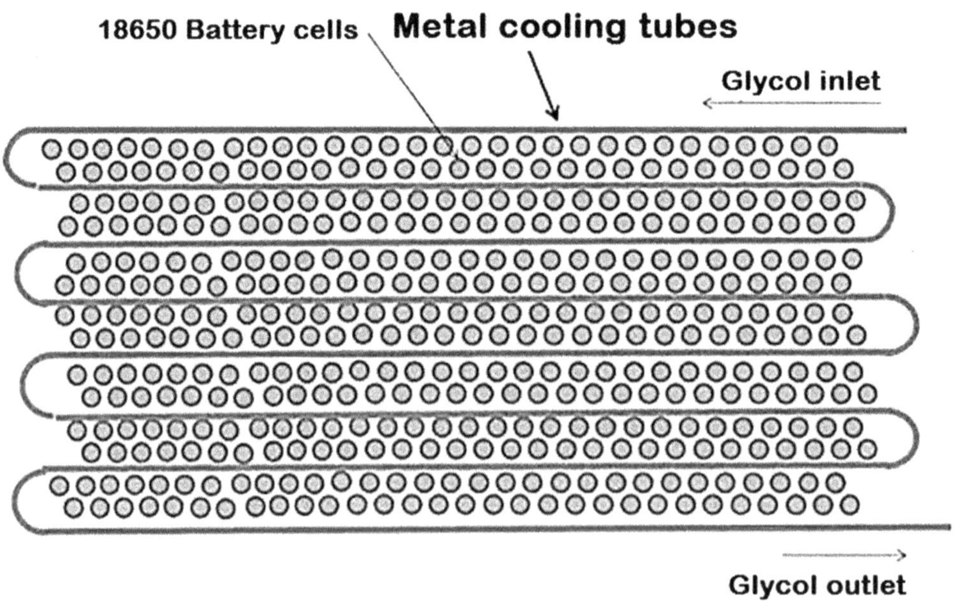

FIGURE 7.2 Ribbed cooling paths for a battery pack in electric vehicles.

Micro-vascular plate liquid cooling using a carbon fiber plate heat exchanger for EV car battery cooling has been reported, however, this design is not suited for cylindrical cells [9]. polymeric hollow-fiber (PHF) heat exchangers are tested with common fluids such as air, water, and cooling oil [10]. PHF heat exchangers utilize hollow flexible polymer fibers as the primary medium for heat exchange, capitalizing on their high surface area contact to facilitate efficient thermal energy transfer. They can be considered micro-heat exchangers because fluid flows within narrow confinements characterized by dimensions typically less than 1 mm. The most common examples of such confinements are microchannel or submillimeter inner diameter holes. The unique structure of these fibers allows for innovative configurations that can significantly enhance performance while maintaining compactness. In comparison to traditional metallic heat exchangers, PHF heat exchangers are particularly well-suited to EV battery cooling systems, where weight, efficiency, and electrical safety are paramount concerns. The high surface area of hollow fibers is a key factor that enables superior heat transfer performance in PHF heat exchangers. When fluids flow through the hollow fibers, the large surface area facilitates efficient thermal energy exchange between the fluids and the surrounding environment. This design also allows for the use of thinner fiber walls, which further enhances heat transfer by reducing thermal resistance. Additionally, the flexibility of polymeric materials enables the production of fibers with customized dimensions and geometries, catering to specific application requirements.

A primary limitation of polymer heat exchangers is the inherently low thermal conductivity of polymer materials. This challenge can be mitigated by minimizing the thickness of the heat exchanger tube. However, a thin-walled tube with a larger diameter tends to have a lower burst pressure. To enhance burst pressure, the tube diameter can be reduced [3]. For example, PHF heat exchangers can be engineered with complex fiber architectures to optimize fluid contact with the heat exchange surface, thereby enhancing thermal performance, as illustrated in Figure 7.3.

In EV battery cooling systems, the advantages of PHF heat exchangers are particularly evident. During operation, batteries produce substantial heat, making temperature regulation essential for optimal performance, lifespan, and safety. Conventional aluminum radiators, widely used in internal combustion engine (ICE) vehicles, offer effective cooling but are not

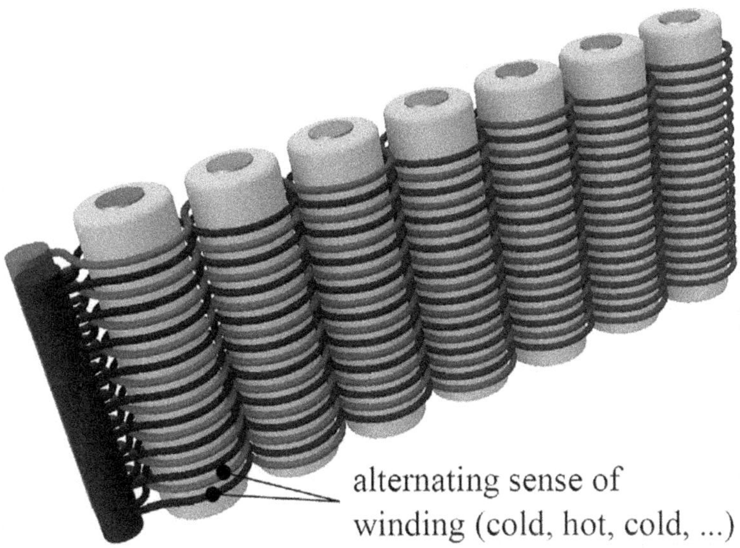

FIGURE 7.3 A schematic representation of hollow fiber heat exchange for cooling of 18650 batteries.

ideally suited to the specific thermal management requirements of EVs. In contrast, PHEs provide superior thermal performance at similar airflow rates, enabling efficient heat dissipation while maintaining advantages in weight reduction and design flexibility. Studies have shown that PHEs can achieve an average weight reduction of 30% compared to aluminum radiators, a critical advantage in the context of EV design where minimizing weight directly contributes to improved energy efficiency and vehicle range. Another noteworthy advantage of PHEs lies in their potential for miniaturization. The compact design enabled by hollow fibers makes PHF heat exchangers ideal for applications where space is a constraint. In EVs, where components must be carefully arranged to optimize functionality and aesthetics, the ability to integrate compact and efficient heat exchangers is invaluable. Moreover, the use of polymeric materials in PHEs allows for seamless integration with other polymer-based components in the vehicle, simplifying assembly processes and reducing overall manufacturing costs.

Figure 7.4 shows the modular heat exchanger design with staggered stacking of cells for improved thermal contact. The corrosion resistance of polymeric materials further enhances the durability and reliability of PHEs. Unlike metals, which are prone to degradation in the presence of moisture, chemicals, or extreme temperatures, polymers exhibit excellent resistance to these factors. This property ensures that PHEs maintain consistent performance over extended periods, reducing maintenance requirements and enhancing the overall lifespan of the system. Additionally, polymers can be engineered to possess specific thermal and mechanical properties, enabling the development of tailored solutions for diverse applications.

While the benefits of PHF heat exchangers are numerous, their adoption is not without challenges. One primary concern is the relatively lower thermal conductivity of polymers compared to metals. However, advances in materials science have led to the development of thermally conductive polymer composites, which bridge this gap by incorporating additives such as carbon nanotubes, graphene, or metal particles into the polymer matrix. These composites enhance the thermal conductivity of the polymer without compromising its other advantageous properties. As a result, PHEs constructed from thermally conductive polymers can achieve performance levels comparable to their metallic counterparts while retaining their inherent benefits. The manufacturing processes for PHEs pose many challenges. Producing hollow polymer fibers with precise dimensions and consistent quality

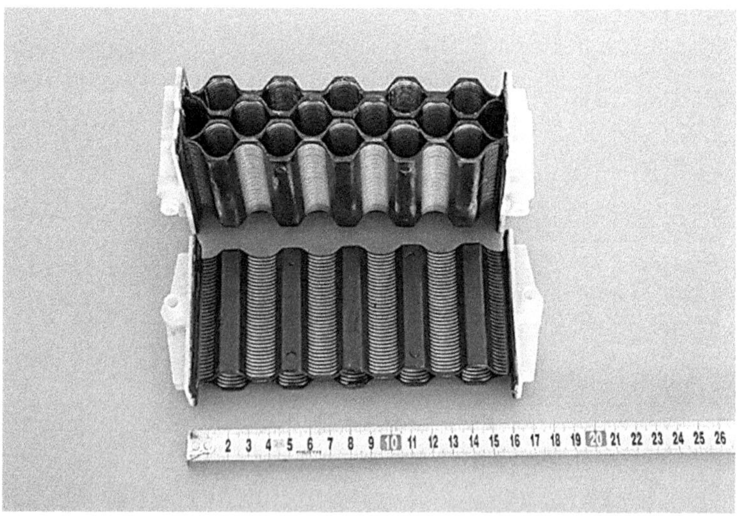

FIGURE 7.4 Image showing the modular cooling PHE device for cylindrical lithium-ion cells.

requires advanced techniques such as extrusion or melt spinning. While these processes are well-established, scaling them for mass production and maintaining cost-effectiveness can be complex. However, ongoing research and development efforts are focused on optimizing manufacturing methods to address these challenges and facilitate wider adoption of PHE technology. In addition to EV battery cooling systems, PHEs have promising applications in other industries. For example, they can be employed in HVAC systems to improve energy efficiency and reduce operational costs. In the medical field, PHEs can be used in devices such as blood oxygenators and dialysis machines, where precise and efficient heat transfer is essential. The versatility of PHEs makes them a valuable technology for addressing a broad spectrum of thermal management challenges. The environmental benefits of PHEs further strengthen their appeal. The lightweight and corrosion-resistant nature of polymeric materials reduces the energy required for manufacturing, transportation, and maintenance. Additionally, the potential for recycling polymeric materials aligns with global efforts to promote sustainability and reduce waste. By replacing traditional metallic heat exchangers with PHEs, industries can achieve significant reductions in their carbon footprint and contribute to a more sustainable future.

7.2.1 Design of Polymer Heat Exchangers

The design of PHEs is similar to any metallic heat exchanger. There are two major types of calculations using log mean temperature difference (LMTD) and effectiveness – number of transfer units (ε – NTU) methods to calculate the area of heat transfer. The overall heat transfer between the hot and cold fluids is given by Eq. (7.2):

$$Q = \dot{m}_t C_{p,t} \left(T_{c,o} - T_{c,i}\right) = \dot{m}_s C_{p,s} \left(T_{h,o} - T_{h,i}\right) \tag{7.2}$$

where $\dot{m}$ is the mass flow rate of the fluid, C_p is the specific heat of the fluids, and T, Q are the temperatures and heat flow rate, respectively. The subscripts h, c, i, o, t, and s stand for hot side, cold side, inlet, outlet, tube side, and shell side, respectively. The overall heat transfer coefficient U is calculated based on the convective heat transfer coefficients on the hot and cold side fluids and the material properties of the tubes. The convective heat transfer coefficient is a function of the Graetz number, which is a dimensionless number used to predict the thermal entrance length in laminar flow, and is given by Eq. (7.3):

$$Gz = \frac{D_h}{L} \mathrm{Re}\,\mathrm{Pr} \tag{7.3}$$

where D_h is the hydraulic diameter, L is the length of the tube, Re is the Reynold's number, and Pr is the Prandtl number. The heat transfer coefficient is calculated using the Nusselt number correlations using the Darcy–Weisbach equation.

The LMTD is calculated using Eq. (7.4):

$$LMTD = \frac{\left(T_{h,i} - T_{c,o}\right) - \left(T_{h,o} - T_{c,i}\right)}{ln\left(\left(T_{h,i} - T_{c,o}\right) \Big/ \left(T_{h,o} - T_{c,i}\right)\right)} \tag{7.4}$$

The heat transfer area is therefore calculated using Eq. (7.5):

$$A = \frac{Q}{\left(U \times LMTD\right)} \tag{7.5}$$

Heat transfer area can also be calculated using NTU and effectiveness calculations. The effectiveness is given by Eq. (7.6):

$$\varepsilon = \frac{q}{q_{max}} = \frac{C_h\left(T_{h,i} - T_{h,o}\right)}{C_{min}\left(T_{h,i} - T_{c,i}\right)} \tag{7.6}$$

where C is the product of mass flow rate and specific heat capacity, with the subscripts the same as previously stated. The NTU method involves the calculation of area of heat transfer using NTU. The expression for NTU is as follows [Eq. (7.7)]:

$$NTU = \frac{UA}{C_{min}} \tag{7.7}$$

The pressure drop across the tube side is calculated using Eq. (7.8):

$$\Delta P = fL\frac{\rho u^2}{2D_h} \tag{7.8}$$

where f is the friction factor and u is the velocity of the fluid in the tube.

7.2.2 Design Configurations of PHF Heat Exchangers

PHF heat exchangers are commonly designed in configurations such as shell-and-tube, parallel flow, counter flow, and crossflow arrangements. Within the shell-and-tube setup, the PHF can be arranged either in a structured pattern or in a randomly distributed manner. The increased tortuosity of randomly oriented PHF can enhance overall heat transfer by introducing greater flow resistance on the shell side. Studies indicate that this random orientation improves heat transfer efficiency, increasing it from 970 $W/m^2{\cdot}K$ to 1091 $W/m^2{\cdot}K$ [11]. Cross-flow heat exchangers are good design alternatives for HVAC applications and automobile radiator designs [12,13]. Since these designs are used for both gas and liquid applications, battery cooling can be achieved using air or with any heat transfer liquid [14]. Additionally, the PHF material being pliant, stretches due to tensile force, therefore it is important to note that the stresses in the PHF do not affect the air side pressure drop in the cross-flow heat exchanger configuration [13]. The tension and resulting stresses are more important in the cross-flow arrangement. Care should be taken in the case of battery cooling as the expansion and contraction of the PHF material may risk loss of good heat transfer contact between the battery and cooling coils.

The fluid flow connections with the PHF are challenging due to their miniature dimensions. One of the methods used for connecting these PHFs is to dip them in a polymer resin. The resin bonding is achieved by first placing the fibers in a designed template such that it makes liquid have a tight

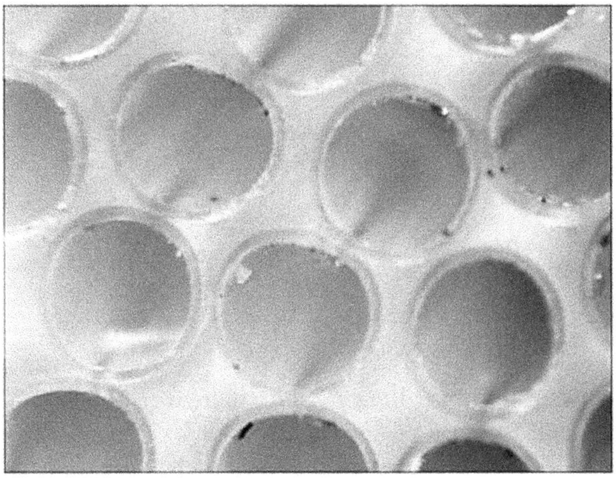

FIGURE 7.5 Image of the polymeric fibers embedded in resin for fluid connections.

contact with the PHF. Resin is carefully poured to bond with the PHF and is released from the template. The bonding achieved by this method has sufficient strength to withstand the flow resistance and pressure drop created by the PHF [15]. A section of the fibers embedded in resin is shown in Figure 7.5.

7.3 THE HEAT EXCHANGE AND FLUID FLOW IN PHF HEAT EXCHANGERS

Hollow fibers are tubular structures with small diameters, typically ranging from 50 to 500 μm, and wall thicknesses of 10–100 μm. The typical dimensions and materials are given in Table 7.1. These dimensions enable a high packing density, which translates into increased surface area for heat transfer. A variety of plastics such as polypropylene [16], fluoroplastic materials such as polyvinylidene fluoride (PVDF) [17], ethylene tetrafluoroethylene (ETFE), and polytetrafluoroethylene (PTFE) [18], are highly suitable materials for hollow fiber heat exchangers due to their robust chemical structures and exceptional thermal stability. These materials exhibit remarkable resistance to a wide range of corrosive agents, including acids, alkalis, oxidizing agents, and organic halogenated solvents. Among them, PVDF has been extensively studied for use in hollow fiber heat exchangers, largely because of its cost-effectiveness and reliable stability in acidic environments. The other common PHF heat exchange materials are polypropylene (PP) [19], polyethersulfone (PES) [19], polyamide (PA) [13,20], and polyetheretherketone (PEEK) [19]. The thermophysical properties of these material and fiber dimensions are given in Table 7.1. Polymers used in PHEs must balance thermal, mechanical, and chemical properties. The key criteria include thermal conductivity, chemical resistance, mechanical strength, and ease of forming. The lower thermal conductivity of polymers compared to that of metals can be neglected because of the small material barrier between the heat exchange fluids [16]. Further, this can be enhanced by suitable additives and fillers.

The thermal conductivity of the material has a direct impact on the overall heat transfer and performance of the heat exchanger. Polymer composites, integrating fillers such as carbon nanotubes or graphene, improve thermal conductivity without significantly compromising workability during processing. Some of the highest reported thermal conductivity values include carbon fiber-reinforced PVDF material with 2.8 W/m^2 K [17]. This material has shown an overall heat transfer coefficient of about 618–900 W/m^2 K for water–water heat exchange [18]. The upper temperature limit of PVDF

TABLE 7.1
Typical Dimensions and Materials of Polymeric Hollow Fibers

	Polypropylene (PP)	Polyethersulfone (PES)	Polyamide (PA)	Polyetheretherketone (PEEK)
ID (μm)	425 [19]	700 [19]	1300 [24]	150, 420 [19]
Wall thickness (μm)	75	150	150	105, 75
Working temperature (°C)	80 [21]	80 [21]	80 [21]	130 [21]
Burst pressure (MPa)	4		6	6

is about 150°C but for PTFE the upper limit range is as high as 200°C [18]. PTFE also has anti-fouling properties that lead to reduced maintenance and longer life of heat exchanger.

Polymer hollow fibers are fabricated using methods such as extrusion, phase inversion, melt spinning, and electro-spinning. Additional surface modifications can be achieved using plasma or chemical grafting. Extrusion is a common method for producing hollow fibers, involving the continuous forcing of molten polymer through a die with a mandrel. Phase inversion involves casting a polymer solution into a coagulating bath, forming hollow fibers with controlled porosity. Melt spinning uses molten polymer to produce hollow fibers through a spinneret, which is ideal for thermoplastics. Electrospinning creates ultrafine fibers with high surface areas but is limited by low production rates and mechanical strength. Surface treatments, such as plasma or chemical grafting, enhance thermal properties, hydrophilicity, and fouling resistance.

The hollow fibers, with an inner diameter of less than 1 mm, exhibit a relative roughness of less than 0.001 mm. As a result, they can be classified as hydrodynamically smooth tubes, where the roughness does not affect heat transfer or fluid flow [21]. Due to the small inner diameter of the tubes, the tube side Reynolds number is in the laminar flow regime for most common fluids. However, the fluid property changes are almost instant due to the high value of the overall heat transfer coefficient (200–400 W/m^2 K approximately [16,20,22]). For example, water has a dynamic viscosity of 1.002 mPa.s at 20°C, and reduces to 0.469 mPa.s at 60°C. These variations in fluid properties have to be accounted for in the design of PHF heat exchangers because they may hinder flow in the tubes. One of the main advantage of a PHF heat exchanger is the heat transfer bundle, composed of small hollow fibers, which is highly flexible. This flexibility allows it to adapt and conform to spaces of various arbitrary shapes. This unique characteristic offers a distinct advantage over traditional rigid metal heat exchangers. Due to these advantages, PHF heat exchangers are used in various applications including ICE radiator, HVAC, water desalination, exhaust gas heat removal, and electronic cooling technology. Additionally, the arrangement of layers of these polymer tubes also plays an important role in heat transfer in the shell side heat transfer. Chaotic arrangement of fibers, although less efficient, can provide a quick installation with sufficient performance due to turbulent flow on the outer surface of the hollow fibers and coverage of the heat transfer area [23].

The compact dimensions of hollow fibers offer a significant advantage by enabling high heat transfer coefficients on both their internal and external surfaces. This phenomenon can overcome the low thermal conductivity of polymer materials. Within the fibers, fluid flow is laminar, resulting in a constant Nusselt number [25]. Consequently, the heat transfer coefficient is inversely proportional to the internal diameter of the fibers. A similar phenomenon occurs on the external surface, where a reduction in fiber size corresponds to an enhancement in heat transfer intensity [25].

The small diameters of hollow fibers can result in significant pressure drops, necessitating careful design to optimize flow dynamics. The pressure drop across 300 mm length hydrophobic polypropylene fiber with 0.6 mm outer diameter, 0.48mm inner diameter, membrane thickness of 0.6 mm, having a flow rate of 700 mL/hour and a maximum pressure drop of 100 kPa [26]. The pressure drop can be further reduced by selecting a hydrophilic material or coating provided to the tube. The surface tension of the fluid over the surface of the polymer is critical in the pressure drop estimation.

7.3.1 Fluid Flow and Heat Transfer in PHE

Computational fluid dynamics analysis of PHE has been described in the literature [16,27]. Heat transfer in PHEs involves conduction through the fiber walls and convection within the fluids on either side. The thin walls of the fibers minimize thermal resistance, while the hollow geometry facilitates efficient fluid flow. The chosen PHF material has to match the temperature and heat transfer requirement for the battery cooling application with typical temperatures ranging from 15 to 35°C. A polypropylene-based PHE with tube side water temperature of 40–70°C, in the same range as that of high discharge battery temperature, has been reported in the literature. Tube side flow is laminar with a Reynold's number less than 300. The tube side heat transfer coefficient increases with an increase in fluid velocity. This behavior is mirrored on the shell side heat transfer coefficient [28]. It is reported that the overall heat transfer coefficient for PHE reached a value of 1675 W/m^2 K. This value is higher compared to a metal heat exchanger with a comparable heat transfer area of 1100–1400 W/m^2 K. This behavior can be related to the resistance to the flow shown on the tube side, shell side, and wall. The tube side resistance is the least among the three, indicating the change in flow has little effect in overall heat transfer. The wall friction resistance is less than the shell side resistance. The shell side resistance increases with the number of tube bundles, thus making the shell side tube layout design and the number of tubes the most important factors in improving the overall heat transfer coefficient.

Numerical studies of the PHF heat exchangers have been reported in the literature with an emphasis on the shell side heat transfer [27,29]. The tube side parameter particularly emphasized is the flow angle at the entry of the tube. The shell side heat transfer is influenced by the cooling medium. For an inlet flow angle of 0° and Reynolds number (Re) below 1000, the water-cooling flow regime inside the tube is assumed to be laminar. Consequently, a laminar flow model is reported for this scenario. Heat transfer from the fiber's outer surface to the cooling medium is assumed to occur solely through the thermal conductivity of water. Convection effects are assumed to be negligible.

For flow angles greater than 0°, the Shear–Stress Transport (SST) turbulence model was chosen for water flow in tubes, integrating both the k-epsilon and k-omega models. The k-omega model was applied to capture turbulence in the near-wall region, while the k-epsilon model described turbulence within the main flow volume. More computational resource-intensive Large Eddy Simulations (LES) and Detached Eddy Simulations (DES) models provide a more detailed simulation of flow turbulence, but their use is not reported for this application.

Research findings from these studies suggest that the entry angle significantly influences heat transfer in tube-side water flow. Additionally, the impact of tortuosity on heat transfer is more pronounced for water flow on the shell side compared to air [27]. The packing fraction also has an effect on the overall heat transfer coefficient. the overall heat transfer coefficient increases initially as the packing fraction increases from 0.025 to 0.15, but it tapers down for higher packing fractions till it reaches 0.35. These results would be extremely difficult to study via experimentation.

7.3.1.1 *Fouling*

The polymer microfibers have smooth internal and external surfaces that are fouling resistant to dust in gas and to organic and inorganic pollutants in liquids. Despite these advantages, the shell side fluid particulate matter is prone to get filtered on the outer surface of fibers. Particulate fouling tests have indicated that polymeric fiber surfaces exhibit significant resistance to fouling. Although the fouling rate was initially high, it was offset by a low asymptotic fouling value and weak deposit adhesion. The asymptotic fouling value was estimated to be approximately 15×10^{-5} m^2 K/W, leading to a reduction of about 20% in the overall heat transfer coefficient (HTC) [30]. The rapid accumulation of deposits was primarily attributed to the low liquid velocities used in the experiments. The air bubbling technique is a potential method for removing fouling particles from the liquid flow in heat exchangers. This approach effectively dislodges loosely attached particles and weakly bound aggregates. However, it is not entirely effective for strongly adhered deposits, and particles located near sharp bends remained unaffected. It is important to note that with routine cleaning procedures, up to 90% of fouling particles were successfully removed using this method.

The shell side and air side of the heat exchanger also show fouling, but the dust is easier to clean and maintain [31]. Individual fibers exhibit strong resistance to fouling due to the excellent surface properties of the polymeric material. However, dense fiber arrangements are more susceptible to fouling, primarily due to particle accumulation and clogging in the gaps between them. To mitigate particulate fouling in highly dusty environments, it is recommended in the literature to maintain a fiber spacing of at least 2 mm [31].

7.3.1.2 *Mechanical Integrity and Durability*

Hollow fibers must withstand internal and external pressures without deforming or rupturing. Mechanical durability is tested under cyclic loading and long-term operation [22]. Long-term durability under fluctuating temperatures and pressures is critical, necessitating advanced testing and material development. Fatigue testing of polymer fibers is reported in the literature, with internal pressure loading of 0.35 MPa, for a time period of 1.4 s and total number of cycles of 1.5 million. These tests indicate that the polymer fibers are capable of handling fluid flow conditions of up to 1 MPa at 80°C for PP material [32]. These fiber heat exchangers are reported to be capable of cooling electronic components without any wall rupture [25].

7.3.1.3 *Cost*

While polymers are cost-effective, the fabrication of hollow fibers and composite materials can increase initial costs. Economies of scale and process optimization are required to make PHEs more competitive. With low initial material cost and low maintenance cost, PHEs are very attractive for adoption in the industry. Also, any component failure can be replaced quickly due to the wide availability and low cost.

7.3.2 Future Directions

7.3.2.1 *Material Innovation*

Polymers inherently exhibit low thermal conductivity, approximately one-hundredth that of metals, which significantly restricts heat transfer efficiency. The inherently low thermal conductivity of polymers limits the heat transfer efficiency. Research into high-conductivity composites and advanced fabrication techniques is ongoing. Developing polymers with higher thermal conductivities and better mechanical properties is a priority. Nanocomposites and hybrid materials hold significant promise [9]. Additionally, the surface modifications can make the tubes hydrophilic or hydrophobic, thus reducing the flow restriction and reducing the pressure drop across the bundle. Experimental evaluations of hollow fibers reported in the literature exhibit high heat transfer coefficients, reaching up to 450 W/(m^{2}K) for liquid-to-gas and up to 2100 W/(m^{2}K) for liquid-to-liquid applications. Mixed matrix membranes and hollow fibers with carbon nanotube polymer mixtures are reported to improve the thermal conductivity of the polymer [33].

7.3.2.2 *Advanced Manufacturing*

Techniques such as 3D printing and microfabrication could revolutionize PHE design, enabling complex geometries and integration with other components. Modifications to extrusion heads of the fused deposition modeler can extrude polymer hollow fibers, and with careful placement of these fibers it is possible to develop PHEs that fit snuggly to the heat transfer surface which can reduce the contact heat resistance. This method could accommodate printed circuit boards also in the future.

7.3.2.3 *Multifunctional Heat Exchangers*

Integrating additional functionalities, such as filtration gas separation and catalytic activity, could expand the utility of PHEs in advanced applications. In applications such as exhaust gas cleaning and flue gas cooling PHEs have the capability of providing combinations of these functionalities.

7.3.2.4 *PHE design with Artificial Neural Networks (ANNs)*

Artificial neural networks and machine learning are used to predict the mass flow of water through porous polymer membranes for water filtration processes [34]. On similar lines, the heat transfer behavior of individual polymer fibers and overall performance of PHEs can be studied and performance predictions can be made. This will make the design of PHEs in the future faster and reliable.

7.4 CONCLUSIONS

In conclusion, hollow fiber heat exchangers represent a significant advancement in heat exchanger technology, offering a compelling combination of efficiency, durability, and design flexibility. Their high heat transfer surface area within a low heat exchanger volume, lightweight properties, and corrosion resistance make them particularly well-suited for applications in EV battery cooling systems and beyond. While challenges such as lower thermal conductivity and manufacturing complexities remain, ongoing innovations in materials and processes are steadily addressing these issues. As industries continue to prioritize energy efficiency, sustainability, and performance, PHEs are poised to play a pivotal role in shaping the future of thermal management solutions. This chapter explores the principles and potential of PHEs, emphasizing their advantages, limitations, and future potential. PHEs with hollow fibers represent a transformative approach to heat transfer, offering unparalleled flexibility, corrosion resistance, and adaptability. Despite challenges related to thermal performance, fouling, and mechanical durability, ongoing advancements in materials, fabrication techniques, and design strategies continue to push the boundaries of what PHEs can achieve. As industries increasingly prioritize sustainability and efficiency, PHEs are poised to play a pivotal role in next-generation thermal management solutions.

REFERENCES

[1] M. H. Soliman, "Thermal management systems for li-ion batteries used in electric vehicles," *Assiut Univ. Bull. Environ. Res.*, vol. 27, no. 1, pp. 51–99 2024, doi: 10.21608/auber.2024.288607.1085.

[2] T. Kroulíková, E. Bartuli, and T. Kůdelová, "Polymeric hollow fiber heat exchangers in higher temperatures," In *Proceedings of the 9th International Conference on Fluid Flow, Heat and Mass Transfer (FFHMT'22)* (pp. 1–4). Niagara Falls, Canada. https://doi.org/10.11159/ffhmt22.192

[3] J. Bohacek, M. Raudensky, and E. Karimi-Sibaki, "Polymeric hollow fibers: Uniform temperature of Li-ion cells in battery modules," *Appl. Therm. Eng.*, vol. 159, no. June, p. 113940, 2019, doi: 10.1016/j.applthermaleng.2019.113940.

[4] A. Jinasena *et al.*, "Online internal temperature sensors in lithium-ion batteries: State-of-the-art and future trends," *Front. Chem. Eng.*, vol. 4, no. February, pp. 1–22, 2022, doi: 10.3389/fceng.2022.804704.

[5] M. Kaliaperumal and R. K. Chidambaram, "Thermal management of lithium-ion battery pack using equivalent circuit model," *Vehicles*, vol. 6, no. 3, pp. 1200–1215, 2024, doi: 10.3390/vehicles6030057.

[6] N. You, J. Ham, D. Shin, and H. Cho, "Performance analysis of the liquid cooling system for lithium-ion batteries according to cooling plate parameters," *Batteries*, vol. 9, no. 11, pp. 1–14, 2023, doi: 10.3390/batteries9110538.

[7] M. Macdonald, S. R. Annapragada, A. Sur, R. Mahmoudi, C. Lents, and A. Jain, "Early design stage evaluation of thermal performance of battery heat acquisition system of a hybrid electric aircraft," *Nutr. Today*, vol. 17, no. 2, pp. 1–9, 2020, doi: 10.1115/1.4046159.

[8] P. T. Tennessen, J. C. Weintraub, and W. A. Hermann, "US8758924 Tesla liquid cooling ribbon 2011," vol. 2, no. 12, 2014, [Online]. Available: https://patents.google.com/patent/US8758924B2/en.

[9] S. J. Pety, M. H. Y. Tan, A. R. Najafi, P. R. Barnett, P. H. Geubelle, and S. R. White, "Carbon fiber composites with 2D microvascular networks for battery cooling," *Int. J. Heat Mass Transf.*, vol. 115, pp. 513–522, 2017, doi: 10.1016/j.ijheatmasstransfer.2017.07.047.

[10] F. Lang, "Polymer hollow fiber oil cooler for a racing car," In *Proceedings of the 10th International Conference on Fluid Flow, Heat and Mass Transfer (FFHMT'23)*, Paper No. 177, pp. 1–8. Canada. https://doi.org/10.11159/ffhmt23.177

[11] E. Bartuli, "The influence of hollow fibers orientation inside the polymeric hollow fiber heat exchanger on the heat transfer intensity," *Eng. Mech.*, vol. 24, *2018*, pp. 61–64, 2018, doi: 10.21495/91-8-61.

[12] K. Mraz *et al.*, "Case study of liquid cooling of automotive headlights with hollow fiber heat exchanger," *Case Stud. Therm. Eng.*, vol. 28, no. December, p. 101689, 2021, doi: 10.1016/j.csite.2021.101689.

[13] T. Kroulikova, K. Mraz, J. Hvozda, and J. Bohacek, "Crossflow polymeric hollow fiber heat exchanger: Fiber tension effects on heat transfer and airside pressure drop," *J. Therm. Anal. Calorim.*, vol. 149, no. 8, pp. 3155–3164, 2024, doi: 10.1007/s10973-024-12956-5.

[14] H. S. Lau *et al.*, "State-of-the-art organic-and inorganic-based hollow fiber membranes in liquid and gas applications: Looking back and beyond," *Membranes (Basel).*, vol. 12, no. 5, pp. 1–69, 2022, doi: 10.3390/membranes12050539.

[15] M. Raudensky, *Polymeric Heat Exchangers with Hollow Fibres*. Brno University of Technology, Faculty of Mechanical Engineering, 2021, [Online]. Available: www.heatlab.cz/research/polymeric-heat-exchangers-with-hollow-fibres/#:~:text=Polymeric hollow fibre heat exchangers,of 0.7–1.5 mm.

[16] I. Astrouski, M. Raudensky, and M. Dohnal, "Polymeric hollow fiber heat exchangers," *21st Int. Congr. Chem. Process Eng. CHISA 2014 17th Conf. Process Integr. Model. Optim. Energy Sav. Pollut. Reduction, PRES 2014*, vol. 4, no. Ht, p. 2481, 2014, doi: 10.2495/ht160101.

[17] G. Yi, J. Li, L. C. Henderson, W. Lei, L. Du, and S. Zhao, "Enhancing thermal conductivity of polyvinylidene fluoride composites by carbon fiber: Length effect of the filler," *Polymers (Basel).*, vol. 14, no. 21, 2022, doi: 10.3390/polym14214599.

[18] G. Yi, X. Tang, L. Du, X. Li, and S. Zhao, "Hollow fiber polytetrafluoroethylene membrane heat exchanger with anti-corrosion properties," *J. Memb. Sci.*, vol. 678, no. April, p. 121664, 2023, doi: 10.1016/j.memsci.2023.121664.

[19] L. Song, B. Li, D. Zarkadas, S. Christian, and K. K. Sirkar, "Polymeric hollow-fiber heat exchangers for thermal desalination processes," *Ind. Eng. Chem. Res.*, vol. 49, no. 23, pp. 11961–11977, 2010, doi: 10.1021/ie100375b.

[20] J. Hvozda, K. Mraz, M. Raudensky, A. Vakhrushev, E. Karimi-Sibaki, and J. Bohacek, "Case study on thermal management of planar elements with various polymeric heat exchangers: Experiment and simulation," *J. Therm. Anal. Calorim.*, vol. 149, no. 11, pp. 5229–5238, 2024, doi: 10.1007/s10973-024-13172-x.

[21] T. Kudelova, M. Raudensky, and E. Bartuli, "Limits and use of polymeric hollow fibers as material for heat transfer surfaces," *Proc. 3rd Int. Congr. Adv. Mater. Sci. Eng. AMSE 2022*, no. February 2023, 2022, doi: 10.1109/AMSE51862.2022.10036675.

[22] M. Raudensky, T. Kudelova, E. Bartuli, T. Kroulikova, and I. Astrouski, "Cooling of flue gas by cascade of polymeric hollow fiber heat exchangers," *Case Stud. Therm. Eng.*, vol. 36, no. March, p. 102220, 2022, doi: 10.1016/j.csite.2022.102220.

[23] T. Kroulíková, I. Astrouski, and M. Raudenský, "Chaotised polymeric hollow fibre bundle as a cross-flow heat exchanger in air-water application," *Acta Polytech.*, vol. 60, no. 4, pp. 318–323, 2020, doi: 10.14311/AP.2020.60.0318.

[24] E. Bartuli and T. Kroulikova, "Testing of polymeric hollow fibre heat exchanger with crossed hollow fibres," (Paper No. HTFF 143). In *Proceedings of the 5th World Congress on Mechanical, Chemical, and Material Engineering (MCM'19)*. Lisbon, Portugal. Paper No. HTFF 143, pp. 2-9 https://doi.org/10.11159/htff19.143

[25] M. Raudensky, I. Astrouski, T. Brozova, and E. Bartuli, "Flexible polymeric hollow fiber heat exchangers for electronic systems," *Proc. 15th Intersoc. Conf. Therm. Thermomechanical Phenom. Electron. Syst. ITherm 2016*, no. May, pp. 1143–1147, 2016, doi: 10.1109/ITHERM.2016.7517677.

[26] T. Kůdelová, E. Bartuli, A. Strunga, J. Hvožd'a, and M. Dohnal, "Fully polymeric distillation unit based on polypropylene hollow fibers," *Polymers (Basel).*, vol. 13, no. 7, pp. 1–18, 2021, doi: 10.3390/polym13071031.

[27] E. Bartuli and M. Raudensky, "Numerical investigation of heat transfer on the outer surface of polymeric hollow fibers," *Mater. Tehnol.*, vol. 52, no. 4, pp. 459–463, 2018, doi: 10.17222/mit.2016.221.

[28] X. Chen *et al.*, "Performance analysis and design implementation of a novel polymer hollow fiber liquid desiccant dehumidifier with aqueous potassium formate," *Therm. Sci. Eng. Prog.*, vol. 13, pp. 1–16, 2019, doi: 10.1016/j.tsep.2019.100366.

[29] J. Zhao, B. Li, X. Li, Y. Qin, C. Li, and S. Wang, "Numerical simulation of novel polypropylene hollow fiber heat exchanger and analysis of its characteristics," *Appl. Therm. Eng.*, vol. 59, no. 1–2, pp. 134–141, 2013, doi: 10.1016/j.applthermaleng.2013.05.025.

[30] I. Astrouski, M. Raudenský, and M. Dohnal, "Particulate fouling of polymeric hollow fiber heat exchanger," *Proc. Int. Conf. Heat Exch. Fouling Clean.*, vol. 2013, no. 3, pp. 233–239, 2013.

[31] I. Astrouski, M. Raudensky, T. Kudelova, and T. Kroulikova, "Fouling of polymeric hollow fiber heat exchangers by air dust," *Materials (Basel).*, vol. 13, no. 21, pp. 1–12, 2020, doi: 10.3390/ma13214931.

[32] T. Brožová, T. Luks, I. Astrouski, and M. Raudenský, "Fatigue testing of polymeric hollow fibre heat transfer surfaces by pulsating pressure loads," *Appl. Mech. Mater.*, vol. 821, pp. 3–9, 2016, doi: 10.4028/www.scientific.net/amm.821.3.

[33] C. Y. Feng, K. C. Khulbe, T. Matsuura, and A. F. Ismail, "Recent progresses in polymeric hollow fiber membrane preparation, characterization and applications," *Sep. Purif. Technol.*, vol. 111, pp. 43–71, 2013, doi: 10.1016/j.seppur.2013.03.017.

[34] X. L. Nguyen, N. Van Trinh, Y. Kim, and S. Yu, "A correlation of overall mass transfer coefficient of water transport in a hollow-fiber membrane module via an artificial neural network approach," *Membranes (Basel).*, vol. 13, no. 1, pp. 1–11, 2023, doi: 10.3390/membranes13010008.

8 Advanced Composites and Sustainable Materials for Lightweight Electric Vehicle Components

Mohamed Azlan Suhot, Nicholas Kuan Hoo Tien, Mohamad Zaki Hassan, and Indran Suyambulingam

8.1 INTRODUCTION

Electric vehicles (EVs) and hybrid electric vehicles (HEVs) are pushing the car industry toward sustainability [1]. With increasing worries about carbon emissions and shrinking resources, EVs and HEVs provide an alternative to internal combustion engine-powered cars [2]. The development of such vehicles seeks to use less energy, reduce harm to the environment, as well as improve performance, while meeting safety standards.

Analyzing new EV registrations in the European Union from November 2023 to November 2024 (Figure 8.1) shows a big shift toward electrification—HEVs climbed from 27.5% to 33.2% of registrations, battery electric vehicles (BEVs) increased to 15%, while plug-in hybrid electric vehicles (PHEVs) took 7.6%. The drop in petrol (30.6%) and diesel vehicles (10.6%) adds pressure on automakers to adopt lightweight materials that improve energy efficiency and range [3]. Lightweight materials such as natural fiber composites help meet these targets because they reduce weight yet keep the integrity of the structure [1,4–7].

Lightweight materials in EVs are important because they improve performance and extend range [8]. They also reduce vehicle weight. This saves energy and uses less power and solves the small battery capacity problem. They meet emerging safety standards by adding strength, providing protection in collisions [9]. Research indicates that heavier EVs consume more energy, leading to a significant reduction in travel range due to higher energy demands. The increased vehicle weight also extends the braking distance, necessitating enhanced braking systems to mitigate road safety risks. According to recent findings [9], passenger fatality risk increases by 40–50% when struck by a vehicle that is 1000 lbs. heavier than their own. Additionally, excessive weight-related tire wear contributes to shortened tire lifespans, requiring new tire designs to handle the added stress. Road infrastructure also suffers, as heavier EVs cause increased surface cracking and deformation, accelerating road degradation compared to internal combustion engine (ICE) vehicles. These concerns emphasize the necessity of lightweight materials to maintain vehicle efficiency, safety, and long-term sustainability, making them critical for the next generation of EV designs [9].

This chapter examines the role of advanced lightweight materials, including traditional natural fiber materials and innovative composites such as bio-enhanced composites and sustainable matrices in EVs. It highlights their applications, performance benefits, and contributions to sustainability, while exploring their life cycle and potential to align with circular economy goals. The evaluation of environmental benefits not only reaffirms hybrid polymers' ecological advantages, but also their ability to contribute to the larger objective of sustainable vehicle manufacture [1,9,10].

DOI: 10.1201/9781003596097-8

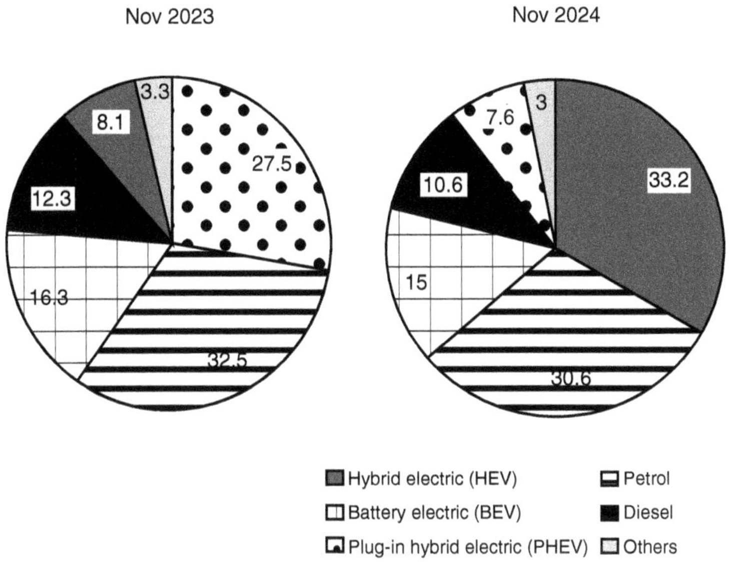

FIGURE 8.1 Comparison of new vehicle registrations by power source in the European Union for November 2023 and November 2024. Figure created by the author. (Data from ACEA [11].)

8.2 THE ROLE OF LIGHTWEIGHT MATERIALS IN EV PERFORMANCE

Lightweight materials significantly enhance vehicle dynamics by improving handling, acceleration, and braking performance. The reduced mass allows for more responsive steering and shorter stopping distances—critical factors for both performance and safety. Recent advancements in composite materials, particularly in polymer matrix composites (PMCs) and hybrid fiber-reinforced composites, have further revolutionized vehicle lightweighting by optimizing strength-to-weight ratios while maintaining crashworthiness [12]. According to Khan et al. [12], composite bumper beams, which are crucial for impact energy absorption, have demonstrated superior performance compared to traditional metallic beams, reducing vehicle weight by up to 30% without compromising safety. Recent studies indicate that glass fiber-reinforced polymer (GFRP) bumper beams offer 40–60% weight savings compared to steel, while maintaining similar energy absorption efficiency. A future concern raised by the automobile industry is the substantial potential for employing natural fiber-reinforced polymer materials and recycled plastics. The growing requirement for natural fiber over time and its outstanding performance will undoubtedly have an impact on the environment and biodiversity [13].

Reducing vehicle weight directly impacts energy consumption, allowing EVs to achieve greater range per charge. A schematic representation of an EV system (Figure 8.2) shows the key components and their relationships, illustrating the role of advanced materials and technologies in improving performance and efficiency. Lightweight composites contribute to better mileage [14] and reduced charging frequency, addressing consumer concerns about range anxiety [3,15]. According to Zhou et al. [11], reducing EV weight by 100 kg increases the travel range of a BEV by 3.6%, while the same reduction in ICE vehicles only yields a 2.4% range improvement. Beyond range extension, weight reduction leads to lower total energy consumption and reduced life cycle greenhouse gas (GHG) emissions. A 100-kg weight reduction in EVs corresponds to fuel savings of 0.3–0.5 L per 100 km and CO_2 emissions reduction of 0.85–1.4 kg per 100 km. Furthermore, the life cycle environmental impact of EVs is highly dependent on electricity generation sources, with coal-based electricity increasing EV life cycle CO_2 emissions by 17–27% compared to gasoline and diesel-powered vehicles. According to Kuan et al. [7], research on the low-velocity impact (LVI)

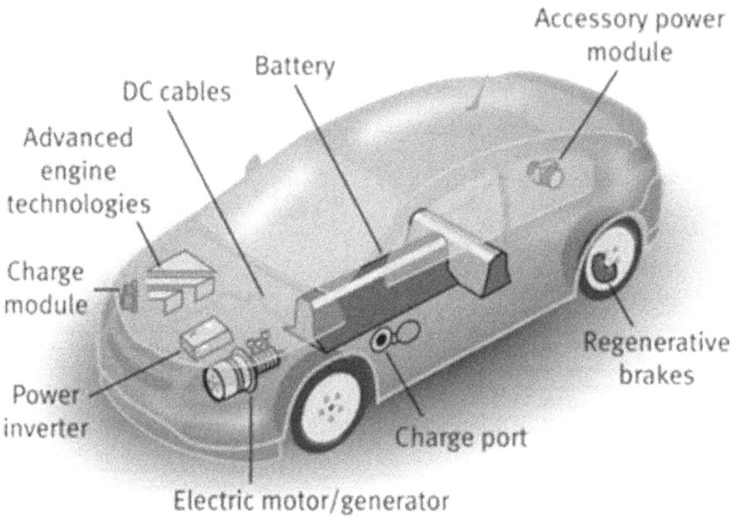

FIGURE 8.2 Schematic representation of key components of an EV [16]. Licensed under CC BY-NC-SA 4.0. https://oli.cmu.edu/courses/electric-vehicle-technology-nsc-stem-pathways/#

response has shown that jute fiber-reinforced composites exhibit superior energy absorption and toughness compared to glass fiber-reinforced composites, which are commonly used in automotive safety applications.

Given the critical role of lightweight materials in enhancing EV performance, material selection becomes a fundamental factor in optimizing vehicle efficiency. Traditional lightweight materials such as carbon fiber- reinforced polymers (CFRPs) and glass fiber composites are used widely because they offer high strength. The next section looks at their applications and limitations in EV design.

8.3 TRADITIONAL LIGHTWEIGHT MATERIALS IN EV APPLICATIONS

According to Kim et al. [17], CFRPs have very good strength properties, so they are ideal in structures for EVs. They have higher strength-to-weight ratio than traditional materials like aluminum or steel. Their density is 1.5 g/cm^3 versus 2.7 g/cm^3 for aluminum or 7.8 g/cm^3 for steel. This lowers weight while keeping a superior structural integrity. They have thermal stability and resist environmental factors [17]. CFRPs are costly to produce and are challenging to recycle, which limits their use for mass production. Iyer et al. [18] state that the embodied energy of CFRPs varies between 183–286 MJ/kg, much more than steel (25–35 MJ/kg) or aluminum (150–200 MJ/kg), which raises the life-cycle costs. Glass fiber composites are used in structural parts such as battery enclosures or chassis components. They provide a cheaper substitute for CFRPs.

While glass fiber-reinforced composites (GFRPs) cost less, their stiffness and strength are inferior to CFRPs, so they are suitable for less demanding applications. Cao et al. [19] demonstrated that modifications in resin systems of glass fiber composites can significantly enhance mechanical performance. According to recent experimental studies, using toughened resins (EPA650-Md) and silane coupling agent (SCA)-treated glass fiber fabrics led to a substantial increase in tensile strength and elastic modulus. The ultimate tensile strength (UTS) of GFRP laminates with high-strength resins showed improvements of 1.9%–2.1% in breaking elongation, enhancing ductility, and structural resilience. The lightweight materials market from 2020 to 2030 is growing as demand climbs for advanced materials like CFRPs and glass fiber composites because they reduce weight and improve energy efficiency in EVs [20]. Polymers and composites lead the market because they are versatile and show superior properties.

Beyond CFRPs and glass fiber composites, other advanced composite materials such as hybrid fiber-reinforced composites and metal matrix composites (MMCs) gain traction in automotive applications because they offer tailored mechanical and thermal properties [21]. Hybrid composites mix carbon, glass, and natural fibers. They offer a balanced combination of mechanical strength, cost efficiency, and sustainability, making them a promising substitute for EV components. MMCs, especially those reinforced with ceramic particles or fibers, provide excellent wear resistance, greater thermal stability, and high strength for demanding structural applications in EVs. As Phiri et al. [21] reported, advancements in manufacturing methods such as vacuum infusion, resin transfer molding (RTM), and additive manufacturing offer feasible production of complex composite structures with fewer defects and optimal material distribution. However, challenges in recyclability, cost reductions, and mass production persist as main concerns. Further research must pursue sustainable composite processing and circular economy methods for EV composite components at the end of their life.

Although CFRPs and glass fiber composites offer high strength-to-weight ratios, their impact on nature and issues with recycling create difficulties for their widespread sustainable use. To solve these problems, sustainable polymer matrices have become a promising alternative. The next section looks at bio-based, recyclable polymers that follow circular economy principles in EV design.

8.4 SUSTAINABLE MATRIX MATERIALS FOR EV APPLICATIONS

Sustainable polymer matrices, such as bio-based and recyclable polymers, solve environmental issues by reducing non-renewable resource reliance. Polylactic acid (PLA) and polyhydroxybutyrate (PHB) are examples of bio-based polymers offering biodegradability and renewable sourcing, making them ideal for lightweight EV applications [22]. According to Wang et al. [22], PLA and PHAs currently constitute over 55.5% of global bio-based plastic production, with PLA accounting for 13.9% and PHAs for 1.2% of total output. The biodegradation rate of PLA in industrial composting conditions reaches 90% within 6 months, while PHAs degrade completely in marine environments within 6–12 months, significantly reducing plastic waste accumulation. The classification of biopolymers based on their sources and chemical nature, as illustrated in Figure 8.3, further emphasizes the potential of these materials in promoting sustainability. Sustainable matrices combine environmental benefits with mechanical performance, supporting industry goals for green manufacturing and product development.

Beyond sustainability, safety remains a key factor in EV material selection. The rising use of lithium-ion batteries calls for advanced materials that lower weight yet offer fire resistance and thermal stability. The next section looks at how fire-retardant polymers enhance EV safety without compromising performance.

8.5 POLYMERS IN EV SAFETY AND PERFORMANCE

Most EVs use lithium-ion batteries, raising the need for materials that resist fire to mitigate thermal hazards. Advanced composites with fire-retardant properties ensure safety without compromising performance or weight [24]. According to Kim et al. [24], incorporating melamine-coated CaAl-layered double hydroxide (CaAl-LDH) into polyphenylene sulfide (PPS)-based composites significantly enhances flame retardancy and thermal conductivity. The PPS/GF/40 m-CaAl-LDH composite achieved a limiting oxygen index (LOI) of 52.74%, a substantial increase from 33% in pure PPS and 45.8% in PPS/GF composites, demonstrating superior fire resistance. Rigorous testing of fire-retardant composites ensures compliance with international safety standards, enhancing consumer confidence. Lightweight composites achieve optimal strength-to-weight ratios, critical for maintaining structural integrity while minimizing weight.

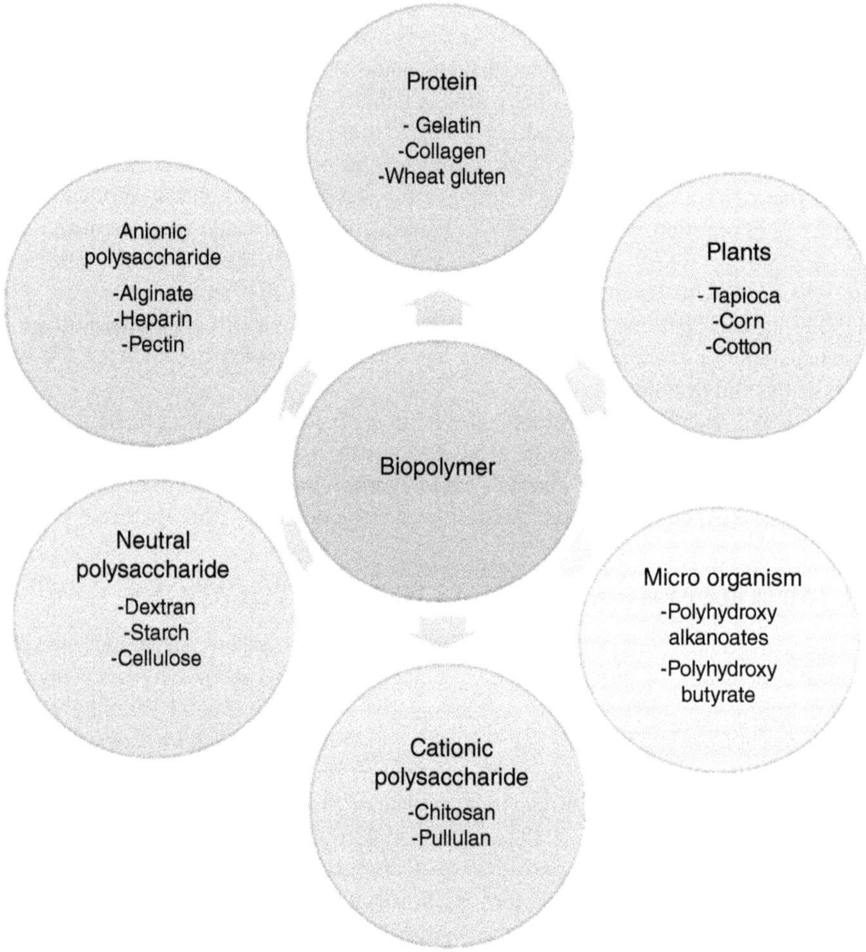

FIGURE 8.3 Classification of biopolymers based on their sources. (Reproduced from Ref. [23] under CC BY license.)

Figure 8.4 shows that modern car bumpers use advanced polymers and hybrid materials. These bumpers meet performance, safety, and environmental requirements. Wolf [25] and Xie et al. [26] provide valuable insights into the advancements in bumper design, focusing on pedestrian safety, energy absorption, and impact resistance. Wolf [25] emphasizes the importance of finite element modeling (FEM) in evaluating pedestrian injuries, highlighting how modern bumpers are designed to comply with Global Technical Regulation (GTR) No. 9 and Euro NCAP pedestrian safety standards. These studies explore how polyurethane composites and energy-absorbing foams contribute to mitigating lower extremity and head injuries by reducing tibia bending moments and head injury criterion (HIC) values. Xie et al. [26] further expand on material innovations by investigating negative Poisson's ratio (NPR) structures, particularly elliptical NPR materials, which offer superior energy absorption and stress redistribution. These materials exhibit transverse expansion upon compression, improving deformation control and impact resilience. The integration of hybrid polymer bumpers with NPR materials enhances both lightweight performance and crashworthiness, addressing key challenges in pedestrian safety and structural durability.

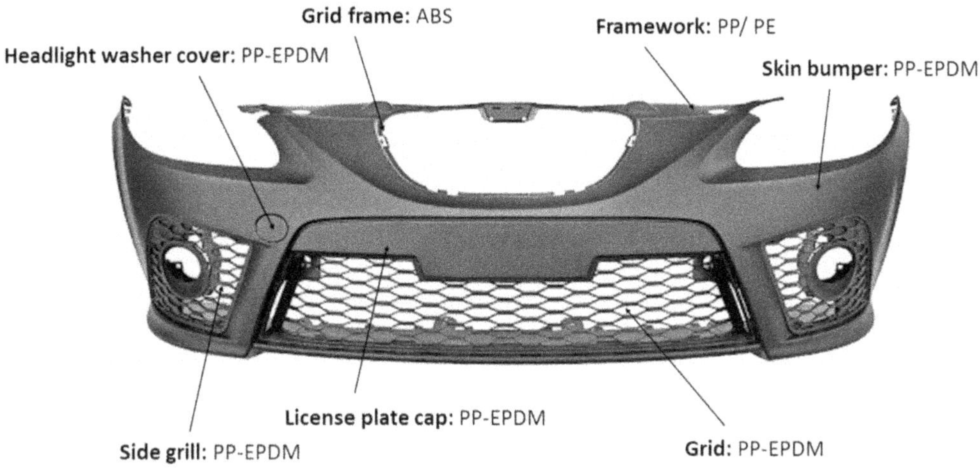

FIGURE 8.4 Material composition and design of a modern automobile bumper. (Reproduced from Ref. [2] under CC BY license.)

Research results help people understand design principles and select materials for energy absorbers in car bumpers. This knowledge plays a big part in improving the safety and performance of hybrid polymer car bumpers. Wu et al. [27] introduced new bi-material multistable auxetic honeycombs (BMAH) that significantly improve energy absorption and impact resistance in automotive applications. These structures show multiple stages of energy dissipation. They shift from elastic to plastic deformation. This change raises bumper effectiveness in low-speed, high-energy crashes. In contrast to conventional bumper absorbers, BMAH structures use two materials: hyperelastic and elastic-plastic elements. This mix optimizes deformation under impact conditions. Wu et al. [27] highlighted NPR materials that expand when compressed. They ensure better force distribution with less stress on the bumper. This innovation meets current pedestrian safety regulations as it lowers injury risks by absorbing impact forces more efficiently. Introducing BMAH structures in hybrid polymer bumpers shows promise. It offers better reusability, light weight, and superior crashworthiness, which support the development of safer, resilient bumpers.

While the integration of fire-retardant polymers improves vehicle safety, the broader sustainability impact of lightweight materials goes beyond their functional lifespan. A full life cycle assessment (LCA) shows how these materials increase resource efficiency, recyclability, and waste reduction in EV manufacturing and disposal. The next section reviews LCA methods and examines their role to advance a circular economy for EV materials.

8.6 LIFE CYCLE ANALYSIS OF LIGHTWEIGHT EV MATERIALS

New ideas with innovative materials in the car industry change how factories work to help them be more environmentally friendly, with greater care for society. This idea called materials innovation for automotive sustainability, tries to make EV manufacturing use materials that are less harmful to nature. It seeks to make recycling easier, use fewer resources, and improve overall efficiency. According to Zahoor et al. [14], the use of recycled aluminum in EV manufacturing reduces CO_2 emissions by 97% compared to new aluminum production, while synthetic SofTex leather generates 85% less CO_2 emissions than genuine leather during processing. Additionally, Ford's initiative to repurpose 1.2 billion plastic bottles annually for vehicle components results in a 50–60% weight reduction and a 37% decrease in CO_2 emissions for new vehicles. Furthermore, recent advancements in bio-based materials such as bamboo, bio-paint, and natural fiber composites have demonstrated the potential to replace petroleum-based materials, significantly reducing carbon footprints in the EV sector.

When focus has been placed on lightweight, recycled, and renewable materials in EV design—though not entirely new—research and development have recently picked up speed. This happens because people care more about how cars affect nature in their life cycle. These efforts study new materials comprehensively to address the car industry's increasing carbon footprint problem and find practical solutions [28]. The idea of a circular economy—seen in Figure 8.5—shows how sustainability integrates into everything: supply chain, material choice, design making, and policy-making. This way encourages the use of renewable and recyclable materials, e.g. bio-based and natural fiber composites, which cuts down on non-renewable resources while allowing recycling or bio-degradability.

A life cycle assessment evaluates materials for various applications and durations, emphasizing the importance of end-of-life considerations. Ideally, automotive materials should be recyclable after their complete life cycle. Consequently, the concepts of renewability and recyclability have become central to sustainable automotive material strategies, supporting a circular economy approach within the industry. Innovative recycling processes for composites support a circular economy, extending the life cycle of EV materials. Lightweight materials enable resource efficiency, reducing the environmental footprint of EV manufacturing and disposal, especially natural fiber composites that have shown significant properties [6,14].

Life cycle analysis highlights the strengths and limitations of current lightweight materials, revealing the need for further advancements in hybrid composites and manufacturing processes. As the automotive industry continues to evolve, new challenges emerge in achieving cost-effective,

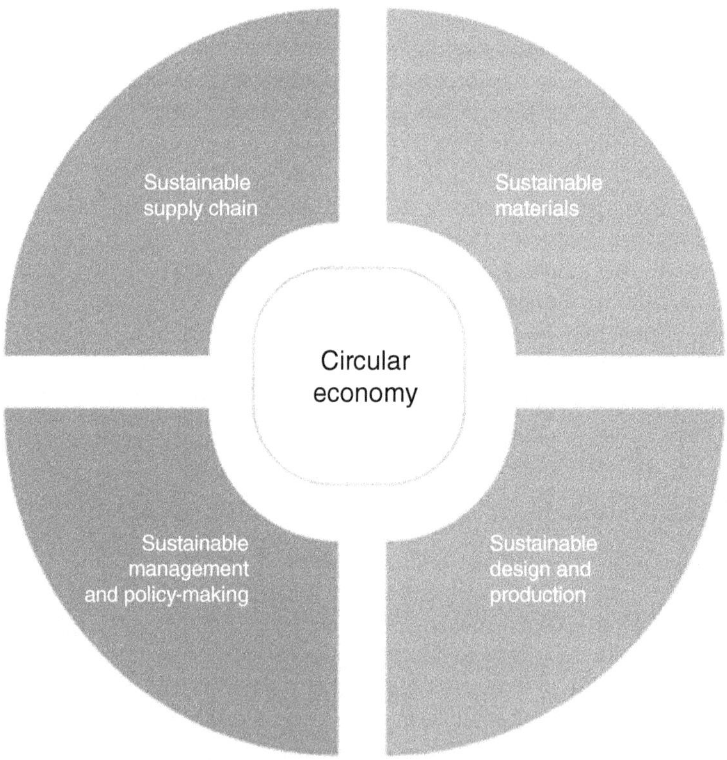

FIGURE 8.5 The circular economy framework, highlighting the integration of sustainable materials, design, production, supply chain management, and policy-making. (Reproduced from Ref. [10] under CC BY license.)

scalable, and high-performance materials. The following section explores ongoing research in hybrid composites and innovative fabrication techniques that aim to bridge this gap.

8.7 FUTURE TRENDS AND CHALLENGES IN LIGHTWEIGHT EV MATERIALS

Research into hybrid composites and advanced biomaterials aims to address performance and sustainability challenges. High production costs and limited scalability remain significant hurdles for advanced composites. Developing materials that achieve an optimal balance between performance, cost, and environmental impact is essential for future innovation. As mentioned by Shekhar et al. [29], the cost distribution for EV production outlines these challenges. Battery costs, including cell manufacturing and packaging, account for a considerable portion of total production costs, reaching up to 50%.

Zhang et al. [30] studied the mechanical performance of CFRP-Al adhesively bonded structures. They showed that hygrothermal conditions affect the strength and durability of these joints. Their results show that long-term exposure to water and temperature fluctuations lowers the bond strength. Tensile capacity decreased by as much as 42.97% after 28 days in hot water. These results show that bond length, thickness, and fiber layup direction are critical for structural integrity and crash resistance. Furthermore, Zhang et al. [30] validated their results using FEM, demonstrating that moisture absorption alters stress distribution and accelerates material degradation, posing challenges for long-term performance in EV applications. This study shows the current state-of-the-art methods and addresses practical insights and future problems in using these hybrid systems for EVs. By giving a full review of the newest joining methods, the research shares useful knowledge that can easily be used in the aerospace and engineering fields.

Hybrid structures combine the lightness of polymers or CFRP with the strong and durable nature of aluminum—they offer immense potential for applications needing better structural efficiency and less total weight. These features are very useful in airplane design and general engineering in the car industry. For instance, these hybrid constructions could play an important role in improving materials for car bumpers—where strength, weight reduction, and performance are very important. According to Keshavarzi et al. [31], adhesively bonded CFRP-Al hybrid structures can achieve significant weight reductions compared to traditional aluminum chassis, while improving specific energy absorption (SEA). Thickness variations played a role, with 1 mm-thick CFRP layers improving energy absorption efficiency, whereas 2-mm layers exhibited adhesive failure and delamination issues. Localized L-shaped reinforcement with an optimal α factor of 12 mm was found to provide 80% of the energy absorption efficiency of a fully reinforced structure while using only one-third of the material, demonstrating an efficient trade-off between material usage and crashworthiness.

To sum up, many research efforts serve as helpful resources for academics, engineers, and professionals trying to develop and use light hybrid materials for engineering tasks in the EV industry. Combining aluminum, polymers, natural fiber [8], and CFRP creates a strong mix of material qualities that can be shaped to meet the tough needs of modern EV engineering. This includes improving fuel efficiency, increasing structure durability, and adjusting overall performance to align with shifting sustainable car design needs.

8.8 SUMMARY

This study explores the role of lightweight materials in advancing EV and HEV performance, efficiency, and sustainability. The research highlights how traditional lightweight materials, such as CFRPs and glass fiber composites, provide significant weight reduction while maintaining structural integrity and safety. Additionally, the study investigates the potential of sustainable and bio-based materials, including natural fiber composites and fire-retardant polymers, in addressing environmental and safety concerns in EV applications.

Through an extensive evaluation of material properties, energy efficiency, and life cycle impact, this work emphasizes the importance of integrating circular economy principles in EV manufacturing. Lightweight materials contribute to energy savings by reducing vehicle weight, thereby extending battery range and lowering greenhouse gas emissions. However, challenges remain in scalability, cost-effectiveness, and recyclability, which necessitate further advancements in hybrid composites, innovative manufacturing processes, and sustainable material development.

The findings reinforce the growing need for high-performance, eco-friendly materials in the EV sector to support global sustainability goals. Future research should focus on improving the mechanical properties of bio-based composites, enhancing recyclability, and optimizing hybrid material systems to create a more efficient, durable, and environmentally responsible automotive industry.

ACKNOWLEDGMENTS

This work was supported by Universiti Teknologi Malaysia under Grant No. Q.K130000.3809.23H19.

REFERENCES

[1] Musa, A.A. and A.P. Onwualu, Potential of lignocellulosic fiber reinforced polymer composites for automobile parts production: Current knowledge, research needs, and future direction. *Heliyon*, 2024.

[2] Zambrano, C., et al., Recycling of plastics in the automotive sector and methods of removing paint for its revalorization: A critical review. *Polymers*, 2024. 16(21): p. 3023.

[3] Togun, H., et al., A review on recent advances on improving fuel economy and performance of a fuel cell hybrid electric vehicle. *International Journal of Hydrogen Energy*, 2024. 89: p. 22–47.

[4] Kuan, H.T.N., et al. The low velocity impact properties of pandanus fiber composites. In Saadah Abdul Rahman, Hao Gong, Kai Li and K.M. Gupta (Eds.) *Materials Science Forum*. 2017. Trans Tech Publ.

[5] Kuan, H.T.N., et al., Mechanical properties study of durio zibethinus skin fiber reinforced polyethylene composite. *J. Sustain. Sci. Manag*, 2022. 17: p. 152–158.

[6] Indor, C.J., et al., Tensile properties of Luffa acutangula reinforced polymer composite. *International Journal of Integrated Engineering*, 2023. 15(5): p. 1–5.

[7] Kuan, H.T.N. and M.Z. Hassan, Low-velocity impact response of lightweight green and cellulosic composite sandwich structures. In M.Y.M. Zuhri and S. M. Sapuan (Eds.) *Biocomposites for Lightweight Sandwich Structures*. 2025, CRC Press. p. 46–60.

[8] Afiq, M.K., H.T.N. Kuan, and M.Z. Hassan. Tensile properties characterisation of hybrid luffa/GCW fiber reinforced polymer composite. In Nisa et al. (Eds.) *Materials Science Forum*. 2023. Trans Tech Publ.

[9] Mohanty, A.K., et al., Sustainable composites for lightweight and flame retardant parts for electric vehicles to boost climate benefits: A perspective. *Composites Part C: Open Access*, 2023. 12: p. 100380.

[10] Kazakova, E. and J. Lee, Sustainable manufacturing for a circular economy. *Sustainability*, 2022. 14(24): p. 17010.

[11] Fuel types of new cars: battery electric 12.1%, hybrid 22.6% and petrol 36.4% market share full-year 2022. [4 January 2025]; Available from: www.acea.auto/fuel-pc/fuel-types-of-new-cars-battery-electric-12-1-hybrid-22-6-and-petrol-36-4-market-share-full-year-2022/.

[12] Khan, M.S., et al., A review of design, materials, and manufacturing techniques in bumper beam system. *Composites Part C: Open Access*, 2024. 14: p. 100496.

[13] Kuan, H.T.N., M.K. Afiq, and S.J. Lumpong. Mechanical properties of luffa fiber reinforced recycled polymer composite. In Emeritus Hideo Nakajima, Danielewski Marek, Eugen Rabkin, S. Linderoth, and Nikolay T. Bagraev (Eds.) *Defect and Diffusion Forum*. 2024. Trans Tech Publ.

[14] Zahoor, A., et al., A systematic study involving patent analysis and theoretical modeling of eco-friendly technologies for electric vehicles and power batteries to ease carbon emission from the transportation industry. *Energy Conversion and Management*, 2024. 321: p. 118996.

[15] Zhou, W., et al., Cost, range anxiety and future electricity supply: A review of how today's technology trends may influence the future uptake of BEVs. *Renewable and Sustainable Energy Reviews*, 2023. 173: p. 113074.
[16] Electric Vehicle Technology — NSC STEM Pathways. Open Learning Initiative (web source) [7 January 2025]; Available from: https://oli.cmu.edu/courses/electric-vehicle-technology-nsc-stem-pathways/#.
[17] Kim, J.H., et al., Experimental investigation of freeze-thaw environmental effects on the fatigue life of CFRP composites. *Polymer Testing*, 2025. 143: p. 108702.
[18] Iyer, K., M.K. Hagnell, and M. Åkermo, The multi-objective optimization framework: A step towards minimizing life-cycle costs and energy consumption of carbon fibre automotive structures. *Composites Part B: Engineering*, 2024. 271: p. 111158.
[19] Cao, Y., et al., Improving tensile properties of glass fiber-reinforced epoxy resin composites based on enhanced multiphase structure: The modification of resin systems and glass fibers. *Materials Today Communications*, 2024. 40: p. 110225.
[20] Research and Markets (2025). Lightweight Materials Market Size, Share & Trends Analysis Report By Product (Aluminum, High Strength Steel, Titanium, Magnesium), By Application (Automotive, Aviation), By Region, And Segment Forecasts, 2025–2030. [7 January 2025]; Available from: www.grandviewresearch.com/industry-analysis/lightweight-materials-market.
[21] Phiri, R., et al., Advances in lightweight composite structures and manufacturing technologies: A comprehensive review. *Heliyon*, 2024. 10(21): pp. 1–24.
[22] Wang, N., et al., Research Hot Spots and development trends of biodegradable plastics. *Process Biochemistry*, 2024. 150: pp. 80–93
[23] Baranwal, J., et al., Biopolymer: A sustainable material for food and medical applications. *Polymers*, 2022. 14(5): p. 983.
[24] Kim, M. and J. Kim, Enhancement of the flame retardant properties of PPS-based composites via the addition of melamine-coated CaAl-LDH fire-retardant filler. *European Polymer Journal*, 2023. 201: p. 112584.
[25] Wolf, J., *Development and Validation of a Vehicle Front Profile Finite Element Model to Evaluate Pedestrian Impacts*. 2024. Marquette University ePublications repository. https://epublications.marquette.edu/theses_open/783/
[26] Xie, M., et al., Research on the mechanical properties of elliptical negative Poisson's ratio structures. *Buildings*, 2024. 14(12): p. 3965.
[27] Wu, X., et al., Bi-material multistable auxetic honeycombs with reusable and enhanced energy-absorbing phases under in-plane crushing. *Thin-Walled Structures*, 2024. 201: p. 111988.
[28] Chen, L. and R. Ma, Clean energy synergy with electric vehicles: Insights into carbon footprint. *Energy Strategy Reviews*, 2024. 53: p. 101394.
[29] Shekhar, H., et al., *Exploring Cost-Reduction Strategies for Electric Vehicle (EV) Batteries*. Indian Council for Research on International Economic Relations. 2020.
[30] Zhang, S., et al., Research on the mechanical properties of CFRP-Al adhesively bonded structures in hygrothermal environments. *International Journal of Adhesion and Adhesives*, 2025. 138: p. 103939.
[31] Keshavarzi, A. and H.S. Googarchin, Local reinforcement of adhesively bonded square shape aluminum energy absorber with CFRP under quasi-static axial loading: An experimental and numerical study. *International Journal of Adhesion and Adhesives*, 2024. 135: p. 103835.

9 Effect of Sustainable Materials to Combat Friction and Wear Behavior in EV and HEV Components using Advanced Surface-Processing Techniques

C. Durga Prasad, Varun K.R., Rashmi K.R., Mohan Kumar S., Mohamad Hamid Hussain, and Adem Abdirkadir Aden

9.1 INTRODUCTION

The idea of electric vehicles (EVs) goes back to the 1820s, when the first crude EVs were constructed. New breakthroughs in batteries and electric motors in the late 1800s meant EVs became widely used in city driving early in the 20th century. However, with the advancements in the internal combustion engine (ICE) and availability of gasoline vehicles early forms of EVs' presence in the market was reduced. A renewed interest in alternative fuel vehicles (AFVs) began in the 1970s' oil crisis and the birth of true hybrid electric vehicles (HEVs) in the late 1990s. HEVs integrate electric and ICE systems with the intention of attaining improved fuel economy but beyond the limits of the battery power. It was in the 21st century that producers concentrated on pure EVs leveraging on lithium-ion battery know-how. Big players such as Tesla have played their part in the EV market by manufacturing high-performing cars. The automotive industry currently views EVs and HEVs as realistic approaches to decreasing the carbon footprint and increasing the use of sustainable energy [1,2].

Environmental concerns and the necessity to decrease greenhouse gas emissions are causing shifts in the automotive sector. The popularity of EVs and HEVs is increasing as cleaner choices for transportation. With the rise in the quantity of such vehicles, there is an increasing need for eco-friendly materials and methods in their production. An important emphasis is placed on creating sustainable materials for tribological parts such as bearings and braking systems, which play a vital role in the performance of vehicles [3]. The unique challenges faced by these parts in EVs and HEVs include high torque and frequent load changes, necessitating materials with improved wear resistance and reduced friction to enhance energy efficiency. New, creative materials are required to fulfill these operational requirements while reducing harm to the environment. A common pair of tribological materials is mainly metal and petroleum-based lubricants, both of which are unsustainable. These materials are often obtained from resources through physically and chemically harsh processes, produce significant amount of waste, and are difficult to recycle at the last stage of the life cycle. As a result, research has broadened to cover environmentally friendly developments such

 DOI: 10.1201/9781003596097-9

as biodegradable polymers, recycled metals, and self-lubricating composites and surface coatings [4]. These materials present attractive attributes that complement the automotive sector's call for sustainable, enhanced-performing components for tribological parts for which the life cycle can be prolonged while at the same time minimizing the impacts on the environment.

The major use of sustainable material in tribological application is to minimize the energy losses and the wear rate, which is proportional to energy loss in vehicles. Small advancements in friction can make radical changes in the range and efficiency of automobiles, particularly in extended-range electric vehicles (EREVs), whose acceptance relies on optimum battery usage and longest possible distance. Conveyor slides, for example, are self-lubricating, and it is a plus that such materials can significantly decrease dependence on oils or require no oil at all, which typically requires replenishment. In the same respect, recyclable/frequently biodegradable polymers provide feasible solutions for components, which are not so crucial in tribological systems, while recycled metals help to decrease the overall CO_2 emission of vehicles [5].

Thermal spray coatings entail a significant deposit on sustainable tribological materials. These coatings improve the wear properties and friction behavior of base materials, reducing the rate of failure of components experiencing high load stresses that would otherwise require a high service frequency due to retreating out of base materials or wearing away of the surface layers. Because of their high hardness and low friction coefficient, these coatings can be applied in the gears and bearings of EVs as wear resistance and to reduce clatter and so are essential for the products. There are various types of thermal spray coatings that are used, one of which is high-velocity oxy-fuel coatings that offer very good corrosive resistance and are extensively used in braking systems because they are wear and heat resistant [6].

The use of these sustainable tribological materials aligns with general objectives of automotive engineering, such as reduced reliance on resources, minimum emissions, and improved recycling. First of all, markings on a product can signal to the consumer that the manufacturing materials are renewable, biodegradable, or highly durable, thus decreasing the detrimental effects of material extraction and processing, as well as disposal. For example, the use of recycled metal saves raw materials in equal measure while reducing energy use during the production process. Furthermore, the use of biodegradable polymers reduces the amount of plastic waste by meeting the challenge of replaceable environmentally friendly material [7]. This chapter reviews the major sustainable materials employed in tribological parts of EVs and HEVs, including properties, uses, and effects on the surroundings. As automobiles become more environmentally progressive, such sustainable tribo-materials will remain central to the automotive industry in delivering both high performance and sustainability in EVs and HEVs. Taking advantage of materials science and engineering the automotive industry can usher the world into a new era of transportation; one that is green, advanced, and not anchored on fossil resources [8].

9.1.1 Critical Tribological Components in EVs and HEVs

In EVs and HEVs, the tribological components are central if performance, efficiency, and durability of the vehicle are to be obtained. The components in contact entail friction and wear, which affect the powering, handling, and durability of components. Figure 9.1 illustrates the classification of components that are crucially affected by friction and wear [9,10].

Below is an overview of the tribological components mentioned, highlighting their functions and tribological challenges.

9.1.2 Motors

Electric motors are essential components in EV and HEV systems, which are used to power the vehicle. In motors, bearings, bushing, and rotor shafts experience friction due to high-speed rotation. The use of technology such as ceramic-coated bearings minimizes wear, friction, and heat

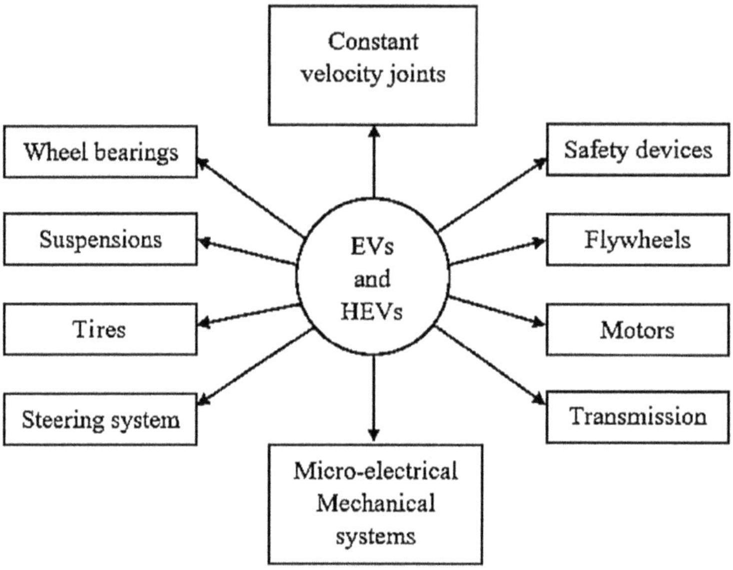

FIGURE 9.1 Classification of crucial tribological parts in typical electric and hybrid vehicles.

generation thus keeping efficiency high in cases where the motor goes at high speeds with fluctuating loads [11].

9.1.3 *Transmission Systems*

Driving EVs and HEVs demands precisely proportioned and uniform tribological components, when talking about multi-speed transmission, which is incorporated into some high-performance models. These systems' gears and shafts are anti-wear, low noise, and efficient through the use of streamlined coatings such as diamond-like carbon (DLC) [12].

9.1.4 *Flywheels*

Energy flow is well regulated using flywheels, which also assist in storing kinetic energy where some forms of hybrids are concerned. From a tribological point of view, flywheels are subjected to rotational force and have resistance to frictional force [13]. High-wear materials and low-friction materials like carbon composites or metals can be anodized to provide long service and energy retention efficiency.

9.1.5 *Wheel Bearings*

Wheel bearings eliminate friction to the wheel radius and provide the supportive structure for the mass of the vehicle body in addition to containing dynamic loads. EVs and HEVs are usually heavier due to the extra power train components of the motor, and therefore, wheel bearings need to be more robust [14]. Anti-wear bearings are often ceramic or polymer coated in order to decrease friction and increase durability in energy applications.

9.1.6 *Tires*

Tires are seen as essential tribological parts of EVs and HEVs as they act as a direct contact between the automobile and the ground. They must withstand the torque outputs of electric motors and have low rolling resistance, which improves range. Sponge and friction materials which are used to decrease wear and enhance adhesion are selected depending on the car weight and driving features, and are often created with regard to the higher weight and usage profiles of the EVs or HEVs [15].

9.1.7 *Suspension Systems*

The complexity of suspension systems in EVs and HEVs necessitates durable and efficient tribological parts owing to the extra load from battery packs. Thus, on several mechanical parts like shock absorbers, bushes, and ball joints, the level of friction should be minimal as well as the parts should be able to absorb shocks and withstand wear. For durability, handling, and ride comfort, bonded counterparts like high-strength polymers and advanced lubricants are used [16].

9.1.8 *Steering Systems*

Steering systems, especially those found in EVs, use electric power steering wherein several tribological parts such as ball bearings, bushing, and gears all undergo friction. All these components need to offer sharp and quick steering with minimal wear patterns on this vital structure. Low-friction materials are very useful in minimizing wear and tear while the same materials can be coated with protective layers to enhance efficiency [17].

9.1.9 *Velocity Joints (Constant Velocity or CV Joints)*

CV joints transmit torque symmetrically from the transaxle to the wheels at constant angular velocity which inflicts torque loads while maintaining smoothness. They are exposed to constant cyclic loads of bending and rotation which leads to high friction and consequently high wearing of the surface. Under loads, their life cycle is enhanced through the use of high-strength steel, heat-treated surfaces, and protection by coating [18].

9.1.10 *Micro-Electro-Mechanical Systems (MEMS)*

MEMS sensors and actuators, including accelerometers, gyroscopes, and pressure sensors, have become widely incorporated into EVs and HEVs. Although these systems are small they contain some moving elements which encounter micro-level tribological problems. MEMS elements also have extremely thin oxide layers, or wear-proof materials, to substantially eliminate wear and increase durability [19].

All these parts have considerable importance with regard to the effectiveness, longevity, and capability of EVs and HEVs. Having access to and incorporation of more exotic materials, coatings, and lubrication systems assists in minimizing wear and unrecovered sources of friction, boosting total energy efficiency, component service longevity, and improving overall driving performance.

9.2 ESSENTIAL MATERIALS FOR COMPONENTS SUBJECTED TO FRICTION AND WEAR

9.2.1 Biodegradable Polymers

A sustainable polymeric material, biodegradable polymers have begun to be incorporated to reduce the environmental impact. Types of biodegradable polymers include polylactic acid (PLA), polyhydroxyalkanoates (PHAs), and polybutylene succinate (PBS), produced from renewable raw materials and have higher degradability than conventional plastics. They can be used for interior components such as the seat cushioning, dashboard, and trim. They are used as light and eco-friendly materials to help to reduce the load of the car and its power consumption. Combining these

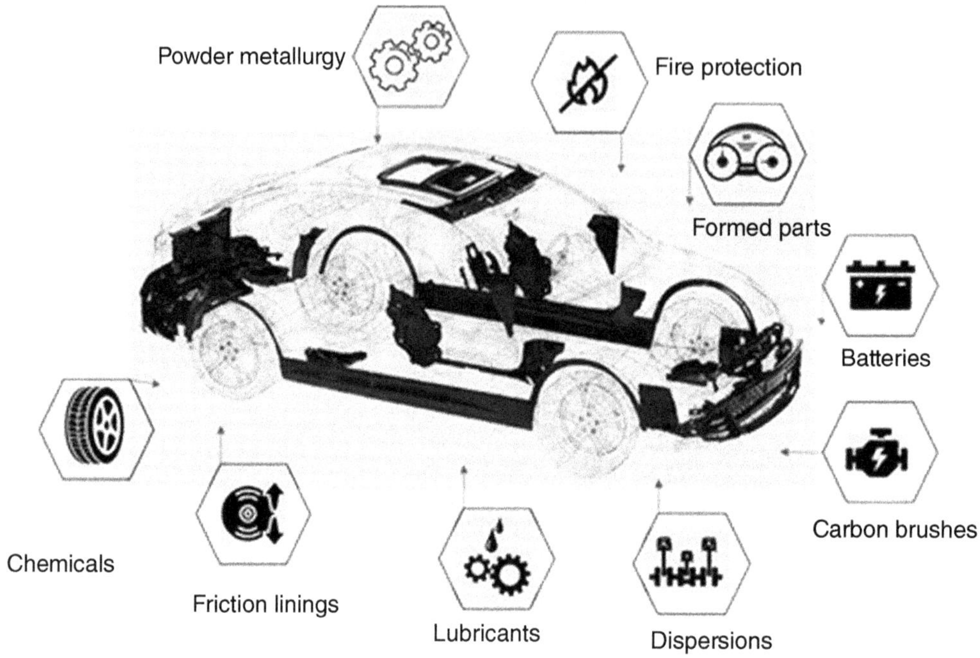

FIGURE 9.2 Biodegradable polymers used in EVs and HEVs.

bio-plastics with natural fibers such as flax or hemp increases the quality of the material overall and its load-bearing capabilities, thus allowing them to be utilized in load-bearing elements of the vehicle cabin [20–23].

Further, in non-structural applications like floor mats and insulation layers, and door trims, cellulose-based bio-plastics and polybutylene adipate terephthalate (PBAT) are used. In EVs and HEVs, weight savings are a big factor for improving battery range, and biopolymers give the same level of strength and environmental friendliness without adding extra weight [24]. They also do away with waste since the materials disintegrate over time, or they can be recycled through an industrial composting process. As biodegradable polymers are used in the construction of some EVs and HEVs, as shown in Figure 9.2, the automotive industry is moving in the direction of energy efficiency and environmentalism.

9.2.2 Recycled Metals

Tribological components very often contain recycled metals such as aluminum and steel. Aluminum alloys have good strength, relatively low weight, and hence a high strength-to-weight ratio, and so they are widely used in the automobile industry for manufacturing the frames of vehicles, suspension systems, brake systems, etc. When produced from scrap materials, these metals reduce the use of raw materials and energy in extraction. Other surface treatments like anodizing of aluminum or thermal spray coatings take the wear resistance to another level, making recycled metals a prize worthy of high-stress parts [25–27].

9.2.3 Self-Lubricating Composites

Self-lubricating composites reduce or eliminate the need for additional lubricants by incorporating materials that provide inherent friction reduction. These composites integrate lubricating fillers such as graphite, molybdenum disulfide (MoS_2), and hexagonal boron nitride (h-BN) into a polymer or metal matrix. This structure helps in reducing wear and friction, particularly in moving parts subjected to high loads, thereby extending component life and reducing maintenance [28–30]. Self-lubricating composites are particularly beneficial for EV and HEV components that are subject to frequent motion and regenerative braking, where traditional lubrication may not be feasible [31].

9.3 ROLE OF THERMAL SPRAY COATINGS IN SUSTAINABLE TRIBOLOGY

9.3.1 Thermal Spray Process and Its Applications

Thermal spray coating is the process to improve or restore a material's surface properties for a given application, as shown in Figure 9.3. This process effectively involves the deposition of a coating on a substrate through creating and launching fine particles of the coating material onto the surface at high velocity with the help of a heat source such as flame, arc, or plasma [32]. In some cases, it can enhance components' durability by protecting or reducing such threats as wear, corrosion, heat, erosion, or abrasion, and therefore is very useful in adverse working conditions.

Thermal spray coatings are unique applications used to develop wear-resistant, anti-corrosive, or thermal barrier coatings. Among them, plasma spraying, high-velocity oxy-fuel (HVOF), cold, flame spraying (flame), and detonation gun (d-gun) are the most popular methods. Consequently, each one provides definite benefits depending on the properties of the coating, operation conditions, and costs involved. These coatings are suitable for high melting point substrates such as ceramics, metals and alloys, copper, aluminum, polymers, and some metals [33–36].

In this special application for EVs and HEVs, advanced coatings have a critical role in boosting and maintaining tribological components. These green coatings result in an improvement of the surface hardness, meaning that wearing is generally reduced and so is friction. It is especially for the case of electric and hybrid powertrains that efficiency and reliability are crucial elements for these forms of vehicles. Advanced coatings are used on the critical components to enhance their dependability in EV and HEV technologies and hence make the technologies sustainable solutions [37–39]. Their application assists in increasing the usage longevity of parts so they seldom need replacement. This reduces costs and the extent of harm caused to the environment to give a better performance in the automotive industry.

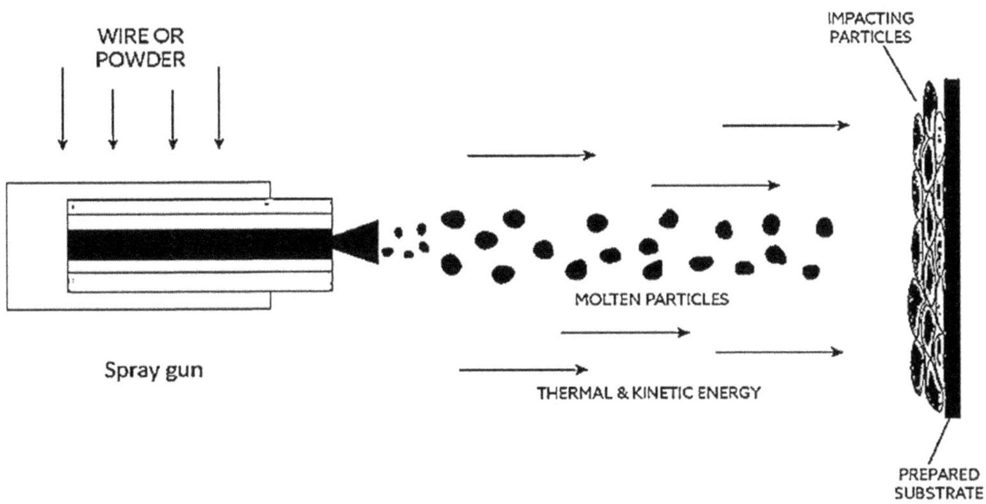

FIGURE 9.3 Method of thermal spraying.

9.3.2 Significance of Sustainable Tribology Components

Thermal spray coatings make a significant contribution to sustainable tribology in EVs and HEVs as they contribute to component durability, minimum energy losses, and improved dependability of the system.

9.3.3 Enhanced Wear Resistance

The rotors, battery enclosure, and gear systems of high-performance EVs and HEVs will expose an engineering material to mechanical stress and wear during operation. Thermal spray coatings, particularly those produced from materials like tungsten carbide (WC) and chromium oxide (Cr_2O_3), provide enormous wear protection in these demanding sites. The desired properties of such coatings are that they form a thin and very hard surface layer on those parts of a component that are exposed to a high level of use, thereby avoiding major wear, which is extremely desirable in areas incurring high costs for replacement or difficult accessibility. Coated components are most likely to last longer than unprotected ones; therefore, fewer replacements are called for, resulting in the conservation of resources, particularly in the reduction of waste and the consequent need for extraction of basic materials. For the manufacturers of EVs and HEVs, this wear resistance has an implication of increased durability, reliability, and therefore an added market advantage within green markets [40–44].

The microstructure of the as-sprayed WC-Co-Cr coating prepared using HVOF spraying is indicated in Figure 9.4. The most common kinds of coating often have WC which provides hardness and wear properties with cobalt-chromium (Co-Cr) because it makes a standard binder and has great resistance to corrosion. The HVOF process delivers a dense coating with low porosity because of the high kinetic energy of the deposited particles. This makes the coating strong and firm and reduces the intrusion of corrosive elements, making it suitable for any surface with abrasive and/or corrosion tasks. The absence of cracks and pores in Figure 9.4, as well as the overall similarity of the visual images and compositions, demonstrates the usefulness of the HVOF technique in generating tough-to-wear coatings [45].

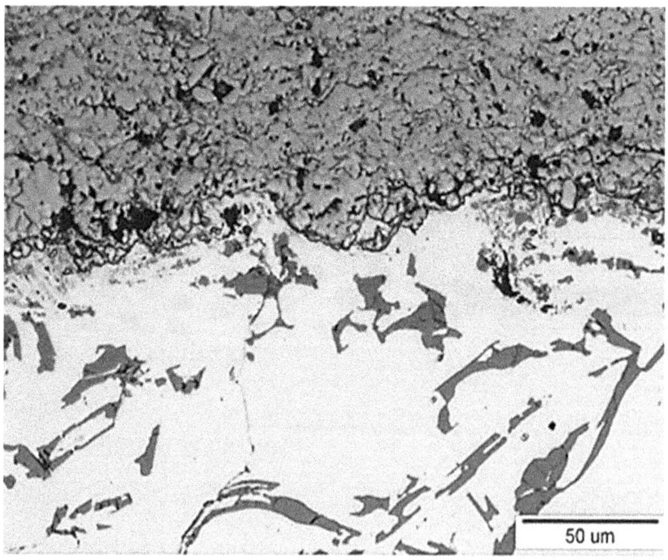

FIGURE 9.4 WC-Co-Cr deposited by HVOF.

9.3.4 Friction Reduction for Efficiency

Efficiency is important in EVs as well as HEVs since battery capacity determines the range and power of vehicles. Molybdenum-based and other low-friction thermal spray coatings also promote this efficiency by reducing friction in critical wearing parts including bearings, gears, and drives. Low friction in turn leads to the loss of less energy during use; hence, the car will need to use less of the available energy to combat internal mechanical drag. This energy conservation is crucial mainly in determination of the battery range and energy efficiency in EVs and HEVs. Reduced wear and heat effects from friction are additional benefits from reduced friction because it is known that friction leads to heat production and hence, additional load applied on the components [46]. In sum, these friction-reducing coatings are valuable in addressing the high energy efficiency requirements of electric and hybrid methodologies for powertrains, thereby supporting sustainability objectives, through battery use lengthening and efficiency improvements.

9.3.5 Corrosion Protection in Diverse Environments

EV and HEV parts are subjected to elements such as road salinity and dampness, various chemicals, and temperature changes. These can cause corrosion and such areas as the chassis, battery trays, and suspensions may be badly affected. The use of thermal spray coats, such as aluminum and zinc, is required to protect parts from corrosion, and to protect them from the environment. The use of such coatings on structural and exterior parts curbs rusting and oxidation, hence increasing their useful lifespan and also decreasing the frequency of their maintenance. For the regions where electric and hybrid vehicles react with salts or urban contaminants, these coatings curb the incidences of component degeneration, and therefore, the subsequent disposal of such parts [47–49]. Protection from corrosion thus plays an enduring role in the vehicle context since it guarantees long life for automobiles in that no component will wear out as a result of exposure to harsh environmental conditions.

Figure 9.5 depicts a plot of corrosion potential by time of steel coated with aluminum. Corrosion potential is an electrochemical value showing the propensities of the coated surface to corrode or oxidize. They show the performance stability of the aluminum coating in minimizing the corrosion vulnerability of steel after some time. A steady or positive shift with time means the material is properly protected, while fluctuations may mean that the coating is wearing out. Studying this trend allows evaluation of the service life of the coating and its resistance to corrosive effects.

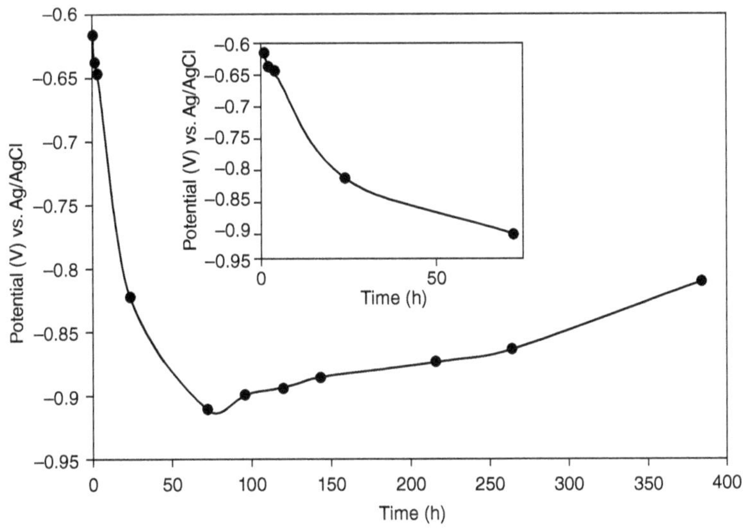

FIGURE 9.5 Corrosion potential–time of Al-coated steel.

9.3.6 Lightweighting with Enhanced Durability

Aluminum and magnesium alloys are preferred more in the production of new-generation EVs and HEVs due to their light weight and enhanced energy efficiency. However, these materials are generally less hard wearing and are prone to corrosion more than conventional steel. Thermal spray coatings are a solution to apply a protective layer on lightweight parts without adding weight to them. This enables automobile manufacturers to incorporate features such as lightweight materials for cars without compromising durability and/or resistance to wear and corrosion. In the context of supporting lightweighting, thermal spray coatings enhance the fuel economy of electric and hybrid vehicles, decrease emissions, and contribute to the development of an environmentally sensitive automotive industry [50–52].

Thermal spray coatings enhance wear resistance, reduce friction, minimize corrosion, and use lightweight materials to support sustainability and operational efficiency of EV and HEV technologies across multiple environments.

9.4 FUTURE DIRECTIONS AND INNOVATIONS IN SUSTAINABLE TRIBOLOGY

Future development trends of sustainable tribology in this field are more inclined toward environmentally friendly materials, lubricants, and coatings that have positive implications for the lifetimes of components and their performance. Development is continuing in the area of biodegradable lubricants that are synthesized from renewable resources including vegetable oils; these substances possess low toxicity and high levels of biocompatibility. There is much effort also in creating solid lubricants and self-lubricating materials for applications where conventional liquid lubricants cannot be used, such as in high-temperature, vacuum, or cryogenic environments. Also, there are newer technologies that are extending to surfaces and hence, new surface treatments like nano-coatings and self-healing materials are coming up as new ways of reducing wear and tear, friction, and hence reducing maintenance and scrap in the long run [53–56].

Another important area is that of low-energy tribosystems which involve a minimum of energy expenditure by virtue of friction. This can involve designing new surface structures and multifunctional coatings that will inherently offer lower frictional force, and the design of components which will have low levels of lubrication requirement while at the same time offering high functionality. Innovations in AM have also been unveiled to enable the realization of designed part surfaces for improved tribology, and this while using minimal resources since AM can reduce the use of materials resulting from the making of multiple parts [57]. Computerization and artificial intelligence are also used for tribological behavior prognostication and enhancement under various functioning states for more environmentally friendly product designs with specific emphases of use. By means of these developments, sustainable tribology aims to achieve a huge contribution toward minimizing the ecological impact of industries depending on tribological systems such as the automobile, aircraft, and manufacturing industries [58–59].

9.5 CONCLUSIONS

Sustainable materials are critical for achieving energy-efficient, eco-friendly, and high-performing tribological components in EVs and HEVs. By leveraging biodegradable polymers, recycled metals, self-lubricating composites, and advanced coatings, automotive manufacturers can develop vehicles that meet stringent environmental and durability standards.

- Thermal spray coatings enhance the lifetime and wear properties of tribological parts in EVs and HEVs, which reduces the removal frequency and demands sustainable resource management.
- Using coatings reduces friction, and plays a pivotal role in electric and hybrid vehicles saving energy needed for effective power, which improves the car's range.

- By incorporating low-toxicity composites and biodegradable materials in thermal spray coatings where possible, adverse effects are not only minimized but the thermal spray coatings are able to meet the stringent standards set by the EV and HEV applications.
- Sophisticated coating methods such as HVOF and cold spraying, coupled with the ability to fine-tune the parameters of the process, emphasize the opportunities to create the required tribological surface for existing or future parts of EVs and HEVs via optimized performances of tribological coatings with sustainable production perspectives.

ACKNOWLEDGMENTS

Dr. Durga Prasad C. would like to thank the Science and Engineering Research Board (SERB) for financial support to carry out this research work. Project File no: CRG/2022/004140, under Core Research Grant (CRG) scheme, Government of India.

REFERENCES

[1] Bradu, C. I., and G. Ianuş. "A prospective view on biodegradable lubricants for hybrid and electric vehicles." In I. Doroftei, A. Popescu, and C. Bujoreanu (Eds.) *IOP Conference Series: Materials Science and Engineering*, vol. 1262, no. 1, p. 012001. IOP Publishing, 2022.
[2] Hemanth, G., and B. Suresha. "Hybrid and electric vehicle tribology: A review." *Surface Topography: Metrology and Properties* 9, no. 4 (2021): 043001.
[3] Shah, Raj, Blerim Gashi, and Andreas Rosenkranz. "Latest developments in designing advanced lubricants and greases for electric vehicles—An overview." *Lubrication Science* 34, no. 8 (2022): 515–526.
[4] Tung, Simon C., Mathias Woydt, and Raj Shah. "Global insights on future trends of hybrid/EV driveline lubrication and thermal management." *Frontiers in Mechanical Engineering* 6 (2020): 571786.
[5] Shah, Raj, Blerim Gashi, Sergio González-Poggini, Melanie Colet-Lagrille, and Andreas Rosenkranz. "Recent trends in batteries and lubricants for electric vehicles." *Advances in Mechanical Engineering* 13, no. 5 (2021): 16878140211021730.
[6] Lee, Seungjoo, Julio A. Cao-Romero-Gallegos, Leonardo I. Farfan-Cabrera, and Ali Erdemir. "Tribological challenges for a sustainable e-mobility." In Leonardo I. Farfan-Cabrera, Ali Erdemir (Eds.) *Electric Vehicle Tribology*, pp. 5–21. Elsevier, 2024.
[7] Chen, Yan, Swarn Jha, Ajinkya Raut, Wenyang Zhang, and Hong Liang. "Performance characteristics of lubricants in electric and hybrid vehicles: A review of current and future needs." *Frontiers in Mechanical Engineering* 6 (2020): 571464.
[8] Ahmed Abdalglil Mustafa, Waleed, Fabrice Dassenoy, Maria Sarno, and Adolfo Senatore. "A review on potentials and challenges of nanolubricants as promising lubricants for electric vehicles." *Lubrication Science* 34, no. 1 (2022): 1–29.
[9] Hemanth, G., and B. Suresha. "Hybrid and electric vehicle tribology: A review." *Surface Topography: Metrology and Properties* 9, no. 4 (2021): 043001.
[10] Farooq, Muhammad Umar. "Tribology of electric vehicles–A comprehensive survey of recent developments in components and key future avenues for surfaces and interfaces research." *Available at SSRN* 4726800 (2024): 1–48.
[11] Chen, Yan, Swarn Jha, Ajinkya Raut, Wenyang Zhang, and Hong Liang. "Performance characteristics of lubricants in electric and hybrid vehicles: A review of current and future needs." *Frontiers in Mechanical Engineering* 6 (2020): 571464.
[12] Tung, Simon C., Mathias Woydt, and Raj Shah. "Global insights on future trends of hybrid/EV driveline lubrication and thermal management." *Frontiers in Mechanical Engineering* 6 (2020): 571786.
[13] Senatore, Adolfo. "Special issue 'Automotive tribology'." *Lubricants* 8, no. 4 (2020): 48.
[14] Skonieczna, Daria, Oleksandr Vrublevskyi, Michał Janulin, and Piotr Szczyglak. "Analysis of tribological properties of engine lubricants used in hybrid vehicles." *Materials* 17, no. 21 (2024): 5304.
[15] Cao-Romero-Gallegos, Julio A., Leonardo I. Farfan-Cabrera, Ali Erdemir, and Juan B. Pascual-Francisco. "Lubricated sliding wear of gear material under electrification–A new approach to

understanding of the influence of shaft currents in the wear of EV transmissions." *Wear* 523 (2023): 204782.

[16] Abebe, Biniyam Ayele, Samet Çelebi, and Recep Kılıç. "A review on tribological considerations in the transition from IC engines to electric vehicles." *International Journal of Automotive Science and Technology* 8, no. 3 (2024): 369–380.

[17] Johns-Rahnejat, Patricia M., Ramin Rahmani, and Homer Rahnejat. "Current and future trends in tribological research." *Lubricants* 11, no. 9 (2023): 391.

[18] del Río, José M. Liñeira, Fátima Mariño, Enriqueta R. López, David E. P. Gonçalves, Jorge H. O. Seabra, and Josefa Fernández. "Tribological enhancement of potential electric vehicle lubricants using coated TiO2 nanoparticles as additives." *Journal of Molecular Liquids* 371 (2023): 121097.

[19] Lee, Seungjoo, Julio A. Cao-Romero-Gallegos, Leonardo I. Farfan-Cabrera, and Ali Erdemir. "Tribological challenges for a sustainable e-mobility." In Leonardo I. Farfan-Cabrera and Ali Erdemir (Eds.) *Electric Vehicle Tribology*, pp. 5–21. Elsevier, 2024.

[20] Zhang, Zheng, Ophir Ortiz, Ritu Goyal, and Joachim Kohn. "Biodegradable polymers." In Kayvon Modjarrad and Sina Ebnesajjad (Eds.) *Plastics Design Library*, pp. 303–335. William Andrew Publishing, 2014. https://doi.org/10.1016/B978-0-323-22805-3.00013-X

[21] Gross, Richard A., and Bhanu Kalra. "Biodegradable polymers for the environment." *Science* 297, no. 5582 (2002): 803–807.

[22] Pachence, James M., and Joachim Kohn. "Biodegradable polymers." *Principles of Tissue Engineering* 22 (2000): 263–277.

[23] Ceschini, Lorella, and Stefania Toschi. "Friction and wear of aluminum alloys and composites." In G. E. Totten (Ed.) *ASM Handbook, Volume 18: Friction, Lubrication, and Wear Technology* (pp. 509–532). ASM International, 2017. https://doi.org/10.31399/asm.hb.v18.a0006388.

[24] Parthasarathy, Meera, and Agnel Arul John. "Tribology of biodegradable polymeric systems." In S. C. George, J. T. Haponiuk, & S. Thomas (Eds.) *Tribology of Polymers, Polymer Composites, and Polymer Nanocomposites*, pp. 235–263. Elsevier, 2023.

[25] Holmberg, Kenneth, Päivi Kivikytö-Reponen, Pirita Härkisaari, Kati Valtonen, and Ali Erdemir. "Global energy consumption due to friction and wear in the mining industry." *Tribology International* 115 (2017): 116–139.

[26] Bendikiene, Regita, Antanas Ciuplys, and Lina Kavaliauskiene. "Circular economy practice: From industrial metal waste to production of high wear resistant coatings." *Journal of Cleaner Production* 229 (2019): 1225–1232.

[27] Mang, Theo, Kirsten Bobzin, and Thorsten Bartels. *Industrial Tribology: Tribosystems, Friction, Wear and Surface Engineering, Lubrication*. John Wiley, 2011.

[28] Menezes, Pradeep L., Pradeep K. Rohatgi, and Emad Omrani, eds. *Self-Lubricating Composites*. Springer, 2018.

[29] Sahoo, Seshadev. "Self-lubricating composites with 2D materials as reinforcement: A new perspective." *Reinforced Plastics* 65, no. 2 (2021): 101–103.

[30] Ouyang, Jia-Hu, Yu-Feng Li, Yun-Zhuo Zhang, Ya-Ming Wang, and Yu-Jin Wang. "High-temperature solid lubricants and self-lubricating composites: A critical review." *Lubricants* 10, no. 8 (2022): 177.

[31] Singh, Satnam, Sarbjeet Kaushal, and Michele Forzan. "Materials system for functional properties of metal matrix composites: Self-lubricating, anti-wear, and self-cleaning properties." In S. A. Bansal, V. Khanna, and P. Gupta (Eds.) *Metal Matrix Composites*, pp. 1–26. CRC Press, 2022.

[32] Kumar, S. Suresh, C. Durga Prasad, and Harish Hanumanthappa. "Role of thermal spray coatings on erosion, corrosion, and oxidation in various applications: A review." *Journal of Bio-and Tribo-Corrosion* 10, no. 2 (2024): 22.

[33] Lakkannavar, Virupakshappa, K. B. Yogesha, C. Durga Prasad, M. Mruthunjaya, and R. Suresh. "A review on tribological and corrosion behaviour of thermal spray coatings." *Journal of The Institution of Engineers (India): Series D* 106, no. 2 (2024): 1–17.

[34] Prasad, C. Durga, S. Kollur, M. Nusrathulla, G. Satheesh Babu, M. B. Hanamantraygouda, B. N. Prashanth, and N. Nagabhushana. "Characterisation and wear behaviour of SiC reinforced FeNiCrMo composite coating by HVOF process." *Transactions of the IMF* 102, no. 1 (2024): 22–28.

[35] Sharanabasava, H., C. Durga Prasad, and M. R. Ramesh. "Characterization and wear behavior of NiCrMoSiC microwave cladding." *Journal of Materials Engineering and Performance* 33, no. 2 (2024): 763–775.
[36] Prasad, C. Durga, Sharnappa Joladarashi, M. R. Ramesh, and M. S. Srinath. "Microstructural and tribological resistance of flame-sprayed CoMoCrSi/WC-CrC-Ni and CoMoCrSi/WC-12Co composite coatings remelted by microwave energy." *Journal of Bio-and Tribo-Corrosion* 6 (2020): 1–15.
[37] Lee, Seungjoo, Leonardo Israel Farfan-Cabrera, Diana Berman, Andreas Rosenkranz, and Ali Erdemir. "Tribological coatings for electric vehicle applications." *Frontiers in Coatings, Dyes and Interface Engineering* 2 (2024): 1426164.
[38] Chavan, Santosh, Bhumarapu Venkateswarlu, Mohammad Salman, Jie Liu, Prakash Pawar, Sang Woo Joo, Gyu Sang Choi, and Sung Chul Kim. "Thermal management strategies for lithium-ion batteries in electric vehicles: Fundamentals, recent advances, thermal models, and cooling techniques." *International Journal of Heat and Mass Transfer* 232 (2024): 125918.
[39] Berman, Diana, Leonardo Israel Farfan-Cabrera, Andreas Rosenkranz, and Ali Erdemir. "Advancing the frontiers of EV tribology with 2D materials–A critical perspective." *Materials Science and Engineering: R: Reports* 161 (2024): 100855.
[40] Huang, Bo, Chao Zhang, Ga Zhang, and Hanlin Liao. "Wear and corrosion resistant performance of thermal-sprayed Fe-based amorphous coatings: A review." *Surface and Coatings Technology* 377 (2019): 124896.
[41] Farfan-Cabrera, Leonardo Israel. "Tribology of electric vehicles: A review of critical components, current state and future improvement trends." *Tribology International* 138 (2019): 473–486.
[42] Fauchais, Pierre, and Armelle Vardelle. "Thermal sprayed coatings used against corrosion and corrosive wear." *Advanced Plasma Spray Applications* 10 (2012): 34448.
[43] Nugroho, Agus, Muhammad Kozin, Rizalman Mamat, Zhang Bo, Mohd Fairusham Ghazali, Muhammad Prisla Kamil, Prabowo Puranto et al. "Enhancing tribological performance of electric vehicle lubricants: Nanoparticle-enriched palm oil biolubricants for wear resistance." *Heliyon* 10, no. 22 (2024): 1–30.
[44] Voevodin, Andrey A., Christopher Muratore, and Samir M. Aouadi. "Hard coatings with high temperature adaptive lubrication and contact thermal management." *Surface and Coatings Technology* 257 (2014): 247–265.
[45] Richert, M., Mikułowski, B., Pałka, P., Hotloś, A. and Perek-Nowak, M. The effect of chemical composition and thermal sprayed method on the chromium and tungsten carbides coatings microstructure. *Journal of Surface Engineered Materials and Advanced Technology* 3 (2013): 1–5.
[46] Lee, Seungjoo, Leonardo Israel Farfan-Cabrera, Diana Berman, Andreas Rosenkranz, and Ali Erdemir. "Tribological coatings for electric vehicle applications." *Frontiers in Coatings, Dyes and Interface Engineering* 2 (2024): 1426164.
[47] Fotovvati, Behzad, Navid Namdari, and Amir Dehghanghadikolaei. "On coating techniques for surface protection: A review." *Journal of Manufacturing and Materials Processing* 3, no. 1 (2019): 28.
[48] Sharun, V., M. Rajasekaran, S. Suresh Kumar, Vikas Tripathi, Rajneesh Sharma, G. Puthilibai, M. Sudhakar, and Kassu Negash. "Study on developments in protection coating techniques for steel." *Advances in Materials Science and Engineering* 2022, Article ID: 2843043 (2022): 1–10.
[49] Lee, Han-Seung, Jitendra Kumar Singh, Mohamed A. Ismail, and Chinmoy Bhattacharya. "Corrosion resistance properties of aluminum coating applied by arc thermal metal spray in SAE J2334 solution with exposure periods" *Metals* 6, no. 3 (2016): 55. https://doi.org/10.3390/met6030055
[50] Guo, Honghe, Xiaoguang Zhou, and Zhenyu Liu. "Advanced lightweight structural materials for automobiles: Properties, manipulation, and perspective." *Science of Advanced Materials* 16, no. 5 (2024): 563–580.
[51] Liu, Ming, Yang Guo, Jeff Wang, and Mark Yergin. "Corrosion avoidance in lightweight materials for automotive applications." *NPJ Materials Degradation* 2, no. 1 (2018): 24.
[52] Lovatt, Carl Ryan. "The development of a lightweight electric vehicle chassis and investigation into the suitability of tial for automotive applications." PhD diss., The University of Waikato, 2008.
[53] Peter, Onu, and Charles Mbohwa. "Ecotribology: A review of eco-frictional materials for sustainable engineering." In *Proc. 5th African IEOM Conf., Paper ID 177*, IEOM Society International, 2024.
[54] Malik, Muhammad Ali Ijaz, M. A. Kalam, M. A. Mujtaba, and Fares Almomani. "A review of recent advances in the synthesis of environmentally friendly, sustainable, and nontoxic

bio-lubricants: Recommendations for the future implementations." *Environmental Technology & Innovation* 32 (2023): 1–18.

[55] Awasthi, Mukesh Kumar, Ashwani Kumar, and Nitesh Dutt. "Recent innovation, technological development, and futuristic application of nanofluids." In Mukesh Kumar Awasthi, Ashwani Kumar, and Nitesh Dutt (Eds.) *Nanofluids Technology for Thermal Sciences and Engineering*, pp. 1–25. CRC Press, 2024.

[56] Johns-Rahnejat, P. M., Ramin Rahmani, and Homer Rahnejat. "Current and future trends in tribological research." *Lubricants* 11 (2023): 391.

[57] Luo, Jianbin, Min Liu, and Liran Ma. "Origin of friction and the new frictionless technology—Superlubricity: Advancements and future outlook." *Nano Energy* 86 (2021): 106092.

[58] Sasaki, Shinya. "Environmentally friendly tribology (Eco-tribology)." *Journal of Mechanical Science and Technology* 24 (2010): 67–71.

[59] Andrew, J. Jefferson, and H. N. Dhakal. "Sustainable biobased composites for advanced applications: Recent trends and future opportunities–A critical review." *Composites Part C: Open Access* 7 (2022): 100220.

10 A Deep Learning Approach for Electric Vehicle Battery Capacity Degradation and Health Early Prediction of Battery Remaining Useful Life

Sivakrishna K., Vaishnavi K., Upendra U., Ahash A., and Leela Sai Sri Krishna

10.1 INTRODUCTION

Research focused on predicting lithium-ion battery (LIB) remaining useful life (RUL) accurately holds essential value for energy storage systems because it determines their operational efficiency and reliability and sustainability. LIBs appear in multiple applications ranging from electric vehicles (EVs) to hybrid electric vehicles (HEVs) and renewable energy storage and consumer electronics because they demonstrate high energy density together with extended cycle life and minimal self-discharge behavior. Batteries naturally degrade with time because they experience reduced capacity along with higher internal resistance and deteriorating performance [1]. In order to avoid system performance degradation and to mitigate potential hazards, precisely monitoring and tracking LIB's capacity degradation charge and discharge [2] and RUL estimation is crucial. Forecasting battery RUL at its initial stage becomes crucial for managing maintenance schedules while avoiding unexpected breakdowns and enhancing battery lifespan to maintain operational safety in critical systems. Predicting the RUL of LIBs, especially in the early stages of the LIBs, is a complex task due to the nonlinearity nature of the battery degradation process. In general, the process of capacity degradation or the process of reduction of battery power starts off gradually but after a few cycles reaches a juncture point known as the "knee point", we could also say a "saturation point", at which the battery losses the linearity. Basic traditional approaches in terms of predicting the battery RUL fall into two main categories, namely: physics-based models and data-driven approaches.

Physics-based approaches such as the dual lithium-ion insertion cell [3] and pseudo-two-dimensional (P2D) models are based on the electrochemical principles of LIBs and can simulate the battery's current and voltage characteristics by incorporating kinetic and transport equations. The physics-based models are used for their accuracy and interpretability, but they demand significant domain expertise which is computationally intensive. On the other hand, the data-driven approach includes the machine learning and deep learning techniques, and utilizing the battery cycling data to identify the degradation patterns and they also may lack the physical interpretability of their physics-based counterparts. For instance, Zhang et al., in order to learn the long-term dependencies among

DOI: 10.1201/9781003596097-10

the degraded capacity, trained a long short-term memory (LSTM) [5]. Even though these methods might have attained satisfactory performances, they still have some inevitable limitations.

Therefore, this chapter presents the approach of an innovative hybrid deep learning framework designed to tackle these challenges. This approach also improves the generalization capabilities and boosts prediction accuracy without the need for additional training costs. The proposed framework integrates the combination of domain knowledge-based features—such as degradation capacity, internal resistance, and discharge voltage variance—with latent deep network features which are automatically extracted by deep neural networks. This combination ultimately increases the model's accuracy and generalization capabilities for the early predictions of battery RUL. This framework consists of a deep learning architecture, incorporating one-dimensional (1D) convolutional neural networks (CNNs) to moral dependencies in the data. Furthermore, methods like nonlinear correlation-based feature selection, dropout, and L2 regularization are employed to enhance robustness and minimize overfitting.

10.2 NOVELTY AND CONTRIBUTION

10.2.1 *Hybrid Deep Learning Framework*

Integration of domain knowledge-based features and deep learning-extracted latent features are used to improve the accuracy and generalization of early battery RUL predictions.

This learning approach is utilized for leveraging both handcrafted features and learned features to address the nonlinear degradation challenges in batteries.

10.2.2 *Advanced Feature Selection*

The SelectKBest method is utilized to identify the most relevant features for the task of predicting the RUL. It ranks features based on their statistical significance and selects the top κ features, ensuring that only the most informative data are used for training.

Feature selection is not limited to statistical methods but also incorporates domain expertise to enhance prediction accuracy.

10.2.3 *Attention Mechanism*

The attention mechanism assigns weights to input features, focusing on critical regions such as charge–discharge cycles or voltage fluctuations, enabling the model to capture long-term dependencies and improve prediction accuracy for RUL.

Attention weights provide insights into which features or time steps are most important, improving model transparency and offering valuable knowledge about battery degradation patterns.

10.2.4 *Comprehensive Feature Integration*

Combining local statistical features with domain-knowledge-based features creates a synergistic approach, allowing the model to benefit from both raw measurements and expert-informed insights.

10.2.5 *Improved Generalization Across Datasets*

Validation of the model's ability allows it to perform well not only on datasets with similar distributions as the training data but also on datasets with significantly different distributions (secondary test sets).

10.2.6 State-of-the-Art Performance

Demonstration of superior performance compared to existing methods in terms of root mean squared error (RMSE) and mean absolute percentage error (MAPE) for both primary and secondary test sets.

10.2.7 Efficient Early Prediction

This has the capability to predict battery RUL accurately using only early cycle data (first 100 cycles), addressing the challenges posed by minimal capacity degradation in the early stages of battery life.

10.2.8 Practical Application

This approach has the potential to accelerate the development cycle of batteries, validate the efficacy of new manufacturing processes, and optimize the utilization of batteries across diverse applications, including EVs, HEVs, consumer electronics, and renewable energy systems.

10.3 LITERATURE SURVEY

Various methods have been studied for RUL prediction in batteries over the years, including physical modeling techniques, machine learning approaches, and hybrid combinations of these methods.

The research of Doyle et al. [2,3] in 1993 presented the dual lithium-ion insertion cell as a physics-based modeling solution that delivers highly precise simulations of electrochemical battery operations. The model demands enormous computing power together with specific technical knowledge for effective implementation, although it produces precise results. The P2D models [1] approved by Santhanagopalan et al. in 2006 improved electrochemical modeling precision through refinement, but these advancements required major computational resources that prevented real-time implementation.

The field of machine learning now serves as an alternative method through which algorithms like support vector machines (SVMs) and long short-term memory (LSTM) networks can predict RUL. The research by Patil et al. [4] in 2015 demonstrated SVM implementation through a cross-to-fine structure which faced issues because its dependence on manually created features required difficult feature engineering work. The application of LSTM networks by Zhang et al. [5] in 2018 resulted in effective long-term battery degradation dependence monitoring. These models need well-selected representative features for ensuring both scalability and robustness on different datasets.

Several architectures integrated into hybrid models lead to better predictive results through their collaborative approach. The research team of Ma et al. [7] launched CNN-LSTM architectures during 2019 which combined CNNs with LSTM networks for automatic feature extraction as well as temporal modeling. This methodology succeeded in detecting more precise battery degradation patterns better than earlier methods. The method still faced obstacles when attempting to make predictions that applied to multiple differing datasets. Rouhi Ardeshiri et al. [6] implemented gated recurrent units (GRU) in 2021 as a method to enhance modeling of battery degradation nonlinear patterns. The advantages of GRU-recurrent neural network (RNN) architectures were unable to compensate for their inability to handle dataset variability, which restricted their widespread use.

The RUL prediction depends on the implementation of a feature selection process alongside feature engineering techniques. Severson et al. [8] highlighted the importance of certain features that apply specifically to the domain when they researched it in 2019. The absence of advanced selection methods in these features caused performance fluctuations when processing different datasets. Recent advancements through correlation-based feature selection along with recursive feature elimination have overcome such limitations by enhancing the accuracy and generalization of models.

Wu et al. (in 2016) together with Chen et al. (in 2020) conducted their research on improving model performance exclusively within a single dataset [9]. The statistical characteristics of datasets [10] determine the effectiveness of these approaches because they tend to perform poorly when handling datasets with different statistical distributions. Consistent reliable RUL predictions in different real-world settings remain blocked because of the lack of robustness.

10.4 PROPOSED SYSTEM

The following examples will demonstrate the embodiments of the invention and the outline of its scope (Figure 10.1): the enhanced battery RUL prediction using hybrid deep learning and attention-based CNN models.

Here, the model will be integrated with a hybrid deep learning approach combining with 1D CNNs with the attention-based mechanism, highlighting the critical features and ultimately improving the prediction accuracy for the battery RUL.

This system actually uses the mutual information regression with SelectKBest technique for the process of feature selection and then identifies the most useful and relevant features so that it can improve the model efficiency by reducing the computational complexity and maintaining the prediction accuracy.

Here, the Adam optimizer adapts the learning rate for each parameter ensuring fast and stable convergence. The concept of L2 regularization and dropout regularization facilitates the robustness in model training and then mitigates the overfitting issue without any additional computational costs.

The combination of the attention-based mechanism within the 1D-CNN layers will dynamically prioritize the significant features from the datasets, enhancing the computational efficiency and also maintaining the prediction accuracy of the battery RUL making this hybrid model adaptable for various datasets and battery chemistries, ensuring the model scalability for real-time deployment such as with EVs.

As mentioned in hybrid deep learning and attention-based CNN models, the system's ability to tackle the increasing need for accurate prediction of the RUL of batteries is highlighted. It offers a flexible and scalable solution that empowers users to proactively schedule battery replacements. The system's robustness for real-world applications is ensured through the integration of feature selection, mutual information regression, attention mechanisms, and other techniques.

Here, this enhanced battery RUL prediction approach will utilize the hybrid deep learning networks with the attention-based CNN models, which are most likely adaptable in LIBs. This range of applicability covers various battery technologies, which are mainly performed in consumer electronics, renewable energy systems, and also in industrial machines. The approach of this hybrid architecture includes the 1D-CNNs, with attention-based mechanism, feature selection, and snapshot ensemble learning techniques, which are ultimately suitable for time-series prediction tasks such as forecasting equipment failures and energy usage trends.

10.4.1 Overview

Systems like LIBs exhibit a long-term degradation process spanning thousands of cycles, and for each cycle there are many variables and measurements collected by various sensors. Generally, directly feeding this substantial amount of raw data into deep learning models is impractical. This is due to the presence of noise and redundant information among those data which might lead in slow model convergence of the result. The nature of battery degradation is nonlinear, therefore it is significantly challenging to predict RUL at an early stage when utilizing limited cycle data. Moreover, the inconsistent distribution between training and test data necessitates a better generalization ability for RUL prediction.

To tackle the above challenges, we propose a model that predicts the RUL of batteries accurately at an early stage. This model utilizes SelectKBest and mutual information regression in order to extract the most relevant features. A hybrid model has been developed that combines both local

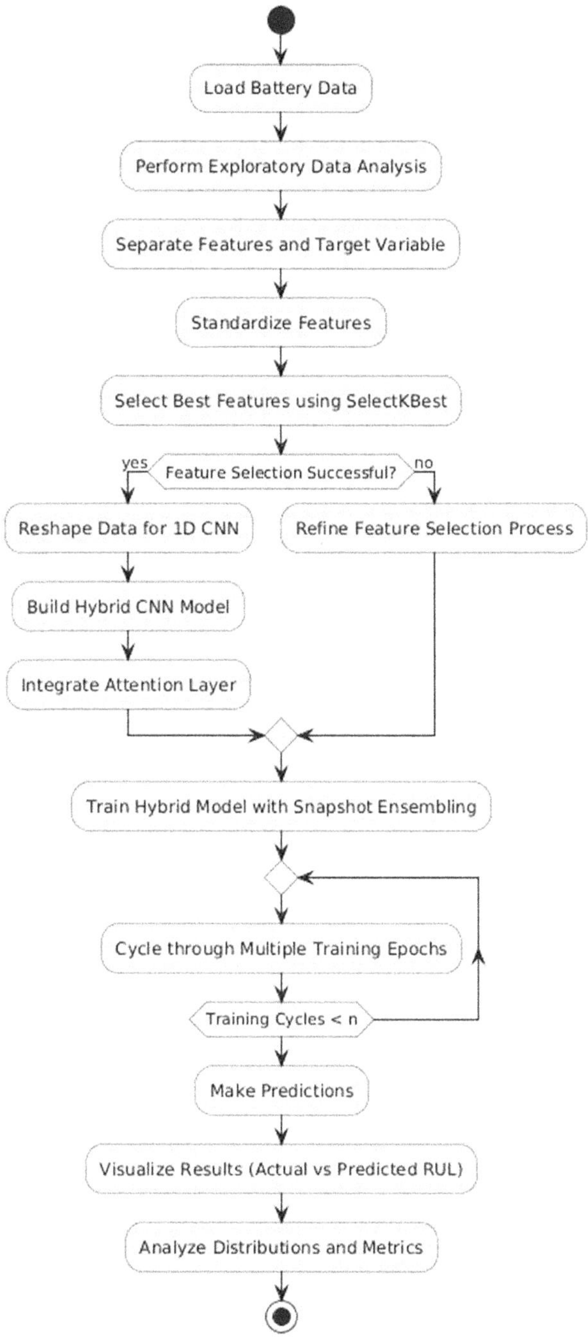

FIGURE 10.1 Flowchart representing the work flow of the system.

statistical features and domain knowledge-based features, as shown in Figure 10.1. Then attention mechanism is integrated with a hybrid 1D CNN model to dynamically prioritize important features or time steps in the sequential data, and some regularization techniques and other techniques are used to overcome the overfitting problem and improve generalization.

10.4.2 Data Collection and Preprocessing

The accurate implementation of prediction models for battery RUL requires proper preprocessing of data inputs. Noisy datasets accompanied by missing values and inconsistencies create performance problems during model implementation. The first stage of data cleaning involves removing both damaged and incomplete records from the dataset. The preprocessing approach for missing data consists of interpolation methods and statistical imputation techniques that use mean, median, and nearest-neighbor interpolation methods. Standard deviation thresholds together with interquartile range (IQR) analysis detect outliers from sensor anomalies and data-logging errors. The methodology decides whether to remove or correct such outliers. The cleaned dataset receives a reformatted structure before deep learning models can utilize it because of its time-series sequence format.

10.4.3 Feature Selection

SelectKBest with mutual information regression (MIR) is used for advanced feature selection.

MIR measures the mutual dependence between input features and the target variable. Features with higher mutual information are retained for further analysis. The mutual information MI (*X*, *Y*) between two variables, input feature *X* and target variable *Y*, measures how much knowing one of these variables reduces uncertainty about the other, and it is given in Eq. (10.1).

$$MI(X,Y) = p(x,y)\log\left(\frac{p(x,y)}{p(x)p(y)}\right)dx\,dy \tag{10.1}$$

where *p*(*x*,*y*) is joint probability distribution of *X* and *Y*, *p*(*x*) is the marginal probability of *X*, and *p*(*y*) is the marginal probability of *Y*.

To standardize these features data standardization is done. The StandardScaler technique is applied. It scales features to ensure uniformity of features. To standardize features *X*, the mean and deviation are used. This ensures that all features are centered around 1 and have a standard deviation of 1, enabling stable model convergence (Eq. 10.2)

$$X_{scaled} = \frac{X-\mu}{\sigma} \tag{10.2}$$

From Eq. (10.2), the mean of feature is given by 'μ' and standard deviation of the feature is given by 'σ'.

10.4.4 Hybrid Deep Learning Framework

In order to capture as much relevant information as possible, we propose a hybrid 1D CNN model. This combines both handcrafted features with domain knowledge and latent features learned by deep networks [11]. The CNN extracts spatial patterns from the time-series data. This operation helps in identifying local patterns.

The input sequence values at position $i+j$ is denoted by $x[i+j]$. The weight of the convolutional kernel (filter) is termed as $w[j]$. Kernel size is given by *k*. The bias term is given by *b*. $y[i]$ gives the output after convolution at position *I* (Eq. 10.3).

$$y[i] = \sum_{j=0}^{k-1} w[j] \,.x[i+j] + b \tag{10.3}$$

The above is the discrete format of the convolutional operation.

The integral format can be written as Eq. (10.4)

$$s(t) = (x * w)(t) = \int_{-\infty}^{\infty} x(T) w(t-T) dT \tag{10.4}$$

where $x(T)$ is the input signal or data, $W(t - T)$ is the kernel function or filter, and $y(t)$ is output after convolution.

Then, the activation function in HybridCNN introduces nonlinearity into the model. This enables the model to learn complex patterns from input data. In this model, the rectified linear unit (ReLU) activation function, i.e. Eq. (10.5) is utilized with CNN layers to extract local statistical features.

$$\text{ReLU}(x) = \max(0,x) \tag{10.5}$$

If $x > 0$, then the output is x, which will be a linear response for positive inputs. If $x <= 0$, then the output will be 0 (zeros out negative inputs).

Now apply max-pooling to reduce dimensionality and retain significant patterns. This is the crucial component for down-sampling the feature maps, generated by the layers. It reduces the spatial dimensions of the input by enhancing computational efficiency and focusing on the most relevant features. This is given by Eq. (10.6).

$$P(i, j) = \max\{X(i + m, j + n)\} \tag{10.6}$$

The feature map output is passed through the pooling layer. For instance, using max pooling with a 2×2 kernel and stride of 2 (Eq. 10.7) can be written as:

$$P(i, j) = \max\{X(2i, 2j), X(2i+1, 2j), X(2i, 2j+1), X(2i+1, 2j+1)\} \tag{10.7}$$

The output is a down-sampled version of the input feature map.

Then add a fully connected (FC) layer to process domain knowledge-based features.

The attention mechanism is integrated with a 1D CNN model to dynamically prioritize important features or time steps in the sequential data. It dynamically assigns weights to input features, focusing on critical areas like charge/discharge cycles or voltage fluctuations and other parameters. Attention weights provide insights about which features or time steps are most significant, enhancing model transparency and providing meaningful knowledge of battery degradation pattens. It typically operates using Query (Q), Key (K), and Value (V) matrices derived from the feature map (Eq. 10.8).

$$Attention(Q,K,V) = softmax\left(\frac{QK^T}{\sqrt{d_k}}\right)V \tag{10.8}$$

During training, the Adam optimizer is employed with this model due to its efficiency and adaptability with a suitable learning rate. It updates the model's parameters (weights and biases) to reduce the loss function. Loss function is calculated during training, which the optimizer minimizes. In this model the RMSE is calculated to measure the loss function, as shown in Eq. 10.9. It penalizes larger errors more heavily, which is critical for precise regression tasks like battery RUL prediction.

$$RMSE = \sqrt{\frac{1}{n}\sum_{i=1}^{n}\left(\widehat{yi} - yi\right)^2} \quad (10.9)$$

Now, the calculated loss function will be optimized. This is given by Eq. (10.10).

$$\theta_t = \theta_{t-1} - \eta \cdot \frac{m_t}{\sqrt{v_t} + \in} \quad (10.10)$$

where the learning rate is denoted by η. ∈ is a small value to prevent division by zero. m_t is the biased first moment estimate (moving average of gradients). v_t is the biased second moment estimate (moving average of squared gradients). It combines momentum and adaptive leaning rate methods. It also automatically adjusts learning rates for each parameter.

The model is trained with early stopping or learning rate scheduling to prevent overfitting. Also, regularization techniques are applied to reduce overfitting. Dropout layers are implemented to reduce overfitting by randomly dropping connections during training. It randomly sets a fraction of neurons to zero during training to reduce reliance on specific neurons. L2 regularization is also utilized to penalize large weights in the network, further improving generalization.

10.5 RESULTS AND DISCUSSION

This section presents the findings and discussion of the introduced scheme. Python was used to carry out the results and discussion of the method.

10.5.1 Dataset Description

The present research used the Kaggle Battery Remaining Useful Life (RUL) dataset to predict the RUL of the LIB. Over 15,000 data points included features such as voltage, current, temperature, charge, discharge cycles, and values of capacity. These parameters are basic for understanding the health of the battery along with the prediction of a lifecycle. This dataset was released into the public domain on the Kaggle website and has been widely used for tasks related to the modeling of battery degradation and RUL estimation. The data were preprocessed to ensure consistency and reliability. Standard-Scaler was used for feature normalization to ensure uniformity in data scaling across different parameters. For feature selection, mutual information regression was utilized to identify and retain the most relevant attributes contributing to the prediction accuracy of the hybrid deep learning model. The dataset was the basis of training and validation of the proposed model to ensure robust and generalizable results. The Kaggle dataset can be accessed at:www.kaggle.com/datasets/ignaciovinuales/battery-remaining-useful-life-rul

The Kaggle Battery RUL dataset was preprocessed by standardizing values and selecting important features through correlation and recursive feature elimination. A hybrid deep learning model with 1D CNN and attention mechanisms was designed, and the performance was checked using mean squared error (MSE) and mean absolute error (MAE).

In this research, we have also used another proprietary dataset of LIBs consisting of battery degradation parameters such as voltage, current, internal resistance, and charge and discharge cycles. The entire dataset is processed to carefully remove the noise, handling the missing values, and normalizing the feature distributions, making sure of optimal model performance.

Making sure of our primary dataset, we have also experimented with an additional dataset from the NASA Battery Dataset (/content/kaggle/working/processed_data.csv). This dataset is widely used in several battery-based studies and mainly contains the battery degradation data of lithium-ion cells under various operational conditions.

The Hawaii Natural Energy Institute dataset, comprising 14 lithium nickel manganese cobalt oxide-lithium cobalt oxide (NMC-LCO) 18650 cells with a nominal capacity of 2.8 Ah, was utilized. These cells were cycled over 1000 times at 25°C using a CC–CV charging protocol at a C/2 rate and a discharge rate of 1.5C.

From that source dataset, we have created features that showcase the voltage and current behaviors over each cycle. Those features can be used to predict the RUL of the batteries. The dataset contained a summary of the 14 batteries.

Variables:
Cycle Index: number of cycles
F1: Discharge Time (s)
F2: Time at 4.15V (s)
F3: Time Constant Current (s)
F4: Decrement 3.6-3.4V (s)
F5: Max. Voltage Discharge (V)
F6: Min. Voltage Charge (V)
F7: Charging Time (s)
Total time (s)
RUL: target

10.5.2 Performance Metrics Overview

The evaluation of each model is calculated by two primary metrics, namely MAE and RMSE. These metrics are crucial for quantifying the accuracy battery RUL predictions, allowing us to assess how closely the predicted values align with actual measurements. The HybridCNN framework functions as the core predictive structure but research was conducted on alternative deep learning models to boost prediction competence and data independence. The following part examines CNN-LSTM along with bidirectional long short-term memory (BiLSTM) and Transformer models through an analysis of their methods combined with their benefits and constraints, which are shown in Figures 10.2–10.5.

The graphs in Figures 10.2–10.5 are the representations of training and validation losses for each model, based on the number of epochs that the model was trained on.

HybridCNN: The HybridCNN's ability to leverage both handcrafted features from domain knowledge and deep learning extracted features significantly enhances the predictive capabilities.

The 1D CNN layers extract the hierarchical feature representations from the battery degradation sequence. The combination of batch normalization with dropout functions was used to stop overfitting through distribution normalization along with neural random deactivation. Fully connected layers serve as the last step to merge learned features for RUL estimation within a predictive framework.

Here, the blue smooth curve lies exactly under the red smooth curve. This implies that the graph demonstrates how closely the predicted RUL values align with the actual values i.e., the prediction is very accurate.

BiLSTM: The BiLSTM model recorded an MSE of 18.3 and an RMSE of 22.1, indicating its effectiveness in capturing the temporal dependencies and also highlighting the limitations in accuracy.

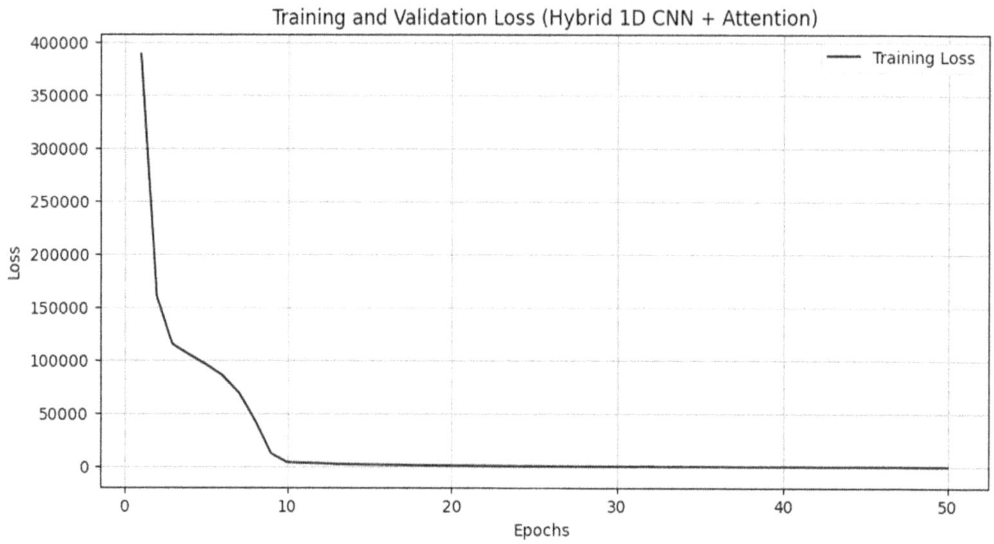

FIGURE 10.2 Training and validation of hybrid 1D CNN.

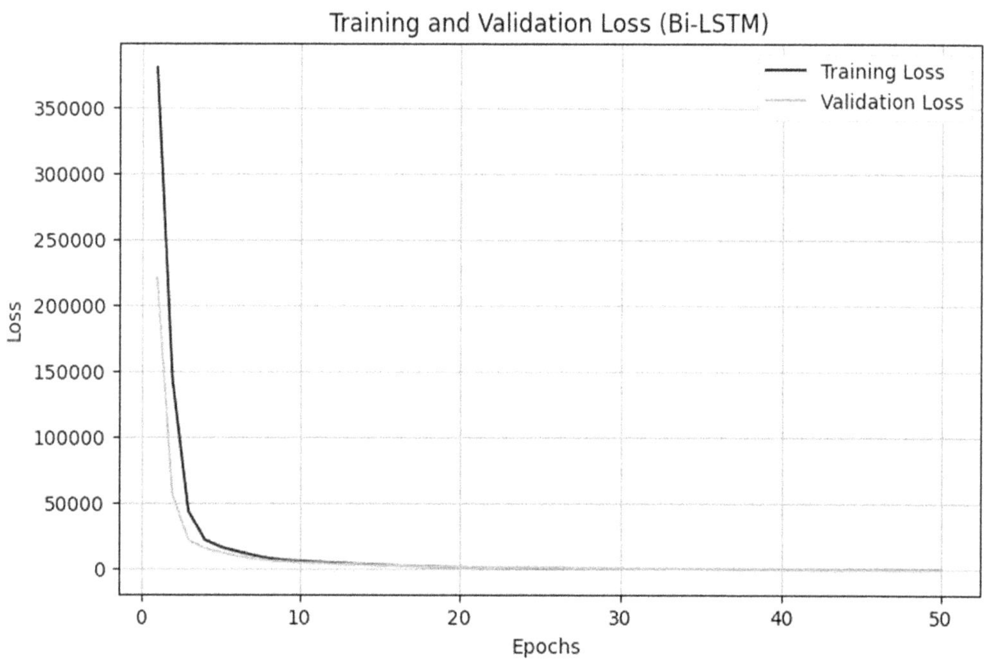

FIGURE 10.3 Training and validation of BiLSTM.

Forward and Backward LSTM Cells: These learn dependencies in both time directions. The concatenation layer combines outputs obtained from both the forward and backward directions. The learned features move through fully connected layers to produce RUL estimation values.

This BiLSTM graph highlights the predictive abilities while revealing the areas where predictions diverge from actual values.

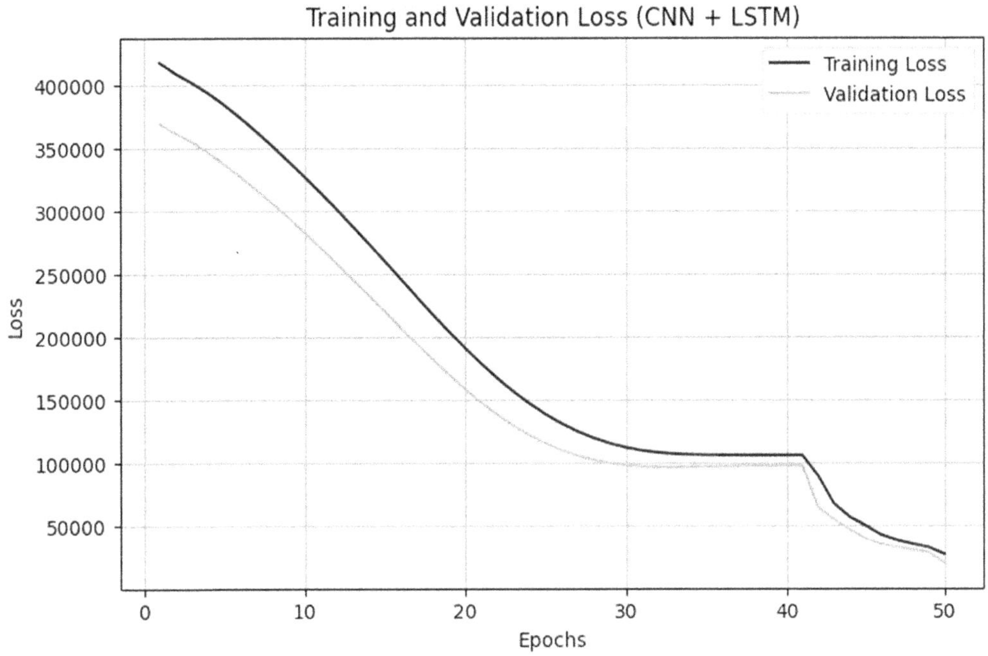

FIGURE 10.4 Training and validation of CNNL-STM.

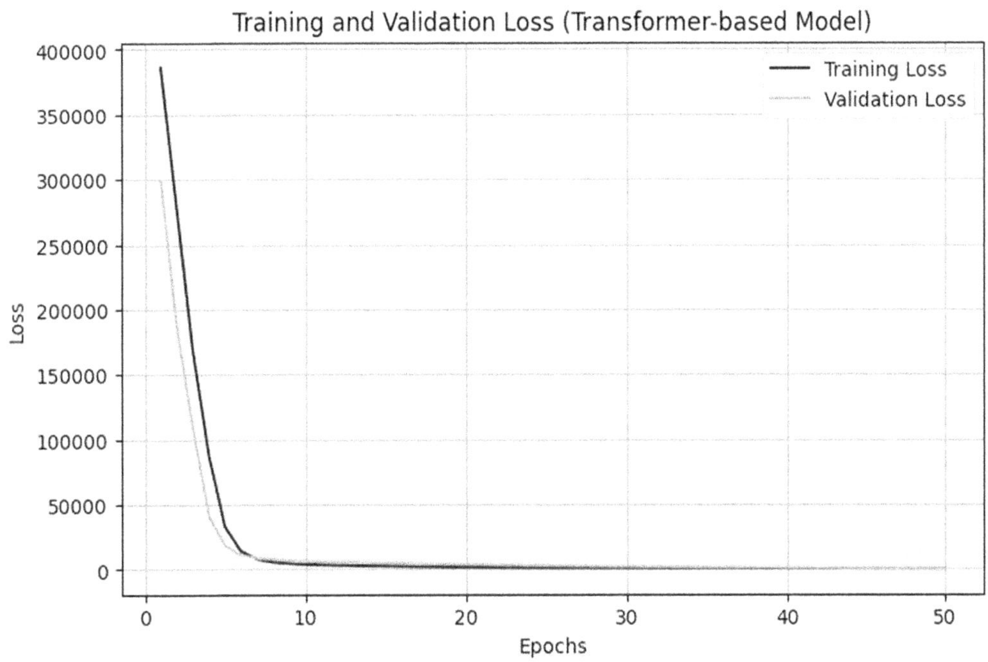

FIGURE 10.5 Training and validation of the Transformer.

CNN-LSTM: CNN-LSTM adds LSTM networks to the HybridCNN framework to detect meaningful relationships between components in battery degradation analysis. The ability of CNNs to extract features successfully does not extend to their understanding of long-term connections between charge–discharge cycles. The model acquires the dual capability to analyze space and time information through the addition of LSTMs that follow CNN layers.

Here, we can see the blue smooth curves clearly at certain points which shows the inaccurate or unsuitable conditions of predictions.

Transformer: The Transformer model has delivered the best performance with an MAE of 14.2 and an RMSE of 19.2, understanding its strength in handling complex patterns through its attention mechanism. The Transformer model, originally designed for natural language processing (NLP), has recently been applied to time-series forecasting. Unlike RNNs which process data sequentially, transformers use self-attention mechanisms to model dependencies across the entire sequence at once.

Multi-Head Self-Attention Mechanism: This assigns different weights to different sequence positions. Positional encoding keeps the order of time-series data. Feedforward layers map attention-weighted features to the final RUL predictions.

These graphs or visual representations enhance understanding and provide a clear view of each model's capabilities in predicting the battery RUL.

10.5.3 Comparative Table

Table 10.1 provides an overview for the different approaches of deep learning methods used in order to predict the battery RUL.

10.5.4 Bar Chart Comparison

The chart contains four models, which are HybridCNN, CNN-LSTM, BiLSTM, and Transformer. Metrics include R square (coefficient of determination), MSE, and MAE. The performances of these four models are analyzed and compared based on these metrics.

10.5.4.1 *R^2 Score Comparison Observation*

All models achieve very high R^2 scores, close to 1, indicating excellent prediction accuracy. The differences between the models are minor, but HybridCNN and BiLSTM seem to perform slightly better. The models are well-tuned, with HybridCNN and BiLSTM marginally outperforming the others (Figure 10.6).

10.5.4.2 *MSE Comparison Observation*

CNN-LSTM has a significantly higher MSE value compared to the others, indicating poor performance in minimizing squared errors. HybridCNN achieves the lowest MSE, followed by Transformer and BiLSTM.

Conclusion: HybridCNN achieves the lowest MSE, followed by Transformer and BiLSTM. HybridCNN is the most accurate in predicting values with minimal deviation, while CNN-LSTM struggles in this metric (Figure 10.7).

TABLE 10.1
Comparative Analysis of Various RUL Prediction Methodologies

Method	MAE	RMSE	Features
Hybrid CNN (Kaggle Dataset)	4.03	0.99947	Combines handcrafted and deep learned features
BiLSTM (Kaggle Dataset)	6.02	0.99911	Bi-directional learning for temporal data
CNN-LSTM (Kaggle Dataset)	33.93	0.959	Merge CNN with LSTM
Transformer (Kaggle Dataset)	7.324	0.9994	Attention mechanism enhancing feature prioritization
HybridCNN (NASA data)	2.4998	0.9995	Handcrafted and deep learned features
CNN-LSTM (NASA data)	200.531	-1.74431	For temporal data
BiLSTM (NASA data)	40.843	0.8868	CNN with LSTM
Transformer (NASA data)	12.4772	0.98344	Attention mechanism with enhanced features

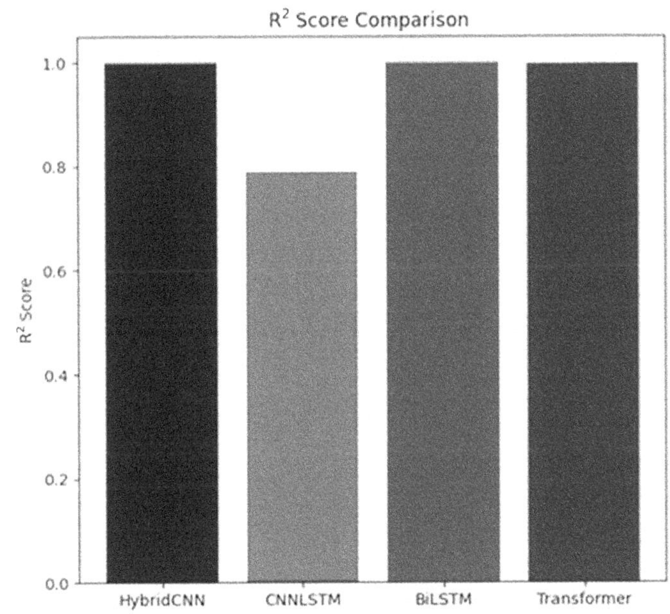

FIGURE 10.6 R^2 score comparison of four models.

10.5.4.3 *MAE Comparison Observation*

CNN-LSTM again has the highest MAE, suggesting it performs poorly in minimizing the average error magnitude. HybridCNN has the lowest MAE, showcasing the best performance, followed by BiLSTM and Transformer. The HybridCNN model is the most effective in providing accurate predictions with minimal errors (Figure 10.8).

10.5.5 Overall Insights

HybridCNN is the best-performing model across all three metrics, making it the most reliable for predicting the RUL of batteries (Figure 10.9).

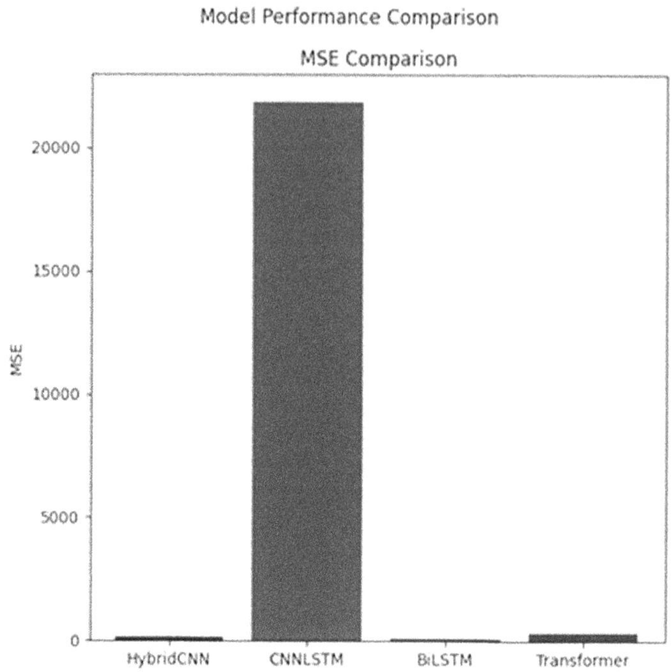

FIGURE 10.7 MSE score comparison of four models.

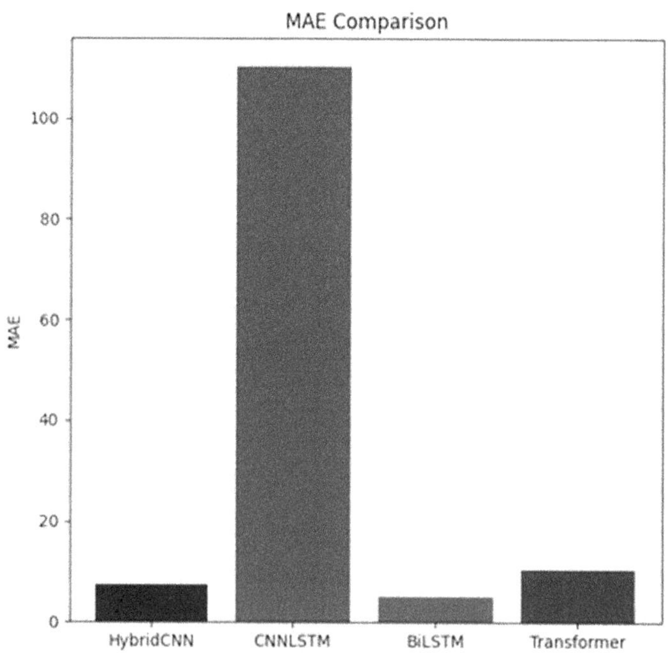

FIGURE 10.8 MAE score comparison of four models.

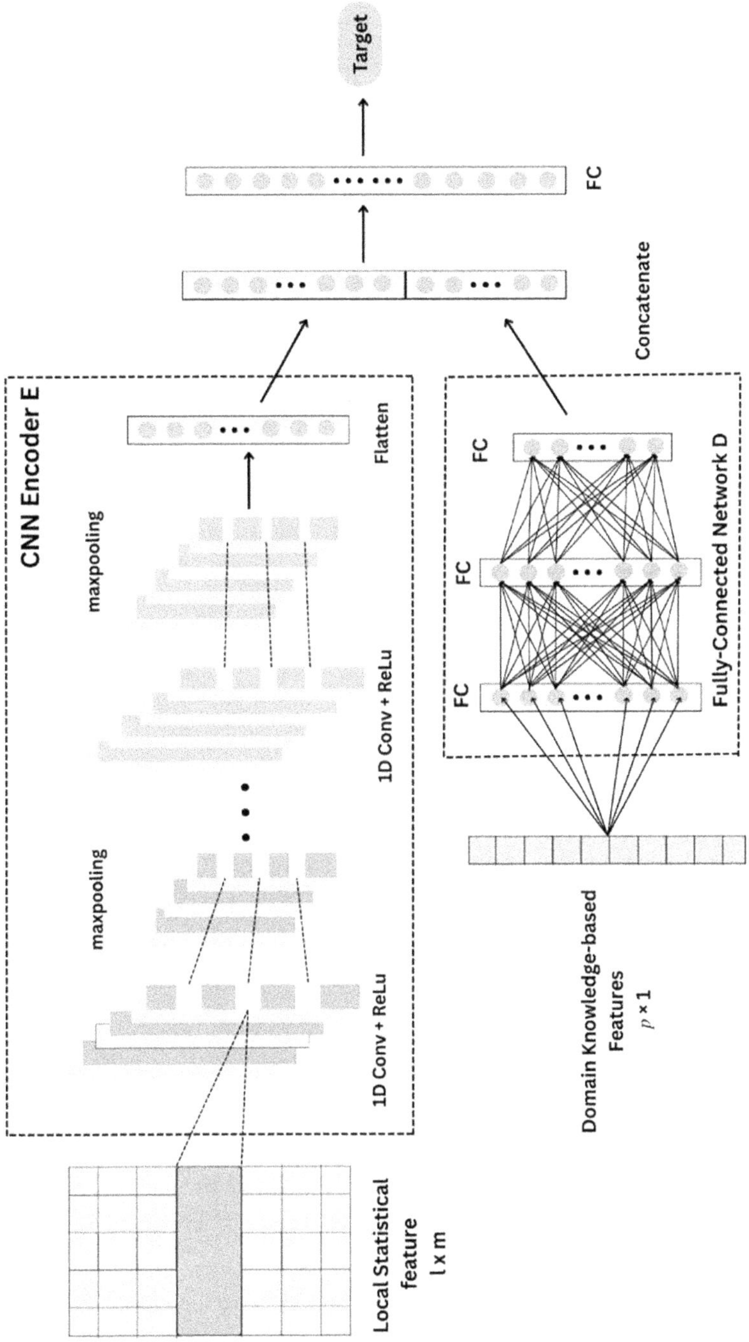

FIGURE 10.9 Working architecture of a hybrid CNN.

CNN-LSTM performs poorly, especially in MSE and MAE, which highlights potential areas for optimization or tuning. BiLSTM and Transformer show comparable performance, but neither surpasses HybridCNN.

10.6 CONCLUSION

In this study, a new proposed hybrid deep learning model with much improved RUL prediction on LIBs is also used to overcome the limitations of conventional methods by integrating CNNs and attention mechanisms and effective feature selection techniques. Analysis shows that early charge and discharge cycles are crucial for predicting the RUL. The model with Snapshot Ensemble Learning has boosted generalization with reduced overfitting and computational requirements. This approach can ensure proactive battery lifecycle management with minimal unexpected failures, thus optimizing scheduling in maintenance. Future work should include real-time implementation and scalability across different chemistries and industries for battery types.

REFERENCES

[1] S. Santhanagopalan, Q. Guo, P. Ramadas, and R. E. White, "Review of models for predicting the cycling performance of lithium ion batteries," *Journal of Power Sources*, vol. 156, no. 2, pp. 620–628, 2006.

[2] M. Doyle, T. F. Fuller, and J. Newman, "Modeling of galvanostatic charge and discharge of the lithium/polymer/insertion cell," *Journal of The Electrochemical Society*, vol. 140, no. 6, pp. 1526–1533, Jun. 1993. [Online]. Available: http://jes.ecsdl.org/content/140/6/1526

[3] T. F. Fuller, M. Doyle, and J. Newman, "Simulation and optimization of the dual lithium ion insertion cell," *Journal of The Electrochemical Society*, vol. 141, no. 1, pp. 1–10, Jan. 1994. [Online]. Available: http://jes.ecsdl.org/content/141/1/1

[4] M. A. Patil, P. Tagade, K. S. Hariharan, S. M. Kolake, T. Song, T. Yeo, and S. Doo, "A novel multistage support vector machine-based approach for li ion battery remaining useful life estimation," *Applied Energy*, vol. 159, pp. 285–297, 2015.

[5] Y. Zhang, R. Xiong, H. He, and M. G. Pecht, "Long short-term memory recurrent neural network for remaining useful life prediction of lithium-ion batteries," *IEEE Transactions on Vehicular Technology*, vol. 67, no. 7, pp. 5695–5705, 2018.

[6] R. Rouhi Ardeshiri and C. Ma, "Multivariate gated recurrent unit for battery remaining useful life prediction: A deep learning approach," *International Journal of Energy Research*, vol. 45, no. 11, pp. 16633–16648, 2021.

[7] G. Ma, Y. Zhang, C. Cheng, B. Zhou, P. Hu, and Y. Yuan, "Remaining useful life prediction of lithium-ion batteries based on false nearest neighbors and a hybrid neural network," *Applied Energy*, vol. 253, p. 113626, 2019.

[8] K. A. Severson, P. M. Attia, N. Jin, N. Perkins, B. Jiang, Z. Yang, M. H. Chen, M. Aykol, P. K. Herring, D. Fraggedakis et al., "Data-driven prediction of battery cycle life before capacity degradation," *Nature Energy*, vol. 4, no. 5, pp. 383–391, 2019.

[9] J. Wu, C. Zhang, and Z. Chen, "An online method for lithium-ion battery remaining useful life estimation using importance sampling and neural networks," *Applied Energy*, vol. 173, pp. 134–140, 2016.

[10] L. Chen, Y. Zhang, Y. Zheng, X. Li, and X. Zheng, "Remaining useful life prediction of lithium-ion battery with optimal input sequence selection and error compensation," *Neurocomputing*, vol. 414, pp. 245–254, 2020.

[11] P. Severson, M. Attia, N. Jin, N. Perkins, B. Jiang, Z. Yang, and M. Z. Bazant, "A hybrid ensemble deep learning approach for early prediction of battery remaining useful life," *IEEE/CAA Journal of Automatica Sinica*, vol. 10, no. 1, pp. 177–187, 2024.

11 Natural Fibre Composites
A Sustainable Lightweight Material for Electric Vehicles

Rahul Murali Cadambi and Sunith Babu L.

11.1 INTRODUCTION

The automotive sector is undergoing tremendous changes to address the challenges and growing pressures from environmental regulations, fuel prices, and stringent specifications from equipment manufacturers. The market is not only demanding lightweight materials but also materials that are recyclable. A 10% reduction in kerb weight is estimated to result in a 6–8% reduction in energy consumption. Usage of fibre-reinforced polymer (FRP) composites based on synthetic fibres such as glass, carbon, and aramid fibres have significantly increased for various engineering applications. Although FRPs have several attractive properties, they are difficult to recycle. This has led to interest in developing sustainable and environment-friendly composites by usage of natural fibre-reinforced composites.

Natural fibre-reinforced composites are made up of natural fibres such as hemp, sisal, jute, kenaf, bamboo, coir, flax, and others reinforced with polymer matrices. With more emphasis towards environmental regulations and sustainability, usage of eco-friendly fibres has been explored by both industries and researchers to replace synthetic fibres. The usage of natural fibres may not only reduce waste disposal issues but also reduce environmental pollution. Natural fibres have advantages including lightweight, energy required for production, availability or sustainable resources, health aspects, recyclability, biodegradability, less abrasiveness to equipment, low cost, and manufacturing flexibility [1–4].

A shift in using "environmentally friendly" natural fibre composites over conventional composites is gaining interests in the automotive, marine, and sporting goods industries [1,4,5] due to several advantages. These advantages include the combination of various mechanical properties with low density [1,6–10]. The use of natural plant fibres as reinforcement in polymer materials is associated with a drawback of water absorption [1,11]. All natural fibres consist of lignin and cellulose. Cellulose contains hydroxyl groups, making it hydrophilic and unsuitable for hydrophobic thermoplastics or thermosets resulting in poor moisture resistance. Plant-based fibre composites' performance depends on the quality of the fibre, poor reproducibility of physical properties of fibres, thermal stability, and water absorption capacity [12–14]. Natural fibres lack compatibility between the fibres and the matrix, resulting in poor final properties, i.e., natural fibres are polar and hydrophilic in nature, whereas all the commercial polymers are non-polar in nature [15].

Due to the incompatibility of natural fibres and polymer matrix, it is essential to modify the surface of fibres for better interfacial properties with the matrix [16]. Modification of the fibre surface can reduce the absorption of moisture [16]. Several studies have demonstrated the appropriate surface modifications for improving compatibility [17]. It is to be noted that natural fibre composites are still in the development stage and have limited industrial applications.

DOI: 10.1201/9781003596097-11

The focus of this chapter is to provide an overview of plant-based fibres, their physical characteristics, chemical composition, and mechanical properties. Different techniques, both physical and chemical treatments, for improving the compatibility of fibres and matrix bonding are discussed. Finally, the potential application of natural fibre composites in the automobile industry is discussed. The chapter will cover an understanding of the characteristics and properties of natural fibre composites for researchers, engineers, and industries to carry out decisions regarding material selection and design, promoting sustainable and environmentally friendly solutions.

11.1.1 Natural Fibres – Physical Characteristics, Chemical Composition, and Mechanical Properties

There are a variety of natural fibres derived from cellulose in the plant species. The morphology and properties of natural fibres depend on the location of fibres in the plant and their function. Plant fibres can be located in bast (e.g., banana, flax, hemp, jute, kenaf, ramie, and rattan), leaf (e.g., banana, pineapple, palm, and sisal), fruit (e.g., coconut, coir, cotton, and rice husk), stalk (e.g., bagasse, bamboo, barley, reed, rice, wheat, and wood), and in the form of agricultural waste such as coir and sugarcane bagasse [1,7,10,18–24]. For composite reinforcement, bast fibres such as flax, hemp, jute, and kenaf are used. Basts are found on the outer portion (bark) of plants and possess excellent stiffness and strength properties [1,7,10,13,25–35].

Plant fibres consist of a strong, stiff cellulose molecule embedded in a matrix consisting of hemicellulose and lignin. Although the fibres are derived from the same plant, the properties of natural fibres are affected by a variety of factors, including geographical conditions, physical characteristics, chemical composition, crystalline cellulose dimensions, microfibrillar angle, defects, structure, and isolation technique. This results in significant variations in the quality and mechanical properties of the fibre [1,7,10,18–24]. The physical properties, chemical composition, and mechanical properties of commonly used natural fibres are tabulated in Tables 11.1, 11.2, and 11.3, respectively.

11.1.2 Mechanical Properties of Natural Fibres

The properties of natural plant fibres are influenced by various factors, such as their cellulose content, which is a high-strength polymer, and the microfibrillar angle, which represents the orientation of cellulose fibrils in the main wall [10]. Fibres with lower microfibrillar angles tend to have higher tensile properties. Table 11.3 summarizes the mechanical characteristics of various natural fibres. The chemical composition, structure, flaws, and cell size are the primary factors that influence

TABLE 11.1
Physical Properties of Selected Natural Fibres

Natural Fibre	Fibre Length (mm)	Fibre Diameter (μm)	Thickness of single cell wall (μm)
Coir	0.3–1.0	12–14	0.06–8
Sugarcane bagasse	0.6–2.7	10–40	1.5–9.5
Bamboo	2–3	14–17.8	3–9
Flax	10–65	5–38	10–20
Hemp	8–55	3–51	8–25
Jute	0.8–6	5–30	5.2–11.3
Kenaf	1.4–11	4–36	1.6–12.6
Sisal	0.8–8	7–47	8–25

TABLE 11.2
Chemical Composition of Selected Natural Fibres

	Chemical Characteristics		
Natural Fibre	**Cellulose (wt.%)**	**Hemicellulose (wt.%)**	**Lignin (wt.%)**
Coir	32–43	0.15–0.25	40–45
Sugarcane bagasse	41.1–55.2	16.8–31.8	22.3–25.3
Bamboo	22–56.7	17.2–43.8	21–31
Flax	64–75	10–21	2–5
Hemp	68–75	15–22.4	3.7–10
Jute	61–71	13–20.4	5–13
Kenaf	31–72	20–33.9	8–21.2
Sisal	60–78	10–38	8–14

TABLE 11.3
Mechanical Properties of Selected Natural Fibres

Properties	Flax	Hemp	Jute	Sisal	Coir	Kenaf	Sugarcane Bagasse	E-Glass	Carbon
Tensile strength (MPa)	700	800	500	600	200	750	230	2000–3500	3400–4800
Young's modulus (MPa)	70	65	30	12	5	45	5	70–75	240–425
% Elongation	3	3	1.8	3	30	3.2	6.5	0.5–3	1.4–1.8
Density (g/cm^3)	1.4–1.5	1.4–1.6	1.3–1.5	1–1.5	1.15–1.5	0.6–1.5	1.1–1.6	2.5–2.6	1.4–1.8
Multifibrillar angle (°)	5–10	2–6	7–10	10–25	30–49	4–8	25–30	–	–

the properties of natural fibres, especially their mechanical capabilities [36] and wear resistances [37,38]. Generally, the tensile strength and Young's modulus of fibres increase with an increase in cellulose content [39].

11.1.3 Factors that Affect the Fibre Properties

Despite the excellent potential shown by plant fibres in terms of their mechanical properties, these are not realized in practice. Fibres show very good tensile properties, but during processing they can be damaged due to the formation of dislocations, thus reducing their properties. These dislocations can be in the form of kink-bands, micro-compressive defects, or nodes. Another significant factor that can affect the properties of plant fibres is moisture. Hydroxyl groups (–OH) present in the cell wall of plant fibres attract water molecules and bind them through hydrogen bonds, rendering the fibres hydroscopic. Hydroxyl groups in the cellulose core of the microfibril form inter- and intra-molecular hydrogen bonds that maintain the crystal structure and provide some lateral stability to the cellulose chains. Since the hydroxyl groups are engaged in inter- and intra-molecular hydrogen bonding, they are unavailable for binding water molecules. However, the hydroxyl groups in lignin and hemicelluloses in the core bind with water molecules. Table 11.4 shows some of the behaviours of natural fibres with respect to moisture uptake.

TABLE 11.4
Behaviours of Some Natural Fibres in | Moisture

Properties	Flax	Jute	Sisal
Absorption regained in % at 65% relative humidity and 20°C	7–12	8.5–17	10–22
Transverse swelling, %	20–25	20–22	18–20
Axial swelling, %	0.05–0.2	0.37	
Volume swelling, %	29.5	44–45	39.5
Water retention, %	50–55	25–35	30–45

11.2 PHYSICAL AND CHEMICAL MODIFICATION TREATMENTS OF NATURAL FIBRES

To reduce the moisture sorption, to improve the dimensional stability, to enhance the fibre–matrix adhesion, and to reduce biodegradation, natural fibres are often modified by chemical or physical means to improve the properties of natural fibre composites.

11.2.1 Physical Treatment

The purpose of physical treatment is to separate natural fibre bundles into individual filaments and modification of the fibres for composite applications. The separation of natural fibre bundles includes chemo-mechanical, water-retting, and dew-retting. The chemo-mechanical treatment produces unacceptable fermentation waste [40–42]. To reduce the fermentation waste, physical separation processes by steam explosion process and thermomechanical process are commonly adapted. Both processes separate the lignocellulosic material into its main components, namely cellulose fibres, amorphous lignin, and hemicellulose. These methods can increase the hydrophobicity of the surface and improve the capability of adhesion with the matrix [40–42].

For the modification of the fibres, composite applications can be divided into thermal and nonthermal methods. For thermal treatments, the method is known as plasma treatment. Plasma treatment brings a physical modification to the surface through roughening of the fibre by the sputtering effect, producing an enlargement of the contact area that increases the friction between the fibre and the polymer. This helps in producing better adhesion characteristics and promotes the action of coupling agents [40–42]. In non-thermal methods, it promotes the polymer surface to enable more wettability and enhance adhesion. Corona discharge and dielectric-barrier discharge are two commonly used non-thermal techniques. Table 11.5 summarizes the physical treatment modification of natural fibres.

11.2.2 Chemical Treatment

Chemical treatment of natural fibres is carried out to improve their compatibility with the polymer matrix and enhance the quality of the resulting composite. Cellulose is the main target for chemical modification in any natural fibres. The amorphous zone of cellulose is more accessible to attack by reagents than the crystalline zone [1,2,19]. The chemical modifications of fibres can decrease the absorption of humidity, thus improving the interfacial bond strengths or adhesion with the polymer matrix [1,13]. The two different strategies for modifying the fibres chemically are by reducing the number of alcohol groups on the fibre's surface to reduce their polarity and increase their compatibility with the polar polymer matrix, or releasing more reactive alcohol sites from the compounds which cover them (waxes, hemicelluloses, and lignins) to make them react with other treatments such as alkaline and silane treatments. Natural fibres are covered with fatty compounds like waxes

TABLE 11.5
Physical Treatment Modification for Natural Fibres

Name of Treatment	Mechanism of treatment	Improvements
Corona	Formation of a high-energy electromagnetic field close to charged thin wires/points induces ionization species (ions, radicals, etc.) and is activated for surface modification through introduction of oxygen-containing functional groups	Existence of hydroperoxide groups that could initiate grafting of the matrix chains leading to significant improvement of interfacial shear strength
Plasma	Similar mechanism to corona. The apparatus required a vacuum chamber and gas feed to maintain the appropriate composition of the gaseous mixture	
Mercerization	Soaking the fibre in NaOH	Improves adhesive characteristics by removing natural and artificial impurities and promotes rough surface topography 1. Fibre fibrillation (breaking down the composites of the fibre bundle into smaller fibres) 2. Increases the effective surface area available for contact with the wet matrix 3. Enhances the reactivity
Heat treatment	When heated the fibre undergoes physical (enthalpy, weight, strength, colour, and crystallinity) and chemical changes (reduction in degree of polymerization by bond scission, creation of free radicals, formation of carbonyl, carboxyl, and peroxide groups)	Increase in yield strength

Source: Ref [1].

and oils, which can hinder the adhesion of fibres to the polymer matrix. Acetone is used to solubilize these fatty compounds [1,40,41]. This subsequently makes the natural fibres more receptive to subsequent chemical treatments aimed at improving their compatibility with the polymer matrix. There are 12 types of chemical treatments: alkaline, saline, acetylation, alkaline hydrogen peroxide, benzoylation, acrylation and acrylonitrile grafting, maleated coupling, permanganate treatment, peroxide treatment, isocyanate treatment, ionic liquid, and thermal decomposition kinetics. Table 11.6 summarizes selective chemical treatments which are more suitable for automotive applications.

11.3 MECHANISM OF FIBRE TREATMENT

The natural fibres have limited thermal stability up to 230°C as usable temperature and exhibit high moisture absorption due to the presence of hydroxyl and other polar groups, which is a major drawback associated with the application of biofibres for reinforcement of polymeric matrices [22,92]. Accordingly, physical and chemical treatments are adapted in the production of natural fibre composites. Physical treatments are used for surface modification and increased mechanical bonding between the fibre and the matrix. Physical treatment does not extensively alter the chemical composition of the fibres or clean them since it does not involve chemicals.

Chemical treatment of natural fibres is carried out to improve their compatibility with the polymer matrix and enhance the quality of the resulting composite. Cellulose is chemically modified in any

TABLE 11.6
Chemical Treatment Modification for Natural Fibres

Chemical Treatment	Fibre	Chemical Reagents Used	Method	Structure Improvement	Application	References
Alkaline	Hemp Jute Sisal	NaOH	Treated fibre with NaOH at 20°C for 48 h and washed using distilled water and acetic acid to neutralize the excess of NaOH	Better fibre–matrix adhesion led to an increase in interfacial energy and thus enhancing the thermal and mechanical properties of the composites	Polymer reinforcements	22,69,94
	Kenaf	NaOH	Treated kenaf fibre with 6% of NaOH solution for 24 h. Then, kenaf fibres were rinsed and immersed into a solution that contained distilled water and 1% acetic acid to neutralize the remaining NaOH. After washing, the kenaf short fibres were dried in an oven for 24 h	Better physical, morphological, and mechanical properties because of the compatibility of kenaf with polypropylene composites	Automotive	22,69,95
Acrylation and acrylonitrile grafting	Flax	Acrylic acid solution	Fibres immersed in NaOH solution for 0.5 h and then soaked in acrylic acid solution at 50°C for 1 h, washed with distilled water, and dried	Improving the physical and mechanical properties	Plastic, automobile, and packaging industries	22,69,99
Maleate coupling	Jute	Maleic anhydride polypropylene (MAPP)	Fibres immersed in MAPP solution in toluene at 100 °C	Increase in mechanical strength	Industrial applications with cost-effective and value-added composite material	104
Permanganate treatment	Sisal	Potassium permanganate	Fibres soaked carefully in a solution of $KMnO_4$/acetone with a concentration of 0.02% for 3 min. After that, the fibres were taken out, washed many times with distilled water, and dried	Improves fibre strength and fibre–matrix adhesion	Industrial applications	101
Peroxide treatment	Sisal	Benzoyl peroxide from acetone solution	Fibres were coated with benzoyl peroxide from acetone solution after alkali pre-treatment. A saturated solution of the peroxide in acetone was used. Fibres were then dried	Enhanced tensile properties	Substitute for wood	105

Source: Adapted from Ref. [1].

natural fibres. The chemical modifications of fibres can decrease the absorption of humidity and improve the interfacial bond strengths or adhesion with polymer matrix [18].

11.4 PROCESSING OF NATURAL FIBRE COMPOSITES

Natural fibre composites are processed by the wet layup or hand layup method, resin infusion process, resin transfer moulding, compression moulding, and injection moulding. Natural fibres are non-abrasive to mixing and moulding equipment, which can contribute to a significant cost reduction [8]. Natural fibres have lower density than glass fibres and so can reduce the mass of the component. The energy needed for production is lower by 80% [2,5].

The major disadvantage of natural fibre-reinforced composites is the inherent polar and hydrophilic nature of the lignocellulosic fibres and the nonpolar characteristics of most thermoplastics, resulting in compounding difficulties and leading to non-uniform dispersion of fibres within the matrix, which impairs the efficiency of the composite. Another problem is that the processing temperature of composites is restricted to less than 200°C as natural fibres undergo degradation at higher temperatures [31]. The result of prolonged high-temperature exposure may be discoloration, volatile release, poor interfacial adhesion, or embrittlement of cellulose components [28,30].

It is quite clear that the advantages outweigh the disadvantages and most of the shortcomings have remedial measures in the form of chemical treatment [8]. However, several major technical considerations must be addressed before the engineering, scientific, and commercial communities gain the confidence to enable wide-scale acceptance of vegetable fibres, particularly in exterior parts where a class A surface finish is required. Challenges include the homogenization of the fibre's properties and a full understanding of the degree of polymerization and crystallization, adhesion between the fibre and matrix, moisture repellence, and flame-retardant properties [3]. Care must be taken so as not to erode the environmental advantage of vegetable fibres with chemical treatments.

11.4.1 Property Comparison between Natural Fibre and Synthetic Fibre Composites

Composites made from natural fibres, such as flax, hemp, jute, and bamboo, have some distinct characteristics compared to composites made from synthetic fibres, such as glass and carbon fibres. In terms of mechanical strength, composites based on synthetic fibres generally have higher strength than composites based on natural fibres. Synthetic fibres, especially carbon fibre, offer exceptional strength and high stiffness, making them preferred choices in applications requiring high structural strength [1,2,42]. However, composites based on natural fibres often have a lower specific weight than composites based on synthetic fibres. This can be advantageous in automotive industry applications [42]. From a cost point of view, composites based on natural fibres generally have a lower cost than composites based on synthetic fibres.

Natural fibres are often cheaper to produce and process, making them more economical for applications. When it comes to durability, composites made from synthetic fibres tend to have better resistance to chemical breakdown, moisture, and mould. Composites made from natural fibres can be more sensitive to these environmental factors and sometimes require special treatments to improve their durability. In terms of environmental impact, composites based on natural fibres are often considered more environmentally friendly than composites based on synthetic fibres due to their renewable origin and their faster degradation at the end of their life. It is important to note that the properties of composites can vary depending on the type of natural fibre, the manufacturing process, and the matrix used. Therefore, it is essential to carry out specific studies and consider the requirements of each application to determine the optimal choice between composites based on natural and synthetic fibres.

11.5 APPLICATIONS IN THE AUTOMOTIVE INDUSTRY

The automotive market is undergoing a massive change. Under growing pressure from environmental regulations, automotive builders are tightening specifications for equipment manufacturers. To reduce CO_2 emissions and shrink the environmental footprint of transportation, the market is demanding more lightweight materials that are recyclable at their end of life. One of the objectives set by European directive is to be recycling 85% and reusing 95% of automotive components soon. The CO_2 emissions are to be gradually reduced to 100 g/km versus the current 140 g/km.

Natural fibre composites coupled with an electric vehicle can help in meeting these new requirements. The low weight of natural fibres is a determining factor in optimizing energy consumption during the parts production process and provides attractive energy savings during use as well. With natural fibre composites, 5–20 kg of total weight can be accounted for. The microstructural morphology and high modulus of elasticity of hollow natural fibres impart good sound- and vibration-dampening properties. Fibres can absorb vibrations without deteriorating the mechanical properties. The automotive parts that are already being produced or under development are hidden interior parts such as door panels, rear seat shells, sound insulation for bulkheads, rear window shelves, and dashboards. The structural parts include floors, under-the-bonnet parts, degas tank caps, air ducts for centre and side ventilators, and rear-view mirror support frames [1–4,18,26,31,32,36,37]. Table 11.7 lists the different applications, types of fibre, and manufacturer models.

TABLE 11.7
List of Automotive Applications

Manufacturer	Model	Application	Natural Fibre
Opel	Vectra, Astra, Zafira	Door panels, headliner panels, instrumental panels	Flax, kenaf
Volkswagen	Passat Variant, Golf, A4, Bora	Seat back, chair backs, door panels, boot liners, boot lid finish panels	Flax, sisal
Audi	A2, A3, A4, A4 Avant, A6, A8, Roadstar, Coupe	Boot-liner, spare tyre-lining, side and back door panels, seat back, and hat rack	Flax, sisal
Daimler Chrysler	A, C, E, and S class, EvoBus (exterior)	Door panels, floor panels, trunk panels, dashboards, pillar cover panels, seat backrests, insolation	Flax, sisal, coir, wood, banana, cotton
BMW	3, 5, and 7 series and others	Driver seat back, headliner panel, boot-lining, door panels, noise insulation panels, and moulded foot well linings	Flax, sisal, cotton, wood, hemp
Fiat	Punto, Brava, Marea, Alfa Romeo 146, 156, 159	Door panel	Beaver
General Motors	Cadillac De Ville, Chevrolet TrailBlazer	Seat backs, cargo area floor mat	Cotton, flax, wood, kenaf, hemp
Peugeot	406 new model	Doors' front and back sections, backs of chairs, and a parcel shelf	Hemp, sisal
Toyota	ES3	Floor mats, spare tyre covers, door panels and seat backs, luggage compartments	Kenaf, sugarcane, bamboo, beaver
Toyota	Raum, Brevis, Harrier, Celsior	Floormats, spare tyre cover, door panels, and seat backs	Kenaf, sugarcane, bamboo
Ford	Mondeo CD 162, Focus	Floor trays, door inserts, door panels, B-pillar, and boot-liner	Kenaf, wheat, beaver
Mercedes Benz	Classes: S, E and A	Door panels, glove box, instrument panel support	Flax, sisal, cotton, wood fibres with epoxy resin

11.6 CONCLUSIONS

Although natural fibre composites may not fully replace their synthetic counterparts like carbon and glass fibres in structural applications due to poor mechanical properties, hybridization presents a potential solution. The hybridization process can promote a balance between environmental benefits and mechanical efficiency. CO_2 emissions from conventional internal combustion engine fossil fuels are resulting in many climatic changes. Electric vehicles can reduce these CO_2 emissions significantly. The amalgamation of both lightweight natural fibre composites and electric vehicles can reduce the environmental effects caused by automotive vehicles. Lightweight materials in electric vehicles are very attractive because they enhance the performance and manufacturers can increase the battery size containment due to a reduction in the mass of structures. Natural fibre composites can become a potential candidate for electric vehicles in decreasing the vehicle weight with low cost and more sustainable alternative lightweight materials.

REFERENCES

[1] S.H. Kamarudin, M.S. Mohd Basri, M. Rayung, F. Abu, S. Ahmad, M.N. Norizan, S. Osman, N. Sarifuddin, M.S.Z.M. Desa, U.H. Abdullah, A review on natural fiber reinforced polymer composites (NFRPC) for sustainable industrial applications, *Polymers* 14 (2022) 3698.

[2] S.H. Kamarudin, L.C. Abdullah, M.M. Aung, C.T. Ratnam, A study of mechanical and morphological properties of PLA based biocomposites prepared with EJO vegetable oil based plasticiser and kenaf fibres, *IOP Conf. Ser. Mater. Sci. Eng.* 368 (2018) 85314.

[3] T.H. Mokhothu, M.J. John, Review on hygroscopic aging of cellulose fibres and their biocomposites, *Carbohydr. Polym.* 131 (2015) 337–354.

[4] N.M. Nurazzi, M.N.F. Norrrahim, F.A. Sabaruddin, S.S. Shazleen, R.A. Ilyas, S.H. Lee, F.N.M. Padzil, G. Aizat, H.A. Aisyah, N.A. Mohidem, Mechanical performance evaluation of bamboo fibre reinforced polymer composites and its applications: A review, *Funct. Compos. Struct.* 4 (2021) 1–25.

[5] A.S.B.F. Yousif, In state of art: Mechanical and tribological behaviour of polymeric composites based on natural fibres, *Mater. Des.* 48 (2013) 14–24.

[6] M. Habibi, S. Selmi, L. Laperri`ere, H. Mahi, S. Kelouwani, Post-impact compression behavior of natural flax fiber composites, *J. Nat. Fibers* 17 (11) (2020) 1683–1691.

[7] H.H.B.F. Abu-Sharkh, Hamid, degradation study of date palm fiber/polypropylene composites in natural and artificial weathering: Mechanical and thermal analysis, *Poly. Deg. Stab.* 85 (2004) 967–973.

[8] A.K.B.J. Gassan, Composite reinforced with cellulose based fibers, *Prog Pol. Sci.* 24 (1999) 221–274.

[9] P. Pradhan, A computational investigation for the impact of particle size on the mechanical and thermal properties of teak wood dust (TWD) filled polyester composites, *Mater. Today: Proc.* 63 (2022) 756–763.

[10] E. Sassoni, S. Manzi, A. Motori, M. Montecchi, M. Canti, Novel sustainable hemp based composites for application in the building industry: Physical, thermal and mechanical characterization, *Energy Build.* 77 (2014) 219–226.

[11] W. Wang, M. Sain, P.A. Cooper, Study of moisture absorption in natural fiber composites, *Comp, Sci. Tech.* 66 (2006) 379–386.

[12] A.K. Mohanty, M. Misra, G. Hinrichsen, Biofibers, biodegradable polymers and biocomposites, *Macro, Mater. Eng.* 276 (277) (2000) 1–24.

[13] Y. Li, Y.W. Mai, L. Ye, "Sisal fibre and its composites: A review of recent developments, *Compos. Sci. Technol.* 60 (11) (2000) 2037–2055.

[14] R.M.V. K, Y.S. Negi, Recent development in natural fiber reinforced polypropylene composites, *J. Reinforc. Plast. Compos.* 28 (10) (2009) 1169–1189.

[15] F. Ahmad, H.S. Choi, M.K. Park, A review: Natural fiber composites selection in view of mechanical, light weight, and economic properties, *Macromol. Mater. Eng.* 300 (2015) 10–24.

[16] R.K. Sharma, S. Obrai, A. Sharma, Chemical modifications of natural fiber for composite material. Der Chemica Sinica, 2 (4) (January 2011) 219–228.

[17] O.M. Omari, *Sustainable Biocomposites: Challenges, Potential and Barriers for Development*, vol. 11, Chapter Green Energy Technol, 2017.

[18] S.S. Kamatha, D. Sampathkumarb, B. Bennehallia, A review on natural areca fibre reinforced polymer composite materials. *Ciência & Tecnologia dos Materiais*, 29 (3) (2017) 106–128. DOI: 10.1016/j.ctmat.2017.10.001.

[19] J. Biagiotti, D. Puglia, J.M. Kenny, A review on natural fibre-based composites - Part I: Structure, processing and properties of vegetable fibres. *J. Nat. Fibers* 1 (2) (2008), 37–68. DOI: 10.1300/J395v01n02_04.

[20] R.M. Rowell, A.R. Sanadi, D.F. Caulfield, R.E. Jacobsen, in: A.L. L, F.X. Carvalho, E. Frollini (Eds.), *Utilization of Natural Fibers in Plastic Composite: Problems and Opportunities*. Lignocellulosic-Plasti Composites, 1997.

[21] O. Faruk, A.K. Bledzki, S.H.-P. Fink, Biocomposites reinforced with natural fibers, *Prog. Polym. Sci.* 37 (11) (2012) 1552–1596.

[22] M.N.M.A.L. Mohammed, G. Pua, M. Jawaid, M.S. Islam, A review on natural fiber reinforced polymer composite and its applications, *International Journal of Polymer Science* 2015 (2015) 243947.

[23] K. Rohit, S. Dixit, A review – Future aspect of natural fiber reinforced composite (2015, November).

[24] M.P.S.N. Tucker, Natural-fiber composites in structural applications, in: K. Pickering (Ed.), *Properties and Performance of Naturalfibre Composites*, Woodhead, 2008.

[25] S.K. Ramamoorthy, S. M, A. Persson, A review of natural fibers used in biocomposites: plant, animal and regenerated cellulose fibers, *Polym. Rev.* 55 (1) (2015) 107–162.

[26] S.D. K Rohit, A review-future aspect of natural fiber reinforced composite, *Polym. Renew. Resour.* 7 (2) (2016) 43–60.

[27] M. Thiruchitrambalam, A. Athijayamani, S. Sathiyamurthy, A.A. Thaheer, A review on the natural fiber-reinforced polymer composites for the development of Roselle fiber-reinforced polyester composite. *J. Nat. Fibers* 7(4) (2010) 307–323. https://doi.org/10.1080/15440478.2010.529299

[28] Bharath, *Applications of Biocomposite Materials Based on Natural Fibers from Renewable Resources: A Review*, Materials Science, Science and Engineering of Composite Materials, 2015.

[29] V. Chauhan, T. Karki, J. Varis, Review of natural fiber-reinforced engineering plastic composites, their applications in the transportation sector and processing techniques, *J. Thermoplast. Compos. Mater.* 35 (8) (2022) 1169–1209.

[30] J. Holbery, D. Houston, Natural-fiber-reinforced polymer composites in automotive applications, *J. Miner. Met. Mater. Soc.* 58 (11) (2006) 80–86.

[31] H.R.T. Sen, Various industrial applications of hemp, kenaf, flax and ramie natural fibres, *Int. J. Innov. Manage. Technol.* 2 (3) (2011) 192–198.

[32] Y.Y. N Reddy, Biofibers from agricultural byproducts for industrial applications, *Trends Biotechnol.* 23 (1) (2005) 22–27.

[33] H.R.T. Sen, Application of sisal, bamboo, coir and jute natural composites in structural upgradation, *Int. J. Innov. Manage Technol.* 2 (3) (2011) 186–191.

[34] A.S. Singha, V. Thakur, Mechanical properties of natural fiber reinforced polymer composites, *Bull. Mater. Sci.* 31 (2008) 791–799.

[35] W. Hoareau, W.G. Trindade, B. Siegmund, Sugar cane bagasse and curaua linens oxidized by chlorine di oxide and reacted with furfuryl alcohol: characterization and stability, *Polym. Degrad. Stabil.* 86 (2004) 567–657.

[36] P. Pradhan, A. Purohit, I. Mohanty, H. Jena, B.B. Sahoo, Erosion wear analysis of coconut shell dust filled epoxy composites using computational fluid dynamics and Taguchi method, *Proc. IME C J. Mech. Eng. Sci.* 237 (23) (2023), 5653–5662. 09544062231167674.

[37] P. Pradhan, et al., Tribo-performance analysis of an agro-waste-filled epoxy composites using finite element method, *J. Inst. Eng.: Series E* 103 (2) (2022) 339–345.

[38] R.V.A. Krishnamoorthy, Review on natural fiber reinforced composites, *Mater. Today: Proc.*16 (2019) 897–906.

[39] U.S. Bongarde, V. Shinde, Review on natural fiber reinforcement polymer composites, *Int. J. Eng. Sci. Inn. Tech.* 3 (2) (2014) 431–436.

[40] P.Y.V.M. Foruzanmehr, M. Robert, S. Elkoun, The effect of grafting a nano-TiO 2 thin film on physical and mechanical properties of cellulosic natural fibers, *Mater. Des.* 85 (2015) 671–678.

[41] L. Boulos, M.R. Foruzanmehr, A. Tagnit-Hamou, S. Elkoun, M. Robert, Wetting analysis and surface characterization of flax fibers modified with zirconia by solgel method, *Surf. Coat.* 313 (2017) 407–416.

[42] S. Mukhopadhyay, R. Fangueiro, Physical modification of natural fibers and thermoplastic films for composites—A review, *J. Thermoplast. Compos. Mater.* 22 (2009) 135–162.

12 Synergetic Effect of Surface-Functionalized 2D Carbon-Based Polymer Nanocomposite to Enhance the Mechanical and Thermal Properties of Lightweight Materials for Electric Vehicles

Sandeep Kumar Singh, Thingujam Jackson Singh, Prashanth S., Puneet Kumar Sonker, Laishram Dhanabir Singh, and Yogendra Kumar Verma

12.1 INTRODUCTION

Polymer composites reinforced with carbon nanomaterials have attracted considerable interest because of their sustainable nature and ability to form strong interfacial bonds, which are essential for superior properties. Understanding the mechanisms of reinforcement is a key point to optimizing the performance of polymer nanocomposites [1]. To address the inherent restrictions of polymers, various carbon fillers are commonly utilized as reinforced fillers in the polymer matrix [2–4]. The incorporation of nanoscale fillers has revolutionized traditional polymer nanocomposites, improving their thermal, mechanical, and physical performances. These fillers, ranging from isotropic to highly anisotropic morphologies such as sheet-like or needle-like structures, possess at least one characteristic dimension in the nanometer range. The molecular-level interactions between nanoscale fillers and polymer matrices enable the development of polymer nanocomposites with improved mechanical and thermal properties to enhance battery performance, lightweight and high-strength materials, 3D printing and additive manufacturing, and self-healing systems.

The authors are focused on two-dimensional (2D) carbon materials, which are a tetravalent nonmetal, meaning they have four valence electrons capable of forming four covalent bonds. This unique valency enables carbon to exhibit exceptional catenation, allowing it to bond with other carbon atoms or various elements, forming a strong bond with a basis of organic chemistry. Due to the instability of atomic carbon, it naturally organizes into various structural forms, known as allotropes, to achieve stability [5]. Carbon exists in multiple allotropes, distinguished by the type of chemical bonding and hybridization (sp^1, sp^2, sp^3) [6]. These carbon materials include zero-dimensional (0D) forms such as sp^2 fullerenes, carbon dots, and carbon black; one-dimensional (1D) structures such as sp^2 hybridized like carbon nanotubes (CNTs), single-walled CNTs (SWCNTs), and multiwalled CNTs (MWCNTs); 2D materials such as sp^2 graphene with its honeycomb lattice, reduced graphene oxide (rGO), and MXenes; and three-dimensional (3D) structures like sp^3 diamond crystals. Each allotrope exhibits

DOI: 10.1201/9781003596097-12

unique electronic and mechanical properties. Emerging materials like graphene, fullerenes, MXene, and CNTs stand out due to their exceptional mechanical and thermal properties. The remarkable versatility of carbon materials stems from the significant influence of the sp^2 (graphite-like) to sp^3 (diamond-like) bond ratio on their physical characteristics [7]. Various forms of sp^2-bonded carbons, ranging from microcrystalline graphite to glassy carbon, display differing levels of graphitic organization and have been extensively studied for their outstanding mechanical and electronic attributes. Due to their nanoscale dimensions and adjustable structural properties, these nano-reinforcements are becoming a promising approach for creating nanocomposites with outstanding properties to be utilized in automobile parts for high-strength and lightweight materials [8,9].

A recent review on nanocomposites incorporating nano-carbon derivatives in polymer matrices has highlighted their exceptional mechanical and thermal properties. The incorporation of carbon nanofiller reinforcement into the polymer matrix contributes to the substantially improved mechanical and thermal performance of these nanocomposites due to the following reasons: uniform dispersion of nanofiller within a polymer matrix material, better interfacial bonding, hindering the nanofiller agglomeration within the polymer matrix, increase in the cross-linking capacities of polymer composite, and improvement of the mechanical interlocking between filler and matrix material which help suppress free vibration and consequently reduce thermal conductivity.

These features collectively enhanced the mechanical and thermal properties, by which the incorporation of carbon nanofiller into the polymer matrix forms polymer nanocomposites, as discussed in this research work [10,11]. Additionally, further increases in the properties of polymer composites, by surface functionalization of the 2D carbon material using various coupling agents (like (3-Aminopropyl)triethoxysilane [APTES], dopamine, [3-glycidyloxypropyl] trimethoxysilane [GPTMS], etc.), result in enhanced nanoparticle dispersion, stronger interfacial bonding, and reduced agglomeration. In recent years, a variety of surface modification methods have been introduced to enhance the distribution and interfacial interaction of carbon nanofillers with the polymer matrix [12,13]. Generally, during the coupling process, various functional groups of silane molecules, such as ethylene, amine, epoxy, and thiohydroxy, are bonded to the surface of the 2D carbon filler. These groups are known to facilitate covalent bonding between graphene and the polymer matrix, thus enhancing the material's properties. In modern commercial applications of polymers, thermomechanical properties play a vital role in determining the lifespan and performance of polymer composite components in the automobile and sports industries. Understanding the recent advancements and underlying mechanisms governing the thermal and mechanical performance of 2D carbon nanofillers, such as graphene, MXene, CNTs, carbon black (CB), carbon quantum dots CQDs), fullerenes, and metal–organic frameworks (MOFs), is essential. This review highlights the dispersion of these carbon nanomaterials and their interactions with polymer matrices that contribute to enhanced thermomechanical performance. This review chapter aims to present a comparative study on the effects of functionalized and non-functionalized coupling agents in 2D carbon nanofillers incorporated into polymer matrices, focusing on enhancing the thermomechanical properties of polymer nanocomposites for commercial and industrial applications in the automobile, sports industries, medical, and transportation sectors. On the other hand, polymer nanocomposites incorporating 2D carbon nanomaterials are developed to reduce the weight of electric vehicles (EVs) while enhancing other properties such as high strength, heat dissipation, electromagnetic interference (EMI) shielding, self-healing capability, wear resistance, coatings and lubrication, etc. Furthermore, based on the current literature, potential future research directions for improving the performance of nano-carbon-reinforced polymer composites are discussed.

12.2 GRAPHENE POLYMER NANOCOMPOSITES

Graphene and its derivatives have become essential elements in composite technology. Graphene oxide (GO), a structurally significant material, is produced by oxidizing and exfoliating the graphite.

Subsequently, GO can be reduced by hydrazine hydrate to form rGO, which is also used in polymer composites, as the elimination of oxygen-containing groups can result in fewer structural defects [14,15]. The authors highlight and elaborate on the literature available on GO and rGO-based polymer nanocomposites, both with and without functionalization of coupling agents. This research work demonstrated the in situ polymerization process of methyl methacrylate monomer with the incorporation of reduced graphene oxide filler, resulting in the development of poly methyl methacrylate (PMMA) beads embedded with rGO, as well as polymer growth on the rGO surface. The impact of varying reduced graphene oxide content (from 0.1 to 2 wt.%) on the thermomechanical properties of poly methyl methacrylate was investigated. Notably, at 1 wt.% rGO, the poly methyl methacrylate (PMMA)/rGO nanocomposite exhibited a 126% enhancement in thermal conductivity and improvements of 9.9% and 28% in mechanical properties with respect to pristine PMMA [16]. In another study, reduced graphene oxide/polyvinyl alcohol nanocomposites were prepared through the coagulation process using 2-propanol. The resulting composite structure exhibited improved interconnections, primarily due to hydrogen bonding among the polymer and reduced graphene oxide sheets due to this better interfacial interaction between them. With the addition of 1 wt.% rGO, the glass transition temperature (T_g) of the polyvinyl alcohol/rGO composite increased by 25% [17]. A similar study using rGO as a reinforcement material enabled the fabrication of versatile microfibrillated cellulose nanocomposites through a direct water-dispersion approach. The two different types of filler incorporated significantly influenced the properties of the microfibrillated cellulose composites. The Young's modulus (YM) increased from 18 GPa to 25 GPa, while the tensile strength (TS) of microfibrillated cellulose/GO and the micro-fibrillated cellulose/rGO composites showed enhancements of 18% and 23%, respectively, at a reinforcement level of 0.6 wt.%. Furthermore, thermal stability was enhanced by 5% with 5 wt.% rGO and by 2% with 2 wt.% rGO with respect to pure melamine-faced chipboard (MFC) [18].

Additionally, GO was synthesized with effective exfoliation using dual-probe ultrasonication and incorporated into an epoxy matrix, resulting in enhanced mechanical and thermal performance of the nanocomposite. The TS improved by 40.763%, while the maximum tensile modulus (TM) increased by 19.3724% at 1 wt.% and 27.63% at 1.5 wt.% GO60 compared to the neat composite. Fracture toughness (K_{IC}) and fracture energy (G_{IC}) showed peak improvements of 1.256 MPa·m$^{1/2}$ and 0.472 kJ/m^2, correspondingly, at 1 wt.% GO60. Notably, the thermomechanical properties of the nanocomposite displayed optimal values, with a storage modulus of 4066.75 MPa and a damping factor of 0.46 at 1 wt.% GO60 [15]. Furthermore, the introduction of 0.5 wt.% GO in GFRP laminate composites resulted in optimal results, such as TS (242.73 MPa), flexural strength (FS) (317.13 MPa), and short beam strength (40.05 MPa). Additionally, the tensile and flexural moduli reached 8.03 GPa and 13.59 GPa, respectively, due to enhanced interfacial adhesion among the epoxy system and GO fillers with respect to the NE composite [19]. Conversely, to increase the mechanical thermal properties due to functionalized graphene nanosheets, created through covalent grafting with APTES, were synthesized. The silane functionalization ensured uniform distribution without agglomeration in the polymer matrix, even at a 1 wt.% filler. This addition increased the T_{ITD} temperature of the polymer nanocomposite from 314°C to 334°C and enhanced TS and tensile strain at the break by 45% and 133%, correspondingly [20]. Another surface functionalization of GO with polyether amine enhanced the mechanical interlocking and hindered the agglomeration among the GO sheets and the epoxy system through chemical bonding during curing. At a filler loading of 0.50 wt.%, the composites revealed significant enhancements, with tensile strength increasing by 63%, and toughness by 119%, compared to the unfilled epoxy [21]. A comparison of the mechanical performance of polymer nanocomposites containing as-produced GO and silane (GPTMS) surface modification of silane-f-GO demonstrated that silane functionalization improved compatibility with the epoxy matrix by strong covalent bonding between matrix and nanofillers. At a loading of 0.25 wt.% functionalized graphene oxide, the tensile modulus and strength showed improvements of approximately 10% (3.4 GPa) and 48% (78.3 MPa), correspondingly with respect to neat. The highest K_{IC} was observed at 0.10 wt.% silane-f-GO, with K_{IC} and G_{IC} values improving by about 39% (~0.695 MPa·m$^{1/2}$) and 83% (~127.5 J/m^2), respectively, with respect to neat composite [22]. In another study, graphene oxide was used as a nanofiller to enhance the dispersion and compatibility of the polypropylene (PP) matrix with ethylene-propylene-diene monomer after fabricated polymer composite. To achieve this, the introduction of GO into PP and ethylene propylene diene monomer (EPDM) to form PP-g-MA nanocomposite was employed in compatibilizer products. The study aimed to optimize blend polymer composition, evaluating mechanical and thermal properties by using thermogravimetric analysis (TGA) and theoretical analysis. The results revealed that increasing GO and adding PP-g-MA significantly enhanced TS, elongation at break, and YM. The morphological images exhibited reduced EPDM dispersion particle size with higher nanofiller concentrations and the formation of agglomeration between polymer and nanofillers [23]. An overview of recent research on graphene-based polymer nanocomposites is tabulated in Table 12.1.

12.3 2D GRAPHITE NANOPARTICLE (GNP) POLYMER COMPOSITE

The exceptional properties of 2D carbon materials like graphene and its derivative (GO/rGO) have garnered significant interest in enhancing the mechanical and thermal performances of polymer nanocomposites. However, challenges persist in enhancing their dispersion within polymers and increasing their interaction capabilities. Therefore, further research is required to develop another, cost-effective 2D carbon nanoparticle that can consistently reinforce and enhance the K_{IC} of polymer composites. The exceptional properties of 2D carbon materials like graphene highlight their potential to enhance epoxy composites' fracture toughness, but challenges in dispersion and interaction remain. The researcher aimed to develop a cost-effective method for improving the mechanical properties of polymers by identifying alternative 2D nanofillers that can be easily dispersed into the polymer matrix. This study illustrates the conversion of graphite into milled graphitic nanoparticles (micro- to nanoparticles) through a high-energy wet and dry ball milling technique and some other chemical techniques to achieve graphite nanoplatelets (GNPs) from graphite. High-energy wet ball milling is used to convert microparticles into nanoparticles under specific parameters. The resulting powder, incorporated into the polymer matrix as reinforcement materials to fabricate polymer nanocomposites, exhibits varied mechanical and thermal behaviors due to particle size, interfacial interaction between matrix and filler, and homogeneous dispersion. The findings indicate that nanoscale GNP provides superior mechanical properties compared to microparticle-reinforced composites. Subsequently, surface functionalization using a coupling agent has been employed to enhance the dispersion of graphite nanoparticles (GrNPs) within the polymer matrix, significantly improving the mechanical and thermal properties of the nanocomposite. This research aims to explore existing literature on GNPs with and without functionalization to improve their properties. This review highlights the production of ball-milled graphite (BMG) nanoparticles via dry milling for 15 and 48 hours, achieving a significant particle size reduction after 15 hours. Similarly, wet ball milling for just 3 hours achieved a comparable size (258 nm). The surface functionalized by the saline coupling agent of BMGs (SBMG@3) was incorporated into an epoxy matrix, yielding nanocomposites with enhanced K_{IC} (2.47 MPa·m$^{1/2}$) and G_{IC} (1.55 kJ/m^2) at 1wt.% of salinized ball-milled graphite (SBMG) filler [29].

In another study graphite was converted into nanoparticles (GrNPs) via 9 hours of high-energy wet ball milling, reducing their size to 227–217 nm. Subsequently, surface functionalized with amine-based silane (APTES) and incorporated into an epoxy matrix, these GrNPs significantly enhanced the nanocomposite crack resistance, achieving fracture toughness (~5.11 MPa·m$^{1/2}$) and energy (~20.42 kJ/m^2) at 1 wt.% loading [32]. Similarly, synthesizing GO is too difficult, so this issue can be addressed using milled GrNPs, which feature a porous structure composed of stacked, randomly oriented planes with numerous defect sites which due to this easily react with another chemical reagent. These defect sites facilitate rapid oxidation within just 1.5 hours. In this study, cost-effective silanized (APTES) GrNPs were utilized as nanofillers to enhance the

TABLE 12.1
An Overview of Recent Research on Graphene-Based Polymer Nanocomposites

Sl. No.	Polymers	Fillers (wt.%)	Silane Agent	σ_u	E	K_{IC}	T_g/E'	Ref.
1	Epoxy	E-GO (1%)	–	41%	20%	46%	186%	[15]
2	MFC	rGO (0.6%)	–	23%	–	–	5%	[18]
3	Epoxy	GO (1%)	APTES	45%	–	–	23%	[20]
4	Polyether amine	GO (0.5%)	–	63%	–	119%	33%	[21]
5	Epoxy	GO (0.25%)	GPTMS	38%	–	49%	52%	[22]
6	PVA	GO (1%)	Dopamine	68%	97%	–	–	[24]
7	Polyurethane	rGO (1%)	Dopamine	69.6%	58%	–	–	[25]
8	PF	rGO (0.12%)	–	150%	38%	–	12%	[26]
9	PET	rGO (1%)	–	17%	–	–	33%	[27]
10	Polyurethane	GO (1%)	–	–	58%	–	63%	[28]

thermo-mechanical properties of GrNP-epoxy nanocomposites [31, 30]. The epoxy nanocomposite containing 0.5 wt.% silanized GrNPs exhibited remarkable improvements, with a ~758% increase in storage modulus and significant enhancements of ~33% in tensile strength and ~43% in fracture toughness [31]. Another similar study has been done on functionalized GrNPs with epoxy groups via silanization and epichlorohydrin grafting over the GrNPs. Incorporating these GrNPs into epoxy matrices improved nanocomposite tensile strength by ~36% and ~30%, respectively, while carbon fiber-reinforced polymer (CFRP) composites showed ~37% and ~36% enhancements compared to neat composite [30]. These fabricated CFRP laminate composites are used across various sectors, including automotive and EVs, defense and ballistic protection, sports equipment and high-performance gear, and wind energy and renewable applications. A summary of recent research on polymer nanocomposites incorporating 2D graphite nanoparticles is presented in Table 12.2.

12.4 CARBON NANOTUBE-BASED NANOCOMPOSITES

CNTs are ideal reinforcing materials for fabricating nanocomposites, particularly in 1D applications like fibers and wires. Surface functionalization is essential to enhance their properties; however, their chemical inertness poses challenges to achieving complete dispersion. The CNTs produced by Iijima had diameters ranging from 4 to 30 nm and lengths reaching up to 1 μm. The nanotubes varied in structure, consisting of two to 50 layers [34]. Based on Iijima's findings, CNTs are classified by their size, with tube diameters measuring just a few nanometers. Their structure is formed by rolling one or more graphitic layers into a hollow tube, resulting in either SWCNTs or MWCNTs [35]. The molecular dynamics simulations were performed using a CNT with varying densities of linked ethylene-diamine molecules. The effect of amine surface modification of CNT on YM, along with the shear and TS of numerous CNT structures, was analytically examined in polymer nanocomposites. The mechanical behavior of the nanocomposites was assessed using a closed micromechanical interphase model, which incorporated factors such as CNT diameter, interphase thickness, and the mechanical properties of the CNT/polymer system. The results indicated that this significantly enhanced the modulus of the CNT/polymer composite, achieving up to a 108% increase with the optimal CNT addition [36, 37].

CNT-reinforced polymer nanocomposites exhibit notable improvements in mechanical properties. Factors such as CNT distribution, aggregation, polymer characteristics, structural defects, and covalent bonding throughout surface modification significantly influence their performance. A multiscale approach was employed, analyzing CNT dispersion and agglomeration at the microscale and their structure and interfacial phases through nanoscale simulations. This is attributed to enhanced dispersion and reduced clustering of CNTs, based on the pristine polymer's parameters, which contribute to strengthening the CNT/polymer nanocomposite [37]. The surface functionalization process of CNTs increases the TS of the CNT/polymer nanocomposite by 34%, as shown in Figure 12.1a. However, the presence of structural defects in CNTs introduced throughout surface modification leads to a decrease in the YM of the nanocomposites. Additionally, agglomerated CNTs can adversely impact YM by influencing the fracture mechanism. To address this, a nanocomposite comprising surface functionalization of CNTs coated with dopamine was integrated into a polyurethane (PU) matrix. This was achieved using solution casting and in situ polymerization techniques, resulting in polyurethane/polydopamine (PU/PDA)-CNT composites. At a 0.3 wt.% concentration of PDA-CNTs, the nanocomposite exhibited a 40% increase in TS and a 100% enhancement in modulus, as illustrated in Figure 12.1b. Fabricating polymer nanocomposite components through 3D printing in advanced manufacturing may yield better results than solution casting and in situ polymerization for lightweight electric vehicle applications. This is attributed to the interaction at the interface between the matrix and filler which was substantially improved. The incorporation of PDA-CNTs altered the phase separation behavior in the polyurethane matrix,

TABLE 12.2
A Summary of Recent Research on Polymer Nanocomposites Incorporating 2D Graphite Nanoparticles

Sl. No.	Polymers	Fillers (wt.%)	Silane Agent	σ_u	E	KIC	$T_{g/}E'$	Ref.
1	Epoxy	GrNP(1%)	APTES	58%	23%	178%	–	[29]
4	Epoxy	GrNPs (1%)	GPTMS	36%	28%	155%	13%	[30]
3	Epoxy	GrNP(0.5%)	APTES	33%	–	43%	758%	[31]
2	Epoxy	GrNP(1%)	APTES	48%	21%	477%	–	[32]
5	Epoxy	GrNPs (1%)	BMI+FGE	446%	42%	–	–	[33]

resulting in the formation of a distinctive network-like structure that contributed to an enhanced reinforcing effect. Furthermore, the thermal performance of the polyurethane coating improved by 15% when an optimal amount of PDA-CNTs was incorporated [38]. Furthermore, a study examining the impact of CNT surface modification using dodecyl amine (DDA) explored the thermal and mechanical performance of high-density polyethylene (HDPE)-based composites. The surface modification of CNT with DDA improved the dispersion and interfacial bonding with HDPE via non-covalent bonding. Consequently, as a result, thermal and mechanical performance improved by 15% as shown in Figure 12.1c, which was attributed to better CNT distribution and strengthened interfacial adhesion in the HDPE matrix, influenced by its increased hydrophilicity [39].

Another literature on epoxy/CNT nanocomposites showed enhanced mechanical properties with CNTs as nanofillers, surface modified with melamine to improve the interfacial bonding between matrix and filler. Testing of CNT/epoxy and melamine nanocomposites revealed a 64% increase in YM, a 22% improvement in TS, and a 95% increase in fracture toughness (K_{IC}) at 2 wt.% melamine-functionalized carbon nanotubes (M-CNTs). These enhancements were credited to the homogeneous distribution and better interfacial bonding among M-CNT and epoxy matrix, affecting crack propagation and fracture toughness, as shown in Figures 12.1d and 12.1e [40]. Similarly, surface functionalization of CNTs with an HNO_3 solution introduced defects and oxygen groups to improve the dispersion and crystallization nucleation in 3D-printed polylactic acid (PLA)/CNT nanocomposites. These polymer composites were compared to conventional CNTs and surface-modified CNTs resulting in significantly enhanced mechanical strength of the polymer composite, increasing it from 29.4 MPa in polylactic acid/c-carbon-nanotube to 41.6 MPa in polylactic acid/f-carbon-nanotube with just 0.5 wt.% of functionalized carbon nanotube. The interfacial bonding and thermal stability were also improved. Another test performed using dynamic mechanical analysis (DMA) revealed a 43% enhancement in storage modulus (δ) at 37°C, significantly enhancing the thermal and mechanical properties of polylactic acid/CNT polymer composites [42]. Unlike SWCNTs, a polymer matrix nanocomposite containing MWCNTs was created to evaluate its thermal and mechanical characteristics. The results showed that a higher MWCNT content led to improvements in both flow stress and fracture strain with a 37% and 50% increase, correspondingly, at 0.5 wt.% MWCNTs with respect to the neat composite. The MWCNT polymer composite also exhibited more crystalline regions, contributing to a tougher composite [43]. To enhance the interfacial bonding among the polymer matrix and CNTs, the MWCNTs were chemically surface functionalized. This surface functionalization effectively improved the dispersion and interfacial bonding between them, enhancing several physical and chemical properties of these materials. The effect of surface-functionalized reinforcement loading on the mechanical performance of PMMA polymer composites embedded with MWCNTs was investigated. Incorporating the 0.5 wt.% MWCNTs into the PMMA matrix enhanced the tensile properties of the nanocomposite by 16% with respect to pure PMMA composite as shown in Figure 12.1f [41]. Additionally, MWCNTs were surface functionalized with amino-silanes through an aqueous deposition method. The surface functionalization of MWCNTs grafted with irregular siloxane oligomers exhibited weak connections with the matrix, whereas the surface functionalization with linear siloxane brushes demonstrated strong network interactions. The DMA and differential scanning calorimetry (DSC) analyses revealed that ds-MWCNTs reduced network segment mobility, promoted uniform cross-linking, and significantly increased T_g at low loading levels with respect to the neat resin. The enhancements in TS and TM at 0.1wt.% were 21% and 6%, however, thermal properties were enhanced by 10% of T_g as compared to neat composite. The MWCNTs were synthesized on carbon fiber using non-destructive fabrication methods, incorporating both homogeneous and heterogeneous approaches. The effect of different MWCNT surface topologies on fatigue damage and the lifespan of carbon fiber-reinforced polymer composite was evaluated. The results indicated that the fatigue life of the CFRPs' laminate composite was improved by 150% with the patterned growth of MWCNTs filler [44]. A summary of recent work related to

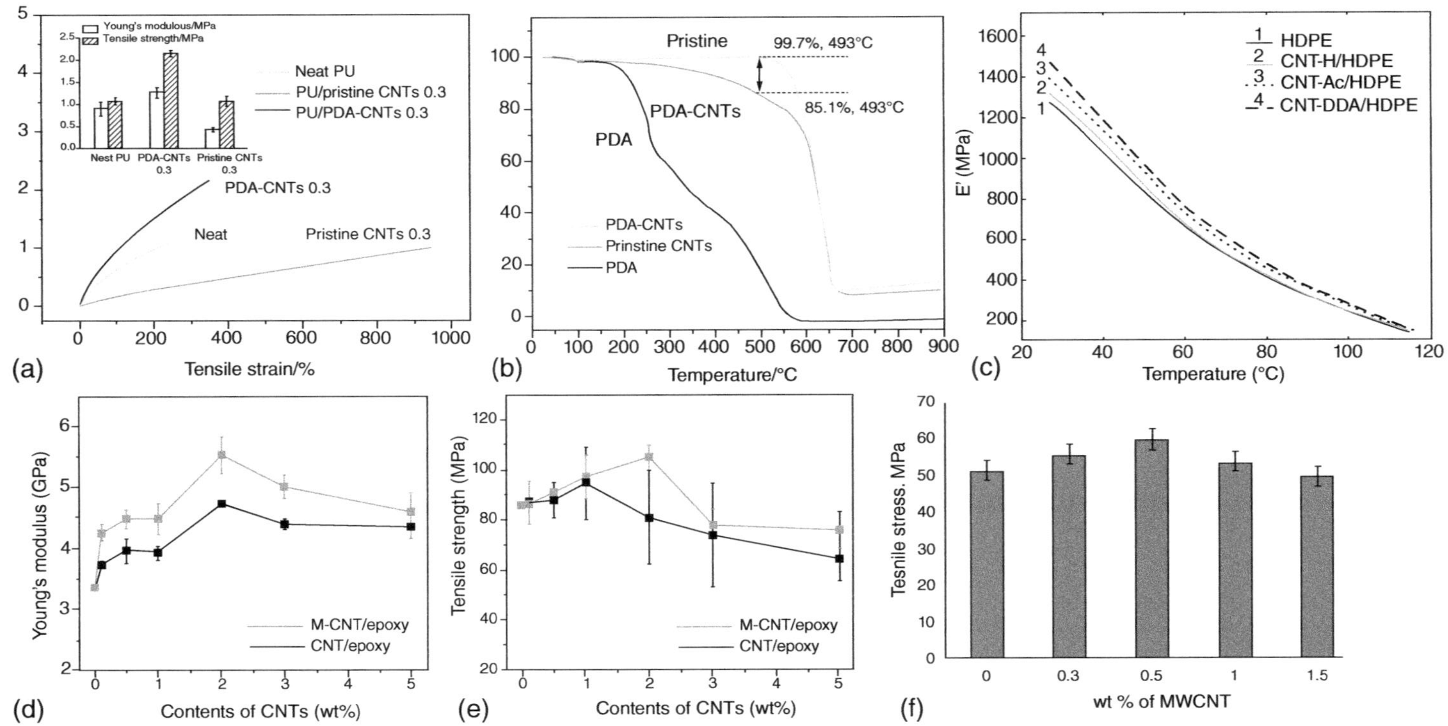

FIGURE 12.1 Mechanical properties: (a) tensile and modulus of CNT/PDA composite, (b) thermal decomposition of PDA/CNT [38], (c) storage modulus of CNT-based polymer composite [39], (d, e) Young's modulus and tensile strength of epoxy/CNT composite [40], and (f) tensile strength [41].

CNT polymer composites with and without functionalization of various coupling agents is provided in Table 12.3.

12.5 MXENE-BASED NANOCOMPOSITE

In the last few years, there has been a considerable increase in interest in 2D materials, driven by the investigation of their properties, including those of graphene, hexagonal boron nitride (Nb_4C_3), metal chalcogenides, and titanium carbide $Ti_3C_2T_x$, i.e. MXene. MXenes are a newly discovered class of 2D nanomaterials known for their remarkable mechanical, thermal, and tribological properties. These materials find applications across several critical areas, such as cancer therapy, energy, and environmental technologies, due to their distinctive attributes, including mechanochemical origins, exceptional mechanical performance, heat resistance, and surface modification capabilities.

Their strong interfacial interactions with macromolecules and terminal groups make them highly promising for use in advanced polymeric composites [49]. The polymer composites reinforced with MXenes have been developed to enhance their mechanical and thermal performance. For example, thermosetting epoxy polymers are commonly used as matrices in fabricating fiber-reinforced composites due to their excellent strength, stiffness, and thermal resistance. However, thermosetting epoxy is inherently brittle and has low fracture toughness, which restricts its various industrial applications. Incorporating micro- and nanoscale MXene particles is a widely adopted approach to enhance the mechanical performance of epoxy composites [50]. Studies on MXene/epoxy composites have revealed that their combined strength and the microscopic mechanisms of failure are influenced by one-directional stress. The specific category of MXene (such as Ti_2CT_x or $Ti_3C_2T_x$) has minimal effect on the bonding forces within the MXene/epoxy system. Reduced hydrogen coverage on the $Ti_3C_2T_x$ surface enhances its compatibility with epoxy, leading to stronger binding between the MXene and the epoxy matrix.

The fabricated polymer nanocomposite, incorporating MXene materials such as Ti_2CT_x or $Ti_3C_2T_x$ into the polymer matrix, enhances the composite's properties, including high electrical conductivity, thermal stability, and mechanical strength. These improvements make them ideal for applications such as lightweight conductive components, EMI shielding, thermal management, and energy storage. In electric vehicles, MXene-based composites contribute to improved battery performance, enhanced structural materials, and efficient thermal dissipation. Additionally, in advanced manufacturing, they enable additive manufacturing (3D printing), flexible electronics, and smart coatings with self-sensing capabilities. A literature review of the stress transfer among the polymer matrix and the MXene particles describes a 16% increase in modulus compared to neat epoxy composite, with the modulus improving as the reinforcement load increases up to 1 wt.%. However, at higher filler concentrations, the modulus increase becomes less pronounced due to filler agglomeration and better interfacial bonding between them. Experimental surface damage analysis revealed the formation of voids around the filler edges in MXene/epoxy composites under strain [50]. A nanocomposite of thermoplastic polyurethane (TPU) and Zr-MXene was developed to evaluate its mechanical performance. The results showed that incorporating 1 wt.% Zr-MXene into the TPU matrix led to a remarkable 20.6% increase in strain, a toughness improvement to 316 MJ/m^3, the maximum documented so far, and a 43.4% enhancement in tensile strength, along with improved fatigue resistance compared to pure TPU composite. Additionally, the synergistic physical and catalytic effects of Zr-MXene significantly reduced the flammability of the TPU matrix material [51]. Another study developed an MXene/PVA/epoxy nanocomposite to assess its tensile strength and thermal degradation performance. The results indicated that aligning high-aspect-ratio nanoparticles significantly enhanced the YM and TS of the nanocomposite. The introduction of 30 vol.% oriented MXene particles into the epoxy matrix represented a 745% increase in YM and a 92% enhancement in TS compared to pure epoxy composite [52]. Another study of ultrathin

TABLE 12.3
Summary of Recent Works Related to CNT Polymer Composites With and Without Functionalization of Various Coupling Agents

Sl. No.	Polymers	Fillers (wt.%)	Silane Agent	σ_u	E	KIC	$T_{g/}E'$	Ref.
1	Epoxy	CNTs (2%)	–	108%	–	–	–	[36]
2	Polyurethane	CNTs (0.3%)	–	100%	40%	–	15%	[38]
3	Polyethylene	CNT (0.8%)	HDPE	47%	–	–	15%	[39]
4	Epoxy	CNTs (2%)	Melamine	22%	64%	95%	–	[40]
5	PMMA	MWCNTs(0.5%)	–	16%	17%	56%	5.2%	[41]
6	PLA	CNTs (3%)	–	34%	–	–	2.8%	[42]
7	Epoxy	MWCNTs(0.5%)	–	37%	–	50%	–	[43]
8	Epoxy	MWCNTs(0.5%)	MPTMS	11%	9%	–	–	[45]
9	Epoxy	MWCNTs(0.2%)	GPTMS	9%	48%	106%	–	[46]
10	Epoxy	MWCNTs(0.2%)	ADMS	17.4%	5.31%	172%	–	[47]
11	Epoxy	MWCNTs(0.2%)	ATMS	20.7%	4%	172%	–	[47]
12	Epoxy	Chopped carbon fiber (0.5%)	APTES	–	–	–	376%	[48]

2D MXene, i.e., $Ti_3C_2T_x$/PP, composites exhibited a 35.3% increase in TS and a remarkable 674.6% improvement in thermal deformability. The initial decomposition temperature rose by 79.1°C, and the storage modulus increased by 102.2% under oxygen-free rapid evaporation. This study highlights a direct approach to creating ductile, thermally stable composites through nano-reinforcement of MXene, paving the way for broader applications of 2D MXenes in polymer matrices [53]. Furthermore, by physically interfacial bonding between melamine cyanurate and titanium carbide nanosheets ($Ti_3C_2T_x$, MXene), a $Ti_3C_2T_x$@MCA (melamine cyanurate) nanohybrid was synthesized and incorporated into a TPU matrix. The result reveals that the polymer nanocomposite achieved a TS of 62 MPa, K_{IC} of 176 MJ/m³, a strain at break of 590%, and a 40% reduction in heat release rate. With the addition of 3.0 wt.% $Ti_3C_2T_x$/melamine cyanurate, it exhibited exceptional mechanical and flame-resistant properties, surpassing previous designs [54]. A recent literature review of MXenes demonstrates significant potential as reinforcements in polymer nanocomposites, with their surface functional groups greatly influencing the mechanical and thermal properties. Incorporating 0.5 wt.% MXene into a TPU matrix enhanced the tensile strength by 47% compared to pure TPU composite. The density functional theory analysis revealed that hydrogen bonding and π-stacking are the primary interaction mechanisms between MXenes and TPU compounds. While the fluorine and hydroxyl terminations group enhanced compatibility with the TPU matrix, oxygen terminations hindered it. These findings emphasize the importance of tailoring MXene surface chemistry to improve their interactions with polymer matrices in nanocomposites [55]. In a separate study, $Ti_3C_2T_x$/silicone rubber nanocomposites were developed to investigate their mechanical and thermal performance. These nanocomposites demonstrated the highest thermal stability to achieve up to 450°C. The incorporation of 2 wt.% $Ti_3C_2T_x$ MXene surpassed previous designs. The result indicated a TS of 430 kPa, elongation at break of 341%, and reduced YM of 402 kPa [56]. Additionally, the properties of $Ti_3C_2T_x$/polyvinyl alcohol nanocomposites were improved using advanced intumescent flame retardants like poly(vinyl phosphonic acid) (PVPA) and polyethylene polyamine (PEPA). MXene reinforcement also increased the thermal stability of the polyvinyl alcohol matrix nanocomposite. Despite the inclusion of PVPA and PEPA, the elongation at the break of the MXene/polyvinyl alcohol nanocomposites peaked at 1.0 wt.% MXene content. Additionally, the TS and YM increased by 25.3% and 64.45%, respectively, with higher MXene concentrations in the polyvinyl alcohol matrix composite [57]. This type of polymer composite is utilized in advanced manufacturing. MXene-based composites enable additive manufacturing (3D printing), flexible electronics, and self-healing material. An overview of recent studies on MXene nanofillers integrated into polymer nanocomposites is presented in Table 12.4.

TABLE 12.4
Overview of Recent Studies on MXene Nanofillers Integrated into Polymer Nanocomposites

Sl. No.	Polymers	Fillers (wt.%)	Technique	σ_u	E	T_g/E′	Ref.
1	Epoxy	$Ti_3C_2T_x$ (1%)	Sonication	16%	–	–	[50]
2	TPU	Zr-MXene (1%)	Stirring	43.4%	33%	86%	[51]
3	PVA/epoxy	MXene (10%)	Mech. Sonic.	91.4%	743%	–	[52]
4	PP	Ti_3C_2Tx (2%)	stirring	35.3%	675%	103%	[53]
5	TPU	$Ti_3C_2T_x$ (3%)	Mech. Stirring	47.5%	588%	40%	[54]
6	TPU	Ti3C2T_x (0.5%)	Mixing	47%	213%	–	[55]
7	SR	$Ti_3C_2T_x$ (1%)	Mixing	430 kPa	402 kPa	–	[56]
8	PVA	$Ti_3C_2T_x$ (1%)	Mech. Mixing	25.3%	64.45%	–	[57]

12.6 FULLERENE-BASED POLYMER NANOCOMPOSITES

Fullerene is a unique carbon-based nanomaterial that has newly garnered attention for use as a reinforcing agent in polymer matrices. Its exceptional properties contribute to enhancing various physical and mechanical properties. Consequently, studies have focused on its design, distribution techniques, and properties, such as thermal stability and tensile strength. For example, research has analyzed the mechanical behavior of both pristine and modified (oxidized) fullerene when incorporated into epoxy matrices [58]. The surface functionalization of C_{60} with the chemical coupling agent APTES was employed to create silanized fullerene, which was used to enhance TPU. The study focused on investigating its thermal flowability and mechanical properties. Historically, the high production costs and low yield of fullerene limited its potential to increase the performance of epoxy nanocomposite materials. However, recent advancements in combustion-based mass production technology have significantly reduced the production cost of fullerene, facilitating its increased application in sports equipment and coating solutions.

12.6.1 Fabrication of Polymer Nanocomposites Using Oxidized and Silanized Fullerene

In this study, fullerenes (C_{60}) were sequentially modified. The oxidation of fullerene was performed by stirring 0.3 g of fullerenes in 20 mL of concentrated HNO_3 at 80°C and 600 rpm for 1.5 hours. The mixture was then washed with DI water, and ethanol until neutral pH was reached, followed by drying at 80°C for 8 hours in a vacuum oven to obtain oxidized fullerenes (O-Fs). To prepare an APTES-derived silane layer and retain the APTES functional group, 0.3 g of oxidized fullerenes (O-Fs) were dispersed in 25 mL of toluene, and bath-sonicated for 1 hour. During sonication, 5 mL of toluene containing 1 wt.% of APTES was added dropwise to the mixture solution. The resulting solution was refluxed at 110 °C with continuous stirring at 600 rpm for 8 hours. The refluxed mixture was then filtered and washed with distilled water and ethanol until a neutral pH was achieved. Finally, the silanized fullerenes (S-Fs) were obtained by drying at 80°C for 8 hours. During silanization, APTES functional groups covalently bonded with oxygenated groups on O-Fs, forming a siloxane layer [59]. Fabrication of epoxy nanocomposites by incorporating pristine and oxidized fullerenes into the epoxy matrix using dual-probe ultrasonic mixing was carried out to investigate their mechanical and thermal properties. Adding 0.5% pristine fullerene improved TS and YM by 13% and 16.5%, while 0.5% oxidized fullerene enhanced them by 23% and 46%. The fracture toughness (K_{IC}) and fracture energy (G_{IC}) for neat epoxy (0.74 MPa$\sqrt{m}$, 457.17 ± 55 J/m²) increased by ~16% and ~18% with pristine fullerene and ~101% and ~85% with oxidized fullerene at 0.5wt.% as compared to neat composite [58]. Fullerenes (C60) are excellent reinforcements for composites, enhancing multifunctional properties.

Fullerene surfaces were modified via nitric acid oxidation and salinization employing APTES to introduce oxygenated and siloxane functional groups. The resulting functionalized fullerene/epoxy nanocomposite demonstrated a substantial improvement in storage modulus (~491%) and a higher impact factor intensity (tanδ peak) with respect to pure epoxy composite [59]. Similar studies at 0.5 wt.% of S-F filler incorporated into an epoxy matrix showed that the TS, YM, and toughness (K_{IC}) were improved by ~ 28%, ~30%, and ~67.5%, respectively, for the epoxy nanocomposites [60]. A similar study has illustrated the eucommia ulmoides rubber (EUR) matrix was subjected to ultrasonic treatment, with varying amounts of fullerene C_{60} incorporated to form nanocomposite films. The thermal and mechanical properties of these polymer composites were extensively studied. The addition of C_{60} enhanced the melting temperature by 12.5% and improved mechanical properties by 67% with respect to pure eucommia

TABLE 12.5
Summary of Current Studies on Polymer Composites Incorporating Fullerene (C60) Nanofillers

Sl. No.	Polymers	Fillers (wt.%)	Functionalization	σ_u	E	T_g/E'	Ref.
1	Epoxy	C60 (0.5%)	O-C60	23%	101%	–	[58]
2	Epoxy	C60 (0.5%)	APTES	–	–	491%	[59]
4	Epoxy	C60 (0.5%)	APTES	28%	67.5%	–	[60]
3	EUR	C60 (2%)	–	67%	–	12.5%	[61]
5	PA	C60 (0.3%)	–	41%	80.1%	12 °C	[62]
6	PC	C60 (0.5%)	–	33%	45%	40 °C	[63]
7	SPEI	C60 (1%)	–	114%	–	46%	[64]
8	PVDF	C60 (0.5%)	–	62%	18%	14%	[65]

ulmoides rubber composite [61]. Considering the numerous research papers published on C_{60}, the remaining studies are compiled in Table 12.5, highlighting the enhancement of mechanical and thermal performance with and without the functionalization of C_{60}. MXene, GNPs, and CNTs have been similarly applied as reinforcement materials in lightweight electric vehicles. A summary of current studies on polymer composites incorporating fullerene (C60) nanofillers is presented in Table 12.5.

12.7 CONCLUSIONS

The incorporation of carbon nanofiller reinforcement into the polymer matrix contributes to the substantially improved mechanical and thermal performance of this nanocomposite due to the following reasons: uniformed dispersion of nanofiller within a polymer matrix material, better interfacial bonding, hindering the nanofiller agglomeration within the polymer matrix, increasing the cross-linking capacities of polymer composite, and enhancing the interfacial interaction between filler and matrix material. These features play a key role in greatly enhancing the mechanical and thermal properties of carbon nanofillers into polymer nanocomposites, as discussed in this review. Incorporating 2D carbon-based materials into polymer nanocomposites enhances electrical conductivity, thermal stability, and mechanical strength in lightweight electric vehicles. These properties enable applications in EMI shielding, thermal management, and energy storage, improving battery performance, structural integrity, and heat dissipation. Additionally, MXene-based composites advance additive manufacturing, flexible electronics, and smart coatings with self-sensing capabilities. A summary of their mechanical and thermal performance across various applications is provided in the tables. The incorporation of nano-carbon materials like graphene, GrNPs, CNTs, MWCNTs, MXenes, and fullerene led to improvements of 150%, 58%, 108%, 100%, 91%, and 114%, and in the mechanical properties of polymer nanocomposites, respectively, at optimal nano-carbon content. The greatest enhancement, 150%, was achieved with the optimal inclusion of rGO. However, regarding thermal stability, the incorporation of graphene, MXenes, CNTs, and fullerene increased thermal stability by 63%, 40%, 20%, 15%, and 40%, respectively, as compared to neat composite, with optimal nano-carbon incorporation into polymer nanocomposite. Notably, thermal stability was enhanced by 63% with the optimal incorporation of graphene.

Future prospect: Although certain carbon nanomaterials enhance the mechanical and thermal performance of polymer nanocomposites, scalable production methods for some remain difficult, and their high production costs limit commercial application. Additionally, there has been limited research on the recyclability of carbon nanocomposites, highlighting the need for further development in recycled nanocomposites. One promising approach may be utilizing waste thermoset-polymer carbon nanocomposites to create high-value carbon nanomaterials.

ACKNOWLEDGMENTS

We sincerely extend our heartfelt gratitude to the editorial teams of all journals for providing us with the opportunity to submit this review chapter. We also deeply appreciate the valuable contributions, insightful discussions, and unwavering dedication of all our co-authors throughout the preparation of this manuscript.

Data Availability: The data can be accessed upon request.

Conflicts of Interest: The author(s) declare no potential conflicts of interest with respect to the research, authorship, and/or publication of this chapter.

REFERENCES

[1] A. Salehabadi, M. F. Umar, A. Ahmad, M. I. Ahmad, N. Ismail, and M. Rafatullah, "Carbon-based nanocomposites in solid-state hydrogen storage technology: An overview," *Int J Energy Res*, vol. 44, no. 14, pp. 11044–11058, Nov. 2020, doi: 10.1002/er.5674.

[2] X. Chen et al., "MXene/polymer nanocomposites: Preparation, properties, and applications," *Polymer Rev*, vol. 61, no. 1, pp. 80–115, Jan. 2021, doi: 10.1080/15583724.2020.1729179.

[3] K. Zeranska-Chudek et al., "Terahertz shielding properties of carbon black based polymer nanocomposites," *Materials*, vol. 14, no. 4, p. 835, Feb. 2021, doi: 10.3390/ma14040835.

[4] A. Kausar, "Epitome of fullerene in conducting polymeric nanocomposite—Fundamentals and beyond," *Poly Plast Technol Mater*, vol. 62, no. 5, pp. 618–631, Mar. 2023, doi: 10.1080/25740881.2022.2121223.

[5] Z. Latif, M. Ali, E.-J. Lee, Z. Zubair, and K. H. Lee, "Thermal and mechanical properties of nano-carbon-reinforced polymeric nanocomposites: A review," *J Composites Sci*, vol. 7, no. 10, p. 441, Oct. 2023, doi: 10.3390/jcs7100441.

[6] V. Singh, D. Joung, L. Zhai, S. Das, S. I. Khondaker, and S. Seal, "Graphene based materials: Past, present and future," *Prog Mater Sci*, vol. 56, no. 8, pp. 1178–1271, Oct. 2011, doi: 10.1016/j.pmatsci.2011.03.003.

[7] R. J. Young, I. A. Kinloch, L. Gong, and K. S. Novoselov, "The mechanics of graphene nanocomposites: A review," *Compos Sci Technol*, vol. 72, no. 12, pp. 1459–1476, Jul. 2012, doi: 10.1016/j.compscitech.2012.05.005.

[8] A. Kausar, "Polymer/carbon-based quantum dot nanocomposite: Forthcoming materials for technical application," *J Macromolecular Sci Part A*, vol. 56, no. 4, pp. 341–356, Apr. 2019, doi: 10.1080/10601325.2019.1578614.

[9] A. Kausar, "Epoxy and quantum dots-based nanocomposites: Achievements and applications," *Mater Res Innovations*, vol. 24, no. 4, pp. 235–243, Jun. 2020, doi: 10.1080/14328917.2019.1636175.

[10] S.-A. Leau, C. Lete, and S. Lupu, "Nanocomposite materials based on metal nanoparticles for the electrochemical sensing of neurotransmitters," *Chemosensors*, vol. 11, no. 3, p. 179, Mar. 2023, doi: 10.3390/chemosensors11030179.

[11] S. Ashique, A. Upadhyay, M. Gulati, D. Singh, P. A. Chawla, and V. Chawla, "One-dimensional polymeric nanocomposites in drug delivery systems," *Curr Nanosci*, vol. 19, no. 6, pp. 825–839, Nov. 2023, doi: 10.2174/1573413719666230110110706.

[12] Y.-J. Wan, L.-X. Gong, L.-C. Tang, L.-B. Wu, and J.-X. Jiang, "Mechanical properties of epoxy composites filled with silane-functionalized graphene oxide," *Compos Part A Appl Sci Manuf*, vol. 64, pp. 79–89, Sep. 2014, doi: 10.1016/j.compositesa.2014.04.023.

[13] C. Bao, Y. Guo, L. Song, Y. Kan, X. Qian, and Y. Hu, "In situ preparation of functionalized graphene oxide/epoxy nanocomposites with effective reinforcements," *J Mater Chem*, vol. 21, no. 35, p. 13290, 2011, doi: 10.1039/c1jm11434d.

[14] S. K. Singh, B. Nayak, T. J. Singh, and S. Halder, "Investigating the role of synthesized reduced graphene oxide and graphite micro-fillers on mechanical and fretting wear performance of glass fiber epoxy-based composite," *High Perform Polym*, vol. 35, no. 9, pp. 946–962, Nov. 2023, doi: 10.1177/09540083231196087.

[15] S. K. Singh, T. J. Singh, B. Nayak, P. K. Sonker, and M. A. Singh, "Analysis of the impact of exfoliated graphene oxide on the mechanical performance and in-plane fracture resistance of epoxy-based nanocomposite," *High Perform Polym*, vol. 36, no. 9–10, pp. 487–507, Dec. 2024, doi: 10.1177/09540083241278209.

[16] S. N. Tripathi, P. Saini, D. Gupta, and V. Choudhary, "Electrical and mechanical properties of PMMA/reduced graphene oxide nanocomposites prepared via in situ polymerization," *J Mater Sci*, vol. 48, no. 18, pp. 6223–6232, Sep. 2013, doi: 10.1007/s10853-013-7420-8.

[17] H. J. Salavagione, G. Martínez, and M. A. Gómez, "Synthesis of poly(vinyl alcohol)/reduced graphite oxide nanocomposites with improved thermal and electrical properties," *J Mater Chem*, vol. 19, no. 28, p. 5027, 2009, doi: 10.1039/b904232f.

[18] J. Phiri, L.-S. Johansson, P. Gane, and T. Maloney, "A comparative study of mechanical, thermal and electrical properties of graphene-, graphene oxide- and reduced graphene oxide-doped microfibrillated cellulose nanocomposites," *Compos B Eng*, vol. 147, pp. 104–113, Aug. 2018, doi: 10.1016/j.compositesb.2018.04.018.

[19] S. K. Singh, T. Jackson Singh, L. D. Singh, P. K. Sonker, and B. Mazumder, "Experimental study on the impact of hybrid GFRP composites with graphene oxide and silicon carbide fillers on mechanical and wear properties," *J Adhes Sci Technol*, vol. 39, no. 7, pp. 1105–1132, Dec. 2024, doi: 10.1080/01694243.2024.2438988.

[20] X. Wang, W. Xing, P. Zhang, L. Song, H. Yang, and Y. Hu, "Covalent functionalization of graphene with organosilane and its use as a reinforcement in epoxy composites," *Compos Sci Technol*, vol. 72, no. 6, pp. 737–743, Mar. 2012, doi: 10.1016/j.compscitech.2012.01.027.

[21] Y.-J. Wan, L.-X. Gong, D. Yan, L.-C. Tang, L.-B. Wu, J.-X. Jiang and G.-Q. Lai Li-Zhi Guan, "Toward effective and tunable interphases in graphene oxide/epoxy composites by grafting different chain lengths of polyetheramine onto graphene oxide," *J Mater Chem A Mater*, vol. 2, no. 15058, 2024.

[22] Y.-J. Wan, L.-X. Gong, L.-C. Tang, L.-B. Wu, and J.-X. Jiang, "Mechanical properties of epoxy composites filled with silane-functionalized graphene oxide," *Compos Part A Appl Sci Manuf*, vol. 64, pp. 79–89, Sep. 2014, doi: 10.1016/j.compositesa.2014.04.023.

[23] M. J. Azizli, M. Barghamadi, K. Rezaeeparto, and S. Parham, "Improvement of mechanical, morphological and thermal properties on PP-enriched graphene oxide/PP-g-MA/EPDM blend compatibilized: PP-g-MA compatibilizer and graphene oxide nanofiller role," *J Poly Res*, vol. 29, no. 8, p. 322, Aug. 2022, doi: 10.1007/s10965-022-03182-4.

[24] S. Zhang et al., "Polydopamine functional reduced graphene oxide for enhanced mechanical and electrical properties of waterborne polyurethane nanocomposites," *J Coat Technol Res*, vol. 15, no. 6, pp. 1333–1341, Nov. 2018, doi: 10.1007/s11998-018-0082-3.

[25] D. Kang, Y. Shin, H. J. Jo, H. Ko, and H. S. Shin, "Mechanical properties of poly(dopamine)-coated graphene oxide and poly(vinyl alcohol) composite fibers coated with reduced graphene oxide and their use for piezoresistive sensing," *Particle Particle Syst Characterization*, vol. 34, no. 9, Article ID: 1600382, pp. 1–9, Sep. 2017, doi: 10.1002/ppsc.201600382.

[26] P. K. Sandhya, M. S. Sreekala, M. Padmanabhan, K. Jesitha, and S. Thomas, "Effect of starch reduced graphene oxide on thermal and mechanical properties of phenol formaldehyde resin nanocomposites," *Compos B Eng*, vol. 167, pp. 83–92, Jun. 2019, doi: 10.1016/j.compositesb.2018.12.009.

[27] B. Ucpinar Durmaz, C. Ozturk, and A. Aytac, "Reduced graphene oxide reinforced PET/PBT nanocomposites: Compatibilization and characterization," *Polym Eng Sci*, vol. 60, no. 10, pp. 2606–2618, Oct. 2020, doi: 10.1002/pen.25498.

[28] Y. Zhang, J. Hu, S. Zhu, T. Qin, and F. Ji, "A 'trampoline' nanocomposite: Tuning the interlayer spacing in graphene oxide/polyurethane to achieve coalesced mechanical and memory properties," *Compos Sci Technol*, vol. 180, pp. 14–22, Aug. 2019, doi: 10.1016/j.compscitech.2019.04.033.

[29] L. N. Borah, S. Halder, and P. Choudhury, "Epoxy composites integrated with functionalized wet-milled graphitic nanoparticles: Featuring enhanced mechanical performance and in-plane fracture resistance," *Polym Compos*, vol. 44, no. 11, pp. 7985–7996, Nov. 2023, doi: 10.1002/pc.27681.

[30] N. I. Khan, S. Halder, S. Das, and M. S. Goyat, "Graphitic nanoparticles functionalized with epoxy moiety for enhancing the mechanical performance of hybrid carbon fiber reinforced polymer laminated composites," *Polym Compos*, vol. 42, no. 2, pp. 678–692, Feb. 2021, doi: 10.1002/pc.25857.

[31] S. Das, S. Halder, B. Paul, N. I. Khan, and M. S. Goyat, "Impact of silanized milled graphite nanoparticles on thermo-mechanical properties of epoxy nanocomposite," *Mater Chem Phys*, vol. 278, p. 125601, Feb. 2022, doi: 10.1016/j.matchemphys.2021.125601.

[32] S. Halder, L. N. Borah, and P. Choudhury, *Functionalized Milled Graphitic Nanoparticle for Improving In-Plane Crack Propagation Resistance to Enhance Fracture Toughness and Energy of Epoxy Nanocomposites,* 2023. SSRN publisher. doi: 10.2139/ssrn.4399323.

[33] N. I. Khan, S. Halder, and J. Wang, "Diels-Alder based epoxy matrix and interfacial healing of bismaleimide grafted GNP infused hybrid nanocomposites," *Polym Test*, vol. 74, pp. 138–151, Apr. 2019, doi: 10.1016/j.polymertesting.2018.12.021.

[34] J.-V. Lim, S.-T. Bee, L. Tin Sin, C. T. Ratnam, and Z. A. Abdul Hamid, "A review on the synthesis, properties, and utilities of functionalized carbon nanoparticles for polymer nanocomposites," *Polymers* (Basel), vol. 13, no. 20, p. 3547, Oct. 2021, doi: 10.3390/polym13203547.

[35] S. Iijima, "Helical microtubules of graphitic carbon," *Nature*, vol. 354, no. 6348, pp. 56–58, Nov. 1991, doi: 10.1038/354056a0.

[36] M. Tarfaoui, K. Lafdi, and A. El Moumen, "Mechanical properties of carbon nanotubes based polymer composites," *Compos B Eng*, vol. 103, pp. 113–121, Oct. 2016, doi: 10.1016/j.compositesb.2016.08.016.

[37] A. H. Esbati and S. Irani, "Effect of functionalized process and CNTs aggregation on fracture mechanism and mechanical properties of polymer nanocomposite," *Mech Mater*, vol. 118, pp. 106–119, Mar. 2018, doi: 10.1016/j.mechmat.2018.01.001.

[38] C. Mu et al., "Modification of carbon nanotubes by a novel biomimetic approach towards the enhancement of the mechanical properties of polyurethane," *Polymer (Guildf)*, vol. 92, pp. 231–238, Jun. 2016, doi: 10.1016/j.polymer.2016.03.085.

[39] F. V. Ferreira et al., "Dodecylamine functionalization of carbon nanotubes to improve dispersion, thermal and mechanical properties of polyethylene based nanocomposites," *Appl Surf Sci*, vol. 410, pp. 267–277, Jul. 2017, doi: 10.1016/j.apsusc.2017.03.098.

[40] J. Cha, G. H. Jun, J. K. Park, J. C. Kim, H. J. Ryu, and S. H. Hong, "Improvement of modulus, strength and fracture toughness of CNT/Epoxy nanocomposites through the functionalization of carbon nanotubes," *Compos B Eng*, vol. 129, pp. 169–179, Nov. 2017, doi: 10.1016/j.compositesb.2017.07.070.

[41] N. Deep and P. Mishra, "Evaluation of mechanical properties of functionalized carbon nanotube reinforced PMMA polymer nanocomposite," *Karbala Int J Modern Sci*, vol. 4, no. 2, pp. 207–215, Jun. 2018, doi: 10.1016/j.kijoms.2018.02.001.

[42] L. S. De Bortoli, R. de Farias, D. Z. Mezalira, L. M. Schabbach, and M. C. Fredel, "Functionalized carbon nanotubes for 3D-printed PLA-nanocomposites: Effects on thermal and mechanical properties," *Mater Today Commun,* vol. 31, p. 103402, Jun. 2022, doi: 10.1016/j.mtcomm.2022.103402.

[43] C.-K. Yang, Y.-R. Lee, T.-H. Hsieh, T.-H. Chen, and T.-C. Cheng, "Mechanical property of multiwall carbon nanotube reinforced polymer composites," *Polymers Polymer Composites*, vol. 26, no. 1, pp. 99–104, Jan. 2018, doi: 10.1177/096739111802600112.

[44] A. Y. Boroujeni and M. Al-Haik, "Carbon nanotube – Carbon fiber reinforced polymer composites with extended fatigue life," *Compos B Eng*, vol. 164, pp. 537–545, May 2019, doi: 10.1016/j.compositesb.2018.11.056.

[45] S. Shadlou, E. Alishahi, and M. R. Ayatollahi, "Fracture behavior of epoxy nanocomposites reinforced with different carbon nano-reinforcements," *Compos Struct*, vol. 95, pp. 577–581, Jan. 2013, doi: 10.1016/j.compstruct.2012.08.002.

[46] S. Shadlou, E. Alishahi, and M. R. Ayatollahi, "Fracture behavior of epoxy nanocomposites reinforced with different carbon nano-reinforcements," *Compos Struct*, vol. 95, pp. 577–581, Jan. 2013, doi: 10.1016/j.compstruct.2012.08.002.

[47] D. Vennerberg, Z. Rueger, and M. R. Kessler, "Effect of silane structure on the properties of silanized multiwalled carbon nanotube-epoxy nanocomposites," *Polymer* (Guildf), vol. 55, no. 7, pp. 1854–1865, Apr. 2014, doi: 10.1016/j.polymer.2014.02.018.

[48] S. Das, S. Halder, J. Wang, M. S. Goyat, A. Anil Kumar, and Y. Fang, "Amending the thermo-mechanical response and mechanical properties of epoxy composites with silanized chopped carbon fibers," *Compos Part A Appl Sci Manuf*, vol. 102, pp. 347–356, Nov. 2017, doi: 10.1016/j.compositesa.2017.07.026.

[49] M. Malaki and R. S. Varma, "Mechanotribological aspects of MXene-reinforced nanocomposites," *Adv Mater*, vol. 32, no. 38, Sep. 2020, doi: 10.1002/adma.202003154.

[50] Y. Sliozberg, J. Andzelm, C. B. Hatter, B. Anasori, Y. Gogotsi, and A. Hall, "Interface binding and mechanical properties of MXene-epoxy nanocomposites," *Compos Sci Technol*, vol. 192, p. 108124, May 2020, doi: 10.1016/j.compscitech.2020.108124.

[51] L. Liu et al., "Functionalizing MXene towards highly stretchable, ultratough, fatigue- and fire-resistant polymer nanocomposites," *Chem Eng J*, vol. 424, p. 130338, Nov. 2021, doi: 10.1016/j.cej.2021.130338.

[52] G. Monastyreckis, L. Mishnaevsky, C. B. Hatter, A. Aniskevich, Y. Gogotsi, and D. Zeleniakiene, "Micromechanical modeling of MXene-polymer composites," *Carbon N Y*, vol. 162, pp. 402–409, Jun. 2020, doi: 10.1016/j.carbon.2020.02.070.

[53] Y. Shi et al., "Strengthening, toughing and thermally stable ultra-thin MXene nanosheets/polypropylene nanocomposites via nanoconfinement," *Chem Eng J*, vol. 378, p. 122267, Dec. 2019, doi: 10.1016/j.cej.2019.122267.

[54] Y. Shi, C. Liu, Z. Duan, B. Yu, M. Liu, and P. Song, "Interface engineering of MXene towards super-tough and strong polymer nanocomposites with high ductility and excellent fire safety," *Chem Eng J*, vol. 399, p. 125829, Nov. 2020, doi: 10.1016/j.cej.2020.125829.

[55] R. M. Ronchi, C. F. N. Marchiori, C. M. Araujo, J. T. Arantes, and S. F. Santos, "Thermoplastic polyurethane – Ti3C2(Tx) MXene nanocomposite: The influence of functional groups upon the matrix–reinforcement interaction," *Appl Surf Sci*, vol. 528, p. 146526, Oct. 2020, doi: 10.1016/j.apsusc.2020.146526.

[56] Y. Zeng et al., "Significantly improved dielectric and mechanical performance of Ti3C2Tx MXene/silicone rubber nanocomposites," *J Alloys Compd*, vol. 905, p. 164172, Jun. 2022, doi: 10.1016/j.jallcom.2022.164172.

[57] X. Yan, J. Fang, J. Gu, C. Zhu, and D. Qi, "Flame retardancy, thermal and mechanical properties of novel intumescent flame retardant/MXene/poly(vinyl alcohol) nanocomposites," *Nanomaterials*, vol. 12, no. 3, p. 477, Jan. 2022, doi: 10.3390/nano12030477.

[58] S. Das, S. Halder, and N. Islam Khan, "Mechanical properties of oxidized fullerene C60/epoxy nanocomposite," *Mater Today Proc*, vol. 18, pp. 655–659, 2019, doi: 10.1016/j.matpr.2019.06.461.

[59] S. Das, S. Halder, and K. Kumar, "A comprehensive study on step-wise surface modification of C60: Effect of oxidation and silanization on dynamic mechanical and thermal stability of epoxy nanocomposite," *Mater Chem Phys*, vol. 179, pp. 120–128, Aug. 2016, doi: 10.1016/j.matchemphys.2016.04.085.

[60] S. Das, S. Halder, A. Sinha, M. A. Imam, and N. I. Khan, "Assessing nanoscratch behavior of epoxy nanocomposite toughened with silanized fullerene," *ACS Appl Nano Mater*, vol. 1, no. 7, pp. 3653–3662, Jul. 2018, doi: 10.1021/acsanm.8b00763.

[61] X. Wei, X. Yang, A. Ma, J. Zhang, P. Peng, and J. Dong, "Fullerene C60/Eucommia ulmoides rubber nanocomposite films with excellent UV-blocking performance for controlling wheat scab fungus," *Ind Crops Prod*, vol. 162, p. 113301, Apr. 2021, doi: 10.1016/j.indcrop.2021.113301.

[62] J. Chen, K. Luo, J. Zhu, J. Yu, Y. Wang, and Z. Hu, "Reversibly cross-linked fullerene/polyamide composites based on Diels-Alder reaction," *Compos Sci Technol*, vol. 176, pp. 9–16, May 2019, doi: 10.1016/j.compscitech.2019.03.021.

[63] Y. Su, S. Ran, Z. Fang, and Z. Guo, "Fullerene-induced crystallization toward improved mechanical properties of solvent casting polycarbonate films," *Appl Phys A*, vol. 126, no. 4, p. 293, Apr. 2020, doi: 10.1007/s00339-020-03473-x.

[64] M. Rajagopalan, I.-K. Oh, J. Palanivel, S. Sabarathinam, and J. Cifuentes-Faura, "Fabrication and characterizations of electro-mechanical actuators based on fullerene-reinforced biocompatible polymer," *Sens Actuators A Phys*, vol. 339, p. 113510, Jun. 2022, doi: 10.1016/j.sna.2022.113510.
[65] M. Behera, "Study of optical, thermal, mechanical and microstructural properties of fullerene/poly (vinylidene fluoride) polymer nanocomposites," *Biointerface Res Appl Chem*, vol. 13, no. 2, p. 121, Mar. 2022, doi: 10.33263/BRIAC132.121.

13 Emerging Materials for Battery Management Systems

Prashantha S., Omkaresh B.R., and Basir Maleki

13.1 INTRODUCTION

The battery material market is evolving rapidly due to the increased demand for electric vehicles (EVs) and consumer electronics such as smartphones and laptops. The growth drivers related to batteries are portable electronic devices, energy storage systems in industry, and EVs, which are the major drivers in the market for battery materials [1,2]. The battery material market is segmented into various materials, including lead dioxide, lithium cobalt oxide, lithium manganese oxide, nickel cobalt aluminum, nickel manganese cobalt, lithium-iron phosphate, lead, synthetic graphite, and natural graphite.

To keeping up with current trends and to remain competitive, innovation (developing new battery chemistries and materials), sustainability (emphasizing eco-friendly materials and recycling processes), and collaboration (partnering with research institutions and other industries to drive advancements) are the need of the hour. The quest for high-energy-density secondary battery systems remains a top priority in the industry. Achieving higher energy density means long-lasting batteries, which is essential for uses ranging from portable gadgets to electric cars. The focus areas for high-energy-density batteries ranges from lithium-ion batteries (LIB), solid-state batteries (SSBs), silicon anodes, advanced cathode materials, and emerging technologies [3,5]. The pursuit of these advancements is critical for meeting the ever-growing energy demands of modern technology. Whether it's extending the range of EVs or improving the battery life of consumer gadgets, industry is always pushing the boundaries of what's possible. The following are the main parts of a standard battery, each of which is essential to its operation and adds to its total weight.

The cells and the materials they contain typically account for 60–75% of a battery's overall weight, with the metal casing, wires, and thermal and battery management systems making up the remaining 25–40%.

Lead Plates: These are vital to the battery's functioning because they store and release energy. A battery's weight increases with the number of lead plates it contains. Typically, these plates have a lead dioxide coating, which increases their weight.

Electrolyte: Water and sulfuric acid are combined to form the electrolyte; it promotes the energy storage and release involving chemical reactions. The electrolyte contributes a substantial amount of weight to the battery but is not heavy as the lead plates.

Separators: Batteries employ separators to keep the lead plates from coming into contact and shorting out. Fiberglass or other insulating materials are frequently used to make them. They are necessary for the battery's safe operation, even though they don't contribute much weight.

Casing: Everything is held together by the casing. Usually composed of sturdy plastic, it might occasionally be constructed of metal. The enclosure must be lightweight to prevent adding extra weight while still being sturdy enough to protect the interior components [6]. The voltage

 DOI: 10.1201/9781003596097-13

of LIBs is higher, often 3.6 V. Despite its great energy density and lengthy service life, it comes at a comparatively high cost. The use of LIBs is growing, particularly in hybrid and electric cars. They cost more than lead-acid batteries, but they are lighter and more effective. Common lithium-ion, the weight of these batteries ranges from 10 to 15 kg. The utilization of various materials, including lithium and other lightweight metals, accounts for the reduced weight. They function exceptionally well and have a greater energy storage capacity than lead-acid batteries despite being lighter [7]. The positive active ingredient in nickel-metal hydride batteries is nickel hydroxide. The negative active component is metal hydride. A solution of potassium hydroxide serves as the electrolyte. Although nickel metal hydride batteries are inexpensive, have a well-established technology, and last a long time, they are bulky with low energy density [8].

Pb-acid batteries are renowned for their dependability and affordability and have been in business for more than a century. Standard Lead-Acid, these are the batteries you usually see in cars. Their typical weight ranges from 15 to 25 kg. The sulfuric acid electrolyte and the many lead plates are the sources of the weight. Due to their affordability and ability to supply enough power for most cars, these batteries are frequently employed. Lead and its oxides make up most of the electrodes in lead-acid batteries. A sulfuric acid solution serves as the electrolyte, and it is inexpensive and performs well at low temperatures. On the other hand, Pb-acid batteries have a low energy density and are big with a short lifespan. Nowadays, Pb-acid batteries are the most widely used kind in automobiles [9]. Nickel-cadmium batteries, there is a 500 times service life for the 1.2V nickel-cadmium battery. Nickel-cadmium batteries have a discharge temperature of –20°C to 60°C and a charging temperature of 0°C to 45°C. It also resists overcharging well [10]. Lithium nickel cobalt manganite is the cathode material in ternary LIBs. It has a long cycle life, high energy density, and good stability at low temperatures. However, at high temperatures, it is not sufficiently stable [11]. Figure 13.1 presents the specific and volumetric energy density, indicating smaller size and lighter cells.

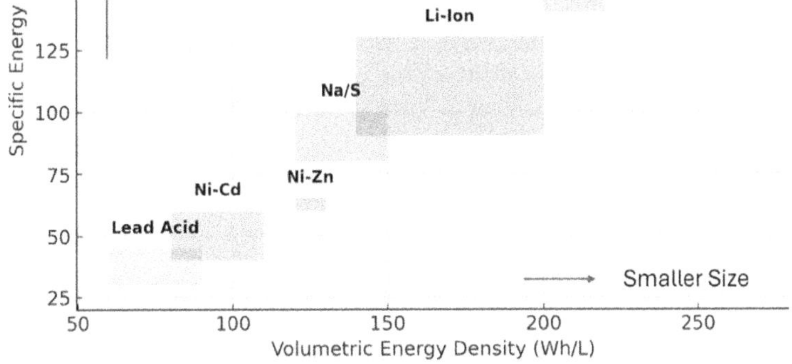

FIGURE 13.1 Energy density comparison by size and weight.

13.2 BATTERY MATERIALS AND HISTORY

The material in the battery is expected to grow its market share by 8% with a value of $750 billion in the years 2020–2025. The trends which influence the material market include solid-state electrolytes, LIBs, ultra-fast carbon electrolytes, sodium-ion batteries, and nanobolt lithium tungsten batteries.

***Lead-Acid Batteries*:** Lead-acid batteries, which had minimal production costs and technology, were the first commercial batteries. The reliability, easy maintenance, and capability of delivering high surge currents made them dominant for over a century [12]. The specific energy is 30–50 Wh kg^{-1} with a specific power of 150–200 W kg^{-1} and a life cycle of 400–800, capable of exhibiting good performance in low and high temperatures, made them popular choices [13].

***Nickel-Based Batteries*:** Various batteries have been developed using nickel oxyhydroxide as the cathode including Ni-Fe and Ni-Cd batteries (invented by Waldemar Junger in 1899) and the Ni-Zn battery which was patented by Thomas Edison (in 1901). In 1967, Ni-MH batteries were invented and in 1970 Ni-H_2 battery was invented at Consmat USA [14]. Among the Ni-based batteries, Ni-Zn has the highest voltage with a high specific energy and is non-toxic. It has better tolerance of overcharge and overdischarge, with a wide operating temperature. Because zinc is soluble in electrolyte, the life cycle is approximately 300. Compared to lead acid batteries, Ni-MH and Ni-H_2 batteries have a greater specific energy and are proven technologies with a specific energy between 35–80 Wh kg^{-1}, specific power of 150–450 W kg^{-1}, and a life cycle of 800–2000, which made them a good choice [15].

***Lithium-based batteries*:** The lithium metal-based primary battery (LMB) in 1976 used metal lithium as the anode with a non-aqueous electrolyte, with low weight, higher energy density, and high specific energy. Varieties of cathode materials have been explored in LMBs, such as iodine, pyrite, manganese dioxide, etc. [16,17]. With the introduction of the $LiCoO_2$ cathode by John B. Goodenough in 1980, and in 1985 the prototype by Akira Yoshino, the first commercial battery by Sony and Asahi Kasei in 1991 made a great leap forward for high-energy and high-power batteries [18]. The implementation of intercalation material by Whittingham was the breakthrough in rechargeable lithium batteries. The LIBs were commercialized in 1991 with $LiCoO_2$ as the cathode and petroleum coke as the anode material, and with further improvements in the anode material using graphite, the energy density was improved [19]. The best cathode material in LIBs was layered lithium nickel cobalt manganese oxide (NCMXXX), which had high capacity and comparatively less cost compared with $LiCoO_2$. By increasing the cathode layered oxides and Ni content, the energy density was further improved. A few battery cells also used $LiMn_2O_4$ (as the cathode) in combination with NMC to match the ratio of power to energy. The specific energy of up to 300 Wh kg^{-1} makes them suitable for hybrid electric vehicles (HEVs) and battery electric vehicles (BEVs) [20,21].

Metal/Air Batteries: These use air as the cathode and a metallic anode. The anode capacity is the limitation of their energy capacity. They can have an energy density of 400 Wh L^{-1} and specific energy of 600 Wh kg^{-1} with varying metallic anodes [22]. The metal/air batteries consist of calcium/air, magnesium/air, zinc/air, iron/air, and aluminum/air. It is more convenient to recycle and replenish mechanically rechargeable batteries, however, the drawbacks are carbonation of alkaline electrolytes and low specific power [23,24].

***Sodium-Beta Batteries*:** The sodium/sulfur (Na-S) and sodium metal chloride (Na/MCl_2) batteries are known for their high energy densities and they must operate at a temperature range of 270–350°C for ionic conductivity [25]. Among the sodium metal chloride batteries, nickel-chloride (Na/$NiCl_2$) and iron-chloride (Na/$FeCl_2$) batteries adopt the transition metal chloride. Sodium-sulfur batteries use a sodium anode, sulfur cathode, and beta alumina ceramic as the electrolyte. Due to internal resistance, the battery degrades and with an incremental depth of discharge this worsens [26,27]. The alternative to Li batteries is sodium-ion, zinc-ion, and magnesium-ion

TABLE 13.1
Quantitative Comparison of Different Batteries

Battery Material	Specific Energy (Wh/kg)	Energy Density (Wh/l)	Specific Power (W/kg)
Ni-Zn	60–65	120–130	150–300
Ni-Cd	40–60	80–110	150–350
Zn-Air	225–235	265–275	100–110
Na-S	95–105	145–155	195–205
Li-polymer	150–160	215–225	310–320
Li-ion	90–130	140–200	250–450

batteries, because of abundant resources of sodium and the working chemistry is similar to that of LIBs.

***Lithium-Air and Lithium-Sulfur Batteries*:** The Li/O_2 batteries are categorized as solid state, hybrid aqueous/non-aqueous, aqueous, and non-aqueous based on the electrolytes. The batteries with non-aqueous electrolytes have a high specific energy of up to 11,700 Wh kg^{-1} [22]. Aqueous electrolyte batteries are prone to burning, but hybrid aqueous/non-aqueous batteries offer the benefit of having two electrolytes. The sulfur exhibits low cost, high gravimetric capacity, but they suffer due to poor electronic conductivity. The formation of soluble lithium polysulfide intermediates and multielectron reactions are drawbacks from the point of view of commercial applications [28].

***Solid-State Batteries (SSBs)*:** The SSB having metal lithium as the anode is a promising technology because of the safety and energy density. The SSB with sulfide electrolyte has a thin layer of silver carbon which increases the electrochemical cyclability. Experimental results show more than 99.8% coulomb efficiency and about 900 Wh L^{-1} energy density [29]. Instead of traction batteries, dual-carbon batteries and dual-ion batteries are preferred for stationary energy storage and, with the progress in new electrolytes and carbon materials, there is promising upgrades in dual-ion batteries (DIBs) and dual-carbon batteries (DCBs) [30,31]. The important parameters to evaluate these batteries are tabulated in Table 13.1.

13.3 MATERIALS IN BATTERY MANAGEMENT SYSTEMS

There is a need for efficient batteries in today's technology to support a wide range of applications including EVs, renewable energy systems, and portable electronic devices. The battery management system (BMS) is the key concept in ensuring the safety, performance, efficiency, and reliability of battery systems making it an integral part [32–35].

Figure 13.2 shows the components of a BMS. It plays a key role in optimization and troubleshooting, and ensures the overall reliability of storage systems by facilitating real-time adjustments and proactive maintenance, ensuring optimal conditions. There have been reports of fires in battery packs due to thermal runaway in recent years [32]. The main challenge is to maintain safety and to ensure the battery is in a safe operating range during charging and discharging (25–40°C) [36]. The conventional method of air-cooling needs extra powered liquid cooling in a battery thermal management system (BTMS) with complex arrangements and does not meet the requirements of new Li-based batteries. Phase change material (PCM) possesses the capability of phase transition temperature within normal operating conditions, relatively low volume change during phase change, and high density with good thermal conductivity [37]. According to their chemical characteristics, PCMs are chemically stable, non-toxic, non-flammable, and non-corrosive. Economically they are reasonable, with availability on a large scale. There are a wide

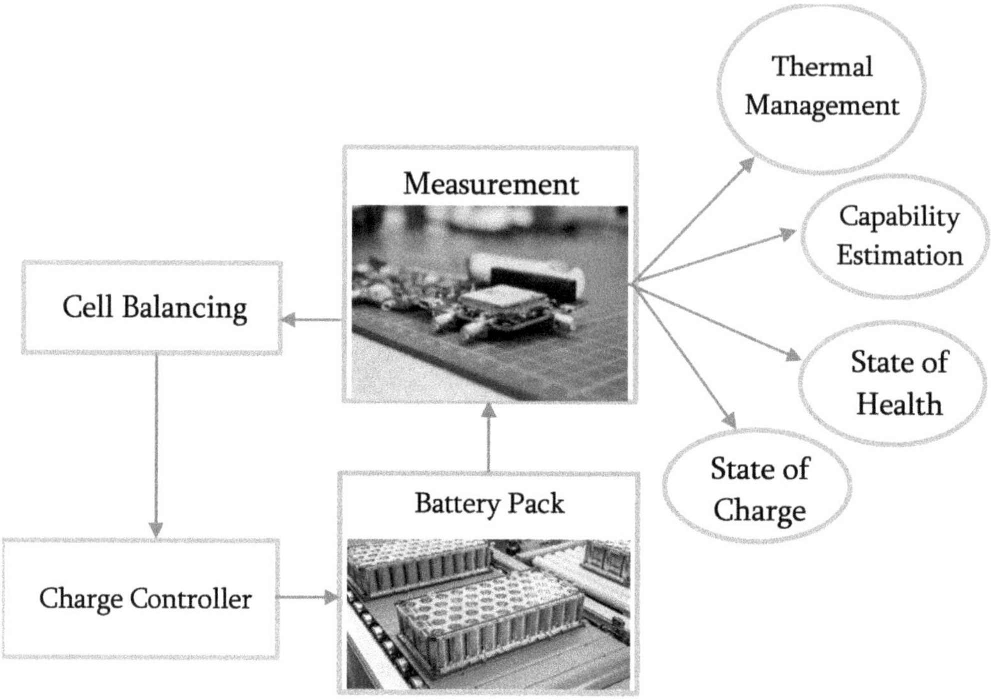

FIGURE 13.2 Battery management system concept.

variety of classification standards for PCMs. These are solid–solid, solid–liquid, solid–gas, and liquid–gas PCMs based on the phase before and after the charging process. Due to the large latent heat capacity, low volume change, and availability the solid–liquid PCMs are in the forefront of applications [38–40]. The common PCMs are organic, inorganic, and eutectics [37]. By combining PCMs with extended graphite, the efficiency of heat dissipation can be improved, adding nanoparticles of carbon nanotubes, silver nanoparticles, graphene nanoparticles, and metallic foam made of nickel/copper/graphite enhances the thermal conductivity [41]. Indeed, simulation-based optimization integrated with BMS while incorporating artificial intelligence (AI) and machine learning (ML) is ongoing. From the theoretical and practical perspective, ML technology plays a crucial role in estimating battery materials' state of health by optimizing energy management systems and thereby accelerating toward achieving carbon neutrality goals.

13.4 ELECTROLYTIC MATERIALS IN BATTERIES

High-voltage LIBs and SSBs are the answers to meet and match the criteria of emerging energy storage systems. Electrolytes play a major role and generate multiple bottleneck issues. Traditionally, electrolytes such as lithium hexafluorophosphate ($LiPF_6$), fluoroethylene carbonate (FEC), and vinylene carbonate have been widely used. However, these are being increasingly replaced or supplemented by high-voltage electrolytes, super-concentrated salts, and ionic liquids. Solid-state electrolytes include gel polymers, solid polymers, ionic liquids, and polyionic liquids [42]. Figure 13.3 shows different types of electrolytes and their relations.

Employing highly concentrated electrolytes facilitates high-energy batteries in organic and aqueous environments. Among non-aqueous solvents, carbonate-based solvents are generally used in commercial batteries. Ether solvent possesses high ionic conductivity, low viscosity, and stability, and include 2-methyltetrahydrofuran polymethyl ethers, tetrahydrofuran, and diethyl ether [42–45].

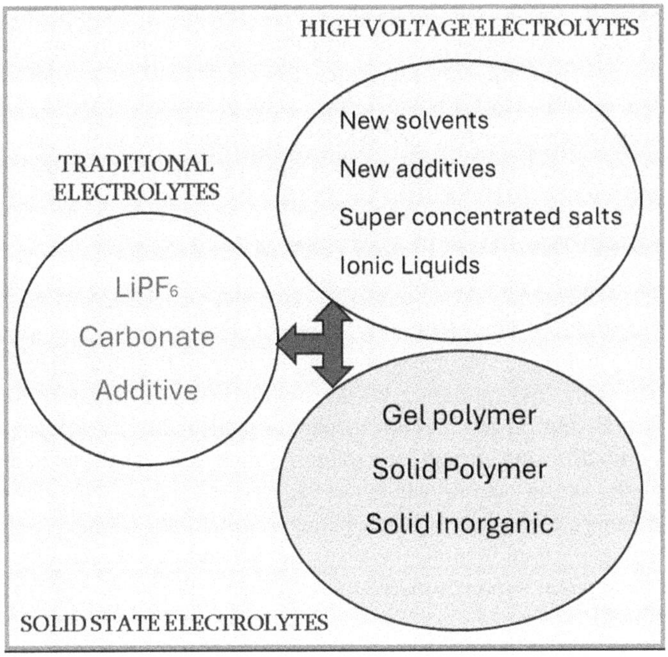

FIGURE 13.3 Different types of electrolytes.

To avoid fire hazards, non-flammable co-solvents such as phosphorus-based materials are used to improve the anodic stability. However, high viscosity and melting point issues restrict its usage. Lithium salts like lithium hexafluorophosphate are the most commonly used in commercial LIB electrolyte systems [46]. $LiPF_6$ is sensitive to moisture at a low ppm level and decomposes at 70°C, hindering its usage due to safety issues. Solid electrolyte interphase layers have been developed with resistive decomposable products of LiF and Li_2CO_3 along with few organic and inorganic byproducts. Dual salt electrolytes were developed to replace $LiPF_6$ [47–49]. The dual blend salt of lithium bis iodine and lithium bis borate is a blend of carbonates as electrolyte additives improve the performance of fuel cells with Li metal (as the anode) and lithium nickel manganese cobalt oxide (NMC) (as the cathode) [50–52].

Additives: To achieve specific properties without affecting the composition, additives are efficient and economical. The additives usually contribute less than 5% of the weight of the electrolyte. During activation cycles they form a protective layer. 1,3-Dioxol-2 or vinylene carbonates (VC) at 1–2% improve the battery life with a reduction in cell impedance and undesirable decomposed elements. LIBs with FEC improve battery metrics by inducing a stable and conductive layer. The cathode electrolyte interface (CEI) formed by adding trimethylsilyl phosphite prevents transition metal dissolution, excessive electrolyte decomposition, and other side reactions. Sulfur-based additives 1,3,2-dioxathiolane-2, tetramethyl sulfone (TMS), and 2-dioxide improve the battery lifetime, suppress gas products, and regulate impedance growth [50,53–57].

***Ionic Liquid Electrolytes*:** These are preferred over carbonate electrolytes because of their non-flammability and low volatility. The main ionic liquid (IL)-based electrolytes are a mixture of IL and lithium salt called ionic electrolytes and a mixture of IL, lithium salt and co-solvents termed IL organic solvents [58–60].

Solid-State Electrolytes: The issues associated with liquid electrolyte-based batteries include fire, explosion, electrolyte leakage, and the formation of dendrites on the anode. Polymer electrolytes composed of polyethylene oxide (PEO) and lithium bis(imide), combined with carboxymethyl cellulose (CMC) binders, have been shown to improve electrochemical performance. The polymer electrolyte and NMC62 (cathode)-containing cell has a specific discharge capacity of about 137 Ah g^{-1} and 100 cycles of Li/NMC 811 at 40°C [61–63].

13.5 CIRCULAR ECONOMY FOR BATTERY MATERIALS

Circular economy is a technique where materials are never converted to waste—it focuses on reusing materials and products. This process includes recycling, maintenance, remanufacture, refurbishment, composting, and reuse. The circular economy confronts global warming and challenges such as waste, biodiversity loss, pollution by decoupling economic activity from the utilization of fixed sources [64]. Execution of a circular economy realistically by repurposing or reutilizing several products is difficult using current technologies due to a lack of established processes [65]. Reutilization is essential as non-operating batteries can also be used for other powerful applications. These procedures are more ecological than extracting new materials [66].

We can decrease resource exploitation, cut waste, and cut carbon emissions by streamlining these procedures. This aligns global interests toward the creation of cleaner energy while also generating economic opportunities [67].

The extraction of rare raw elements like nickel, cobalt, and lithium is the first step in the lifecycle of EV batteries [68]. Batteries are then made using these materials in the manufacturing process. The crucial question is what happens to these batteries when their lifespan of 10–15 years comes to an end [69]. Lead-acid, lithium-ion, nickel-metal hydride, and ultracapacitor batteries are the four primary battery types found in EVs [70]. The EV industry favors LIBs because of their long lifespan, controllable temperature range, and excellent energy efficiency in relation to their physical size. Because of these qualities, they are perfect for a circular economy because they last longer and require fewer replacements, which lowers the demand for materials over the long run [71,72]. LIBs are made from a variety of naturally occurring metals and alloys. The energy required to extract these components is high, which significantly increases the environmental effect of producing LIBs [73].

13.6 CONCLUSIONS

Novel conducting materials need to be found for better electrochemical performance. In order to boost a cell's capacity and voltage, solid-state electrolytes—which are anticipated to drive future trends—are the focus of research on the entire engineering of battery packaging. Market participants for battery materials can improve their capacity to create superior battery materials through investment to boost expertise in cutting-edge graphite technologies for use in EV applications. Research and development initiatives are aimed at creating affordable battery components. Studying new chemistry and effective materials in depth will provide new information and shed light on other crucial energy-related activities that rely on multi-electron and multiphase electrochemical reactions.

REFERENCES

[1] Dunn, B., H. Kamath, and J.-M. Tarascon, Electrical energy storage for the grid: a battery of choices. *Science*. 2011. 334(6058): p. 928–935.

[2] Pang, Q., et al., Advances in lithium–sulfur batteries based on multifunctional cathodes and electrolytes. *Nature Energy*. 2016. 1(9): p. 1–11.

[3] Van Noorden, R., Sulphur back in vogue for batteries. *Nature*. 2013. 498(7455): p. 416–417.

[4] Xia, C., C. Kwok, and L. Nazar, A high-energy-density lithium-oxygen battery based on a reversible four-electron conversion to lithium oxide. *Science*. 2018. 361(6404): p. 777–781.
[5] Liu, J., et al., Pathways for practical high-energy long-cycling lithium metal batteries. *Nature Energy*. 2019. 4(3): p. 180–186.
[6] Borah, R., et al., On battery materials and methods. *Materials Today Advances*. 2020. 6: p. 100046.
[7] Li, Y., P. Zhao, and B. Shen, A review of new technologies for lithium-ion battery treatment. *Science of The Total Environment*. 2024. 951: p. 175459.
[8] Quintero Pulido, D.F., C.F. Covrig, and M. Bruchhausen, On the performance of portable NiMH batteries of general use. *Batteries*. 2025. 11(1): p. 30.
[9] Vangapally, N., et al., Lead-acid batteries and lead–carbon hybrid systems: A review. *Journal of Power Sources*. 2023. 579: p. 233312.
[10] Njema, G.G., R.B.O. Ouma, and J.K. Kibet, A review on the recent advances in battery development and energy storage technologies. *Journal of Renewable Energy*. 2024. 2024(1): p. 2329261.
[11] Nasajpour-Esfahani, N., et al., Comprehensive review of lithium-ion battery materials and development challenges. *Renewable and Sustainable Energy Reviews*. 2024. 203: p. 114783.
[12] Garche, J., et al., *Lead-acid batteries for future automobiles*. 2017: Elsevier.
[13] Wong, Y. and C. Chan, Vehicle energy storage: Batteries, in electric, hybrid, and fuel cell vehicles. 2021: Springer. p. 293–313.
[14] Zeng, Y., et al., An ultrastable and high-performance flexible fiber-shaped Ni–Zn battery based on a Ni–NiO heterostructured nanosheet cathode. *Advanced Materials*. 2017. 29(44): p. 1702698.
[15] Chen, H., et al., Hierarchical micro-nano sheet arrays of nickel–cobalt double hydroxides for high-rate Ni–Zn batteries. *Advanced Sciences*. 2019. 6(8): p. 1802002.
[16] Placke, T., et al., Lithium ion, lithium metal, and alternative rechargeable battery technologies: The odyssey for high energy density. *Journal of Solid State Electrochemistry*. 2017. 21: p. 1939–1964.
[17] Brandt, K., Historical development of secondary lithium batteries. *Solid State Ionics*. 1994. 69(3–4): p. 173–183.
[18] Whittingham, M.S., Electrical energy storage and intercalation chemistry. *Science*. 1976. 192(4244): p. 1126–1127.
[19] Whittingham, M.S., Lithium batteries and cathode materials. *Chemical Reviews*. 2004. 104(10): p. 4271–4302.
[20] Mizushima, K., et al., LixCoO2 (0< x<-1): A new cathode material for batteries of high energy density. *Materials Research Bulletin*. 1980. 15(6): p. 783–789.
[21] Winter, M., et al., Carbonaceous and graphitic anodes. *Lithium Batteries: Science and Technology*. 2003: p. 145–194.
[22] Yu, T., et al., Nonprecious electrocatalysts for Li-air and Zn-air batteries: Fundamentals and recent advances. *IEEE Nanotechnology Magazine*. 2017. 11(3): p. 29–55.
[23] Wang, Q. Battery state of charge estimation based on multi-model fusion. in 2019 *Chinese Automation Congress (CAC)*. 2019: IEEE.
[24] Tan, W.K., et al., Current progress in the development of Fe-air batteries and their prospects for next-generation batteries. *Sustainable Materials for Next Generation Energy Devices*. 2021. 18(14): p. 59–83.
[25] Li, G., et al., Advanced intermediate temperature sodium–nickel chloride batteries with ultra-high energy density. *Nature Communications*. 2016. 7(1): p. 10683.
[26] Wei, S., et al., A stable room-temperature sodium–sulfur battery. *Nature Communications*. 2016. 7(1): p. 11722.
[27] Xu, X., et al., A room-temperature sodium–sulfur battery with high capacity and stable cycling performance. *Nature Communications*. 2018. 9(1): p. 3870.
[28] Huang, G., J. Wang, and X. Zhang, Electrode protection in high-efficiency Li–O2 batteries. *ACS Central Science* 2020. 6(12): p. 2136–2148.
[29] Deng, Z., et al., Sensitivity analysis and joint estimation of parameters and states for all-solid-state batteries. *Australian Business Law Review*. 2021. 7(3): p. 1314–1323.
[30] Chen, S., Q. Kuang, and H. Fan, Dual-carbon batteries: Materials and mechanism. *Small*. 2020. 16(40): p. 2002803.
[31] Chan, C.C. and K. Chau, *Modern electric vehicle technology*. 2001: Oxford University Press.

[32] Ghalkhani, M. and S. Habibi, Review of the Li-ion battery, thermal management, and AI-based battery management system for EV application. *Energies*. 2022. 16(1): p. 185.

[33] Kalghatgi, G., Is it really the end of internal combustion engines and petroleum in transport? *Applied Energy*. 2018. 225: p. 965–974.

[34] Koszela, W., et al., The combined effect of surface texturing and DLC coating on the functional properties of internal combustion engines. *Tribology International*. 2018. 127: p. 470–477.

[35] Waseem, M., et al., Battery technologies and functionality of battery management system for EVs: Current status, key challenges, and future prospectives. *Journal of Power Sources*. 2023. 580: p. 233349.

[36] Shim, J., et al., Electrochemical analysis for cycle performance and capacity fading of a lithium-ion battery cycled at elevated temperature. *Journal of Power Sources*. 2002. 112(1): p. 222–230.

[37] Liu, C., et al., Phase change materials application in battery thermal management system: A review. *Materials*. 2020. 13(20): p. 4622.

[38] Khan, Z., et al., A review of performance enhancement of PCM based latent heat storage system within the context of materials, thermal stability and compatibility. *Australian Business Law Review*. 2016. 115: p. 132–158.

[39] Huang, X., et al., Morphological characterization and applications of phase change materials in thermal energy storage: A review. *Renewable and Sustainable Energy Reviews*. 2017. 72: p. 128–145.

[40] Mohamed, S.A., et al., A review on current status and challenges of inorganic phase change materials for thermal energy storage systems. *Renewable and Sustainable Energy Reviews*. 2017. 70: p. 1072–1089.

[41] Goli, P., et al., Graphene-enhanced hybrid phase change materials for thermal management of Li-ion batteries. *Journal of Power Sources*. 2014. 248: p. 37–43.

[42] Wang, S., et al., Advanced ether-based electrolytes for lithium-ion batteries. *Advanced Energy Materials*. 2024. 14(37): p. 2401526.

[43] Das, D., S. Manna, and S. Puravankara, Electrolytes, additives and binders for NMC cathodes in Li-Ion batteries—A review. *Batteries*. 2023. 9(4): p. 193.

[44] Jiao, L., et al., An advanced lithium ion battery based on a high quality graphitic graphene anode and a Li [Ni0. 6Co0. 2Mn0. 2] O2 cathode. *Electrochimica Acta*. 2018. 259: p. 48–55.

[45] Chen, S., et al., Progress and future prospects of high-voltage and high-safety electrolytes in advanced lithium batteries: From liquid to solid electrolytes. *Journal of Materials Chemistry A*. 2018. 6(25): p. 11631–11663.

[46] Koch, V., et al., Specular lithium deposits from lithium hexafluoroarsenate/diethyl ether electrolytes. *Journal of the Electrochemical Society*. 1982. 129(1): p. 1.

[47] Myung, S.-T., et al., Electrochemical behavior of current collectors for lithium batteries in non-aqueous alkyl carbonate solution and surface analysis by ToF-SIMS. *Electrochimica Acta*. 2009. 55(1): p. 288–297.

[48] Zhang, X. and T.M. Devine, Identity of passive film formed on aluminum in Li-ion battery electrolytes with LiPF6. *Journal of the Electrochemical Society*. 2006. 153(9): p. B344.

[49] Wang, J., et al., Superconcentrated electrolytes for a high-voltage lithium-ion battery. *Nature Comunications*. 2016. 7(1): p. 1–9.

[50] Mai, S., et al., Tris (trimethylsilyl) phosphite as electrolyte additive for high voltage layered lithium nickel cobalt manganese oxide cathode of lithium ion battery. *Electrochimica Acta*. 2014. 147: p. 565–571.

[51] Aurbach, D., et al., On the use of vinylene carbonate (VC) as an additive to electrolyte solutions for Li-ion batteries. *Electrochimica Acta*. 2002. 47(9): p. 1423–1439.

[52] Liu, Y., et al., Understanding the improved high-temperature cycling stability of comprehensive analysis approach utilizing LC-MS and DART-MS. *Journal of Physics and Chemistry C* 2018. 122 : p. 5864–5870.

[53] Wang, D.Y., et al., A systematic study of electrolyte additives in Li [Ni1/3Mn1/3Co1/3] O2 (NMC)/graphite pouch cells. *Journal of the Electrochemical Society*. 2014. 161(12): p. A1818.

[54] Sinha, N.N., J. Burns, and J.R. Dahn, Comparative study of tris (trimethylsilyl) phosphate and tris (trimethylsilyl) phosphite as electrolyte additives for Li-ion cells. *Journal of the Electrochemical Society*. 2014. 161(6): p. A1084.

[55] Xia, J., et al., A comparative study of a family of sulfate electrolyte additives. *Journal of the Electrochemical Society.* 2013. 161(3): p. A264.
[56] Xia, J., et al., Comparative study on prop-1-ene-1, 3-sultone and vinylene carbonate as electrolyte additives for Li (Ni1/3Mn13Co1/3) O2/graphite pouch cells. *Journal of the Electrochemical Society.* 2014. 161(10): p. A1634.
[57] Madec, L., et al., Understanding the role of prop-1-ene-1, 3-sultone and vinylene carbonate in LiNi1/3Mn1/3Co1/3O2/graphite pouch cells: electrochemical, GC-MS and XPS analysis. *Journal of the Electrochemical Society.* 2015. 162(14): p. A2635.
[58] Lewandowski, A. and A. Świderska-Mocek, Ionic liquids as electrolytes for Li-ion batteries—An overview of electrochemical studies. *Journal of Power Sources.* 2009. 194(2): p. 601–609.
[59] Zheng, F., et al., Review on solid electrolytes for all-solid-state lithium-ion batteries. *Journal of Power Sources.* 2018. 389: p. 198–213.
[60] Kato, T., et al., Preparation of thick-film electrode-solid electrolyte composites on Li7La3Zr2O12 and their electrochemical properties. *Journal of Power Sources.* 2016. 303: p. 65–72.
[61] Gupta, H., et al., Development of polymer electrolyte and cathode material for Li-batteries. *Journal of the Electrochemical Society.* 2018. 166(3): p. A5187.
[62] Kaboli, S., et al., Behavior of solid electrolyte in Li-Polymer battery with NMC cathode via in-situ scanning electron microscopy. *Nano Letters.* 2020. 20(3): p. 1607–1613.
[63] Kobayashi, T., et al., Oxidation reaction of polyether-based material and its suppression in lithium rechargeable battery using 4 V class cathode, LiNi1/3Mn1/3Co1/3O2. *ACS Applied Materials & Interfaces.* 2013. 5(23): p. 12387–12393.
[64] Antony Jose, S., et al., Pathways to circular economy for electric vehicle batteries. *Recycling.* 2024. 9(5): p. 76.
[65] Masterson, V and Shine, I. What is the circular economy, and why does it matter that it is shrinking? World Economic Forum, Jun 14, 2022 (white paper). www.weforum.org/stories/2022/06/what-is-the-circular-economy.
[66] Pal, S.S., Energy Grid Optimization- AI & Digital Technologies for Improving Efficiency. 2024. www.cyient.com
[67] EV manufacturers looking to decarbonize—And profit from it. 2023. www.here.com/learn/blog/ev-battery-sustainability-recycling.
[68] The key minerals in an EV battery. https://elements.visualcapitalist.com/the-key-minerals-in-anev-battery/ (accessed on 10 August 2024).
[69] Policy challenges for accessing critical minerals for electric vehicles. www.rff.org/publications/reports/policy-challenges-for-accessing-critical-minerals-for-electric-vehicles/ (accessed on 10 August 2024).
[70] Vehicles., A.F.D.C.B.f.E. https://afdc.energy.gov/vehicles/electric-batteries (accessed on 6 August 2024).
[71] Zhang, R. and S. Fujimori, The role of transport electrification in global climate change mitigation scenarios. *Environmental Reasearch Letters.* 2020. 15(3): p. 034019.
[72] Zubi, G., et al., The lithium-ion battery: State of the art and future perspectives. *Renewable and Sustainable Energy Reviews.* 2018. 89: p. 292–308.
[73] Lai, X., et al., Critical review of life cycle assessment of lithium-ion batteries for electric vehicles: A lifespan perspective. *Etransportation.* 2022. 12: p. 100169.

14 Perovskites for Next-Generation Energy Storage Applications

Innovations and Challenges

Kiran M.S., Hareesha N., Yuvarajagouda P., Niyaz Ahamed M.B., and Veerabhadraswamy B.N.

14.1 INTRODUCTION

Energy is the basic need of mankind that ensures quality of life and is a key component to steer the modern economy. The current energy demand is at a high rate due to population growth and industrial expansions that have created an urgent need for alternative energy solutions. To meet potential demand, scientists have been looking for sustainable and more environmentally friendly ways to generate energy [1–3]. Fossil fuels, which have traditionally dominated the energy landscape, are not only limited but also major contributors to environmental greenhouse gases. The burning of coal, oil, and natural gas has led to an alarming rise in greenhouse gas emissions, accompanied by climate change and other damage to the ecosystem and human health [4,5]. As conventional resources are depleting, geopolitical instability with fluctuating prices further complicates their long-term viability. The world is turning to renewable energy sources such as solar, wind, and hydroelectric power to address the emerging energy crisis. However, the intermittent nature of renewables poses significant challenges to consistent energy supply, particularly when coupled with the limitations of existing energy storage technologies [6,7]. Addressing these challenges requires innovative materials that can enhance energy storage systems, ensuring reliability and efficiency in the transition to a sustainable energy future [8,9].

Current energy storage technologies, including lithium-ion batteries (LIBs) and supercapacitors, play a pivotal role in modern energy systems. LIBs are highly effective for portable electronics and electric applications due to their high energy density and long cycle life. However, they rely on lithium and cobalt materials, and their high cost in addition to environmental concerns related to extraction and recycling raise questions about their long-term sustainability. Similarly, supercapacitors offer exceptional power density and rapid charge–discharge cycles but suffer from low energy densities, limiting their utility in large-scale energy storage applications. Moreover, the high cost of electrode materials, such as activated carbon or metal oxides, poses an economic barrier to widespread adoption. These challenges have prompted researchers to search for alternative materials that can overcome limitations in desired specifications such as cost-effectiveness, efficiency, scalability, and environmental impact [10–14].

Among the available materials for next-generation energy materials, perovskites have gathered significant attention due to their excellent and versatile properties. Perovskites are a class of materials with the general formula ABX_3, where A and B are cations of different sizes, and X is an anion, generally oxygen or halide. Named after the mineral $CaTiO_3$ discovered in 1839 by Gustav Rose and later named after the Russian mineralogist Lev Perovski, perovskites display exceptional tunable physical and chemical properties due to their unique crystal structure [15–18]. Ease and

DOI: 10.1201/9781003596097-14

inexpensive production characteristics positioned perovskite materials as promising for diverse energy applications, including photovoltaics, electrocatalysis, and energy storage. Initially explored for their photovoltaic abilities, especially in solar cells, perovskites have achieved exceptional advancements in power conversion proficiency, exceeding 25% within a decade of active research. This rapid development can be attributed to their strong light absorption, larger carrier diffusion lengths, and high defect tolerance. Apart from photovoltaics, perovskites demonstrated substantial capabilities in energy storage applications, owing to their ionic conductivity, charge storage capacity, and electrochemical stability. Halide perovskites have shown capability in photo-rechargeable batteries and supercapacitors, integrating solar energy conversion and storage into a single device. Similarly, perovskite oxides have been utilized in LIBs and supercapacitors because of their excellent catalytic activity and structural stability [19,20]. The above advancements highlight the versatility of perovskites and their ability to address critical challenges in energy storage. The ability to tailor perovskite materials at the atomic ranges enables precise performance optimization for specific applications. Properties such as ionic conductivity, electronic band gap, and mechanical stability can be fine-tuned through manipulation of A, B, and X site composition in their crystal lattice. The incorporation of transition metals into perovskite oxides can enhance their catalytic activity, making them appropriate for oxygen reduction and evolution reactions in fuel cell applications. Likewise, halide perovskites can be engineered to achieve a balance between ionic mobility and structural integrity, critical for their performance in batteries and supercapacitors. Further, halide perovskites can be engineered to achieve a balance between ionic mobility and structural stability, which is essential for the effective performance of batteries and supercapacitors. Advancements in nanostructure technologies have further enriched the functionality of perovskites, facilitating the design of materials with high surface area, controlled porosity, and improved charge transport characteristics [21–23].

Despite their significant potential, the practical enactment of perovskites in energy storage systems is quite a challenge in terms of their stability, particularly for halide perovskites because of their sensitivity toward environmental factors such as moisture, temperature, and UV radiation [24,25]. Degradation under operational conditions can lead to a loss of performance and limit the long-term viability. Furthermore, the toxicity of certain perovskite constituents, markedly lead, raises ecological and health concerns, and necessitates the development of lead-free alternatives or effective recycling techniques [26]. Scalability of perovskite materials is also a critical issue, as it requires optimization of synthesis and processing of high-quality industrial production without conceding their performance or affordability. Overcoming these challenges will require multidisciplinary approaches, combining advanced materials science, engineering techniques, and environmental policies to explore the potential of perovskites as promising energy materials [27–30].

Recent advancements in the materials field have demonstrated the potential for addressing the barriers. Encapsulation of synthesis techniques and the development of hybrid perovskites ensured stability in operating conditions. Innovative strategies, like doping technology and interface optimization, have further enhanced the durability and efficiency of these materials [31,32]. Currently, research is also focused on the development of nonhazardous substitutes for lead-based perovskites by employing tin or mixed-cation modifications, although these materials entail optimization to achieve comparable performance. Moreover, scalable synthesis methods are being developed to facilitate the commercialization of perovskite-based devices. Improvements in device design, such as tandem structures and integrated storage systems, are contributing to practical viability [33,34]. This chapter discusses the potential advancements in perovskite materials, highlighting the abundance of novel mixtures dependent on A and B site inhabitancy, which leads to a wide scope of materials frameworks with interesting properties and wide applications. It also delves into recent progress in the production of these materials and examines the challenges currently faced by researchers in cutting-edge perovskite material creation to solve the global energy crisis, paving the way for a cleaner and more sustainable future.

14.2 SPECIAL PROPERTIES OF PEROVSKITE MATERIALS

Perovskite materials have gained substantial attention in recent years due to their distinctive properties, which make them applicable for several energy storage applications. This section explores the characteristic properties of perovskite materials, focusing on their tunable band gaps, ionic conductivity (IC), electron mobility (EM), charge storage capacity, and cyclic stability.

14.2.1 Tunable Band Gaps

The band gaps of perovskites can be frequently amended through composition engineering. The "A, B, and X sites" have marked impacts on the band gap of perovskites. For illustration, the band gap for FAPbI3, MAPbI3, and CsPbI3 materials is 1.48, 1.55, and 1.73 eV, respectively. That is, as the ionic radius of the A-site declines, the band gap increases. This is due to the modulation of the Pb-I bond arising from the size impacts of A-site cations [35]. Replacing Pb with Sn will result in a distinctive reduction in the band gap, creating Pb–Sn mixed perovskite excellent rear sub-cells for all-perovskite tandem solar cells (Figure 14.1) [36,37]. The X-site regulation is a widely utilized concept for altering the perovskite material's band gaps. As the X (halogen) ions in MAPbX3 alter from I to Br and Br to Cl, the perovskite band gap will rise from 1.55 to 1.8 and 1.8 to 2.3 eV [38]. Mainly, as documented in prior literature, the MAPbI3-xClx material delivers an equivalent band gap with MAPbI3 material. The rationale for the nominal impact of Cl doping could be the small portion of Cl in the lattice [39]. Fascinatingly, in all-inorganic perovskites or mixed organic–inorganic perovskites rich in bromine ("Br"), chlorine ("Cl") can occupy the "X sites" and broaden the band gap. This is likely because "Cl" can substitute for "Br" but not for iodine ("I"), due to differences in ionic radii [40]. Commonly, for hybrid perovskite materials with mixed halides and diverse "A" and "B" cations, the "X-site" plays a major role in the band gap. For instance, the wide band gap organic and inorganic mixture perovskite materials are generally received via heavy doping of Br. Indeed, if we set the band gap and decrease the content of Br, the content of MA or Cs should be boosted in CsMAFAPbI3-xBrx. In complement to composition, the extension of the crystal structure also dramatically impacts the perovskite materials band gap [41]. It should be remarked that the material band gap rises as the structural extension declines [42]. For instance, in quasi-2D perovskite materials, the band gap grows as the "n" value declines [43].

14.2.2 Ionic Conductivity and Electron Mobility

Perovskites have achieved substantial consideration in the scope of energy storage and production, specifically due to their notable IC and EM, which are essential for improving battery functioning. Current findings have emphasized how these properties can be boosted to durability, total energy storage proficiency, and enhanced battery efficacy.

14.2.3 Ionic Conductivity

Perovskites have appeared as favorable contenders for improving IC in battery functions, specifically in Li-ion and Na-ion systems. Their exclusive crystal structure (ABX_3), permits for substantial tunability in structure and properties, compelling them appropriate for several energy storage applications.

In recent years, the lithium lanthanum titanate perovskite-type materials have been intensively studied due to their elevated Li ion bulk conductivity at room temperature, getting up to 10^{-3} S/cm [44] due to the high concentration of "A site" vacancies that enable Li cation migration via back-ups [45]. Among perovskite materials, $Li_{0.34}La_{0.56}TiO_3$ has been recognized as a material with high IC with a bulk IC of 10^{-3} S/cm [46,47]. Similarly, the $Li_3/8Sr_7/16Ta_3/4Zr_1/4O_3$ perovskite system,

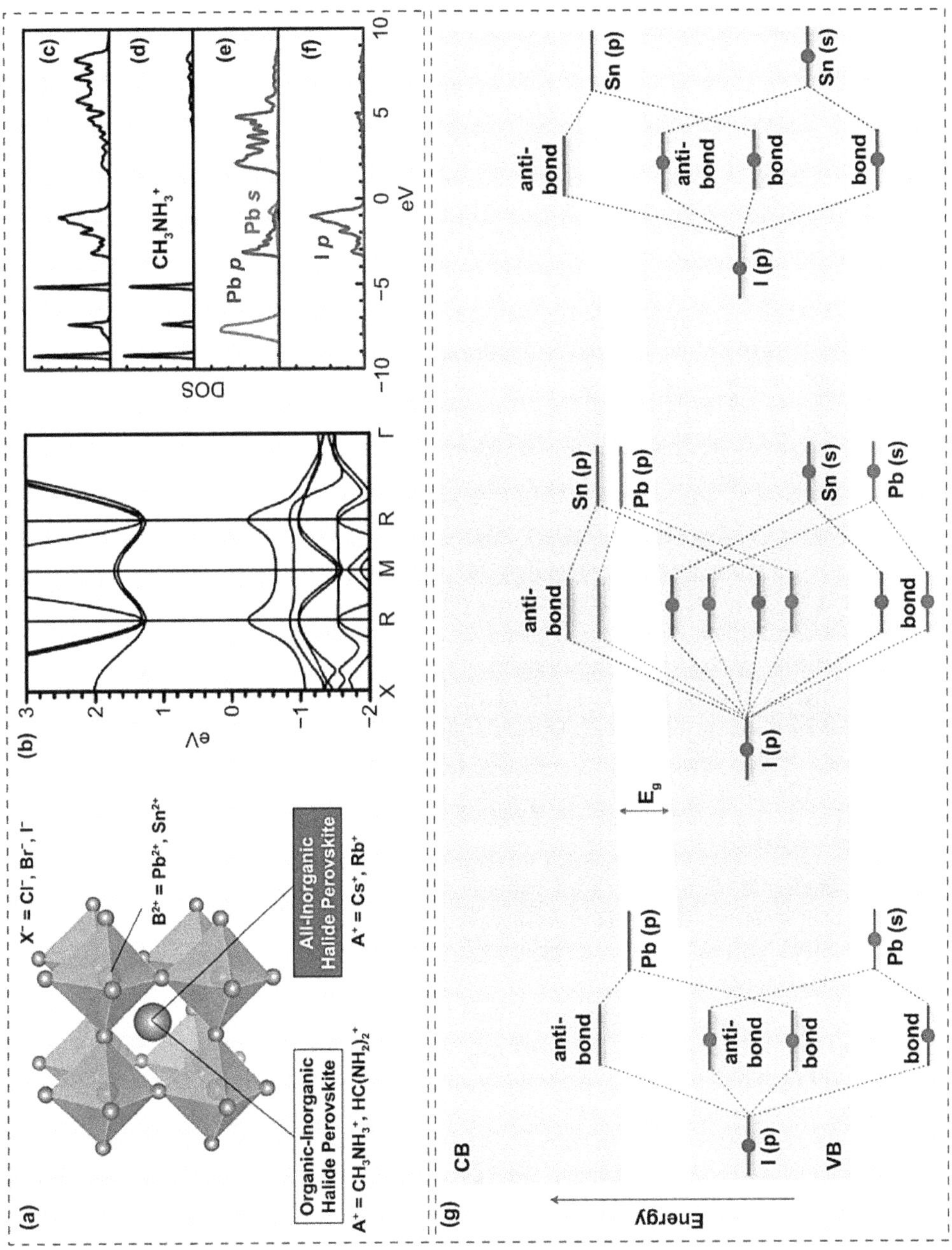

FIGURE 14.1 Schematic of origin of the band gap bowing in $MA(Pb_{1-x}Sn_x)I_3$ [36].

TABLE 14.1
Some Examples of Solid-State Electrolytes Connecting to Ionic Conductivities

Solid-state electrolyte	IC (S/cm)	Reference
Li-I	1.0×10^{-7}	[51]
Li-β-alumina	1.3×10^{-4}	[51]
Na-β-alumina	1.4×10^{-2}	[51]
$Li_7La_3Zr_2O_{12}$	7.7×10^{-4}	[51]
$Na_3Zr_2PSi_2O_{12}$	0.45	[51]
$La_{0.55}Li_{0.36}TiO_3$	1.5×10^{-3}	[51]
$Na_2Zn_2TeO_6$	6.29×10^{-4}	[52]
$Na_{1.9}Zn1.9Ga_{0.1}TeO_6$	1.09×10^{-3}	[52]
$Na2LiFeTeO_6$	4.0×10^{-2}	[53]
$Li_{6.4}La_3Zr_{1.4}Ta_{0.6}O_{12}$	1.0×10^{-3}	[54]
$Li_{1.3}Al_{0.3}Ti_{1.7}(PO_4)_3$	1×10^{-2}	[55]
Na_3PS_4	2.0×10^{-4}	[56]
$Na_{2.9}Sb_{0.9}W_{0.1}S_4$	4.1×10^{-2}	[56]
$Na_2(CB_9H_{10})(CB_{11}H_{12})$	7.0×10^{-2}	[57]
$Li_{5.4}Al_{0.2}PS_5Br$	2.5×10^{-3}	[58]

which has the identical configuration as $SrZrO_3$, has been reported to demonstrate a high Li ion conductivity of 4.1×10^{-4} S/cm at room temperature [48,49]. Unlike the lithium lanthanum titanate, $Na_{0.33}La_{0.55}ZrO_3$ and $Na_{1/3}La_1/3Sr_{1/3}ZrO_3$ encompass Zr, Sr, and La, which are further stable in contrast to Ti ions found in lithium lanthanum titanate. As a result, $Na_{0.33}La_{0.55}ZrO_3$ and $Na_{1/3}La_1/3Sr_{1/3}ZrO_3$ may have a better extent of chemical stability in permutation with Na-metal [50]. These materials present fine ionic conductivities. The perovskite material $Na_{1/3}La_1/3Sr_{1/3}ZrO_3$ shows an elevated IC of 1.025×10^{-5} S/cm at room temperature [50]. Some examples of solid-state electrolytes connecting to ionic conductivities are shown in Table 14.1.

14.2.4 Electron Mobility

Perovskites are essential for Na-ion and Li-ion batteries because of their proficient electron mobility and distinctive crystal structure. Their 3D planning of corner-sharing octahedra enables quick electron and ion movement, portraying them as ideal for high-performance battery functions, particularly in Na-ion and Li-ion batteries.

$La_{0.5}Li_{0.5}TiO_3$ is a promising anode material for Li-ion batteries due to its high specific capacity and excellent cycling stability. Its low band gap of 2.1 eV allows efficient electron transport during lithiation and delithiation processes. Additionally, its ability to undergo a reversible phase transition enhances its electrochemical performance, allowing rapid Li^+ diffusion pathways, and supporting high-rate capabilities during battery operation [59]. Also, $Na_{0.5}Bi_{0.5}TiO_3$ (Figure 14.2) is a perovskite framework in sodium bismuth titanate, which has been explored as a promising anode material for Li-ion batteries due to its structural stability and favorable electronic properties. The presence of oxygen vacancies in the perovskite lattice improves ionic conductivity, facilitating faster charge transport and electron mobility during cycling [60,61]. Further, metal halide perovskites are promising for Na-ion and Li-ion batteries due to their high charge carrier mobility and long diffusion lengths. These materials enable reversible electrochemical intercalation of Li and Na ions, enhancing their conductivity through n-type doping. Li doping in metal halide perovskites

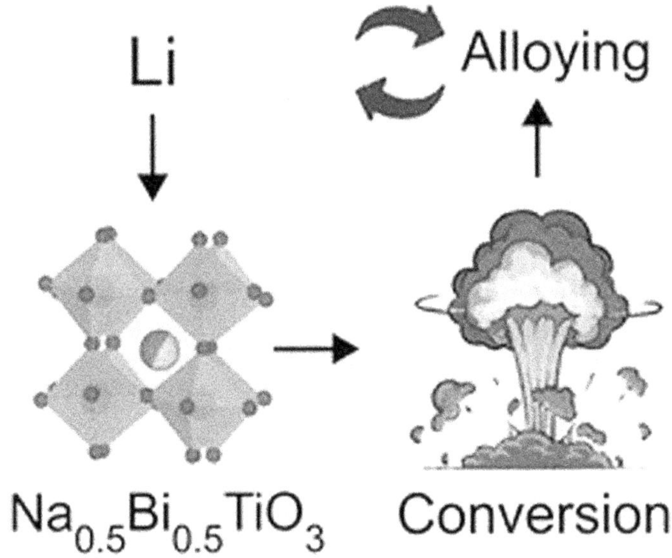

FIGURE 14.2 Pictorial representation of $Na_{0.5}Bi_{0.5}TiO_3$ perovskite material [61].

improves electron mobility and electrochemical performance, enabling higher capacity retention during cycling [62].

14.2.5 Charge Storage Capacity and Cyclic Stability

Perovskites have developed as capable contenders for improving charge storage capacity with good cyclic stability in Li and Na ion batteries. Their exclusive structural properties permit high IC and EM, critical for effective energy storage. In Li-ion batteries, perovskite-based oxide materials like $La_{0.5}Li_{0.5}TiO_3$ demonstrate a high specific capacity of 225 mAh/g and demonstrate exceptional cycling stability, keeping about 79% capacity after 3000 cycles [59]. Further, halide-based perovskite materials have exhibited capacities of 153 mAh/g for 3-D forms and 149 mAh/g for 2-D structures, representing their ability in energy storage systems [63]. For Na-ion batteries, perovskites like $Na_{0.67}Mn_{0.95}Mg_{0.05}O_2$ can serve capacities up to 175 mAh/g, showcasing analogous functioning to commercial Li-ion systems. Furthermore, $Na_{0.5}Bi_{0.5}TiO_3$ material shows capacities of around 220 mAh/g with good cycling stability [61]. The capability to combine several elements in the perovskite structure allows for improved ionic conductivity, charge transfer, tunability, and functioning, crucial for next-generation battery applications.

14.3 APPLICATION OF PEROVSKITES IN BATTERY TECHNOLOGIES

14.3.1 Lithium-Ion Batteries

The growing demand for high-performance LIBs across various applications [64], including portable electronics, electric vehicles, and grid storage, drives the need for innovative battery materials. Research efforts focus on optimizing both organic and inorganic materials for different battery components, with the anode being a key area of development. Graphite and lithium titanate ($Li_4Ti_5O_{12}$, LTO) are among the most widely used anode materials. LTO exhibits excellent high-rate capability but has a higher working voltage (1.55 V vs. Li/Li^+), which reduces energy density in full-cell configurations. Graphite, offering a theoretical capacity of 372 mA h g^{-1}, operates at

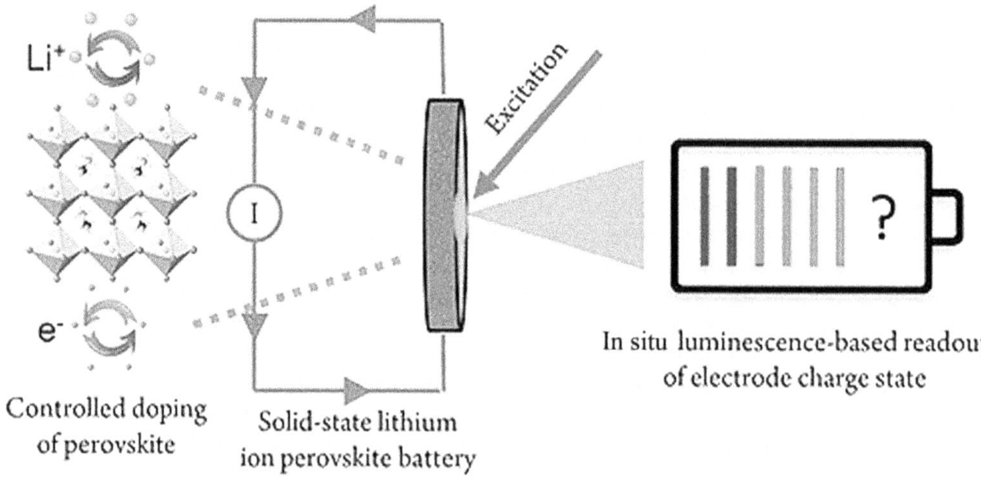

FIGURE 14.3 Schematic representation of the perovskite-based Li-ion battery application [67].

a low potential (0.2 V vs. Li/Li^+), increasing the risk of lithium plating at high currents, which compromises safety. To overcome these limitations, the search for advanced anode materials continues, aiming to achieve a balance between power density, safety, and energy efficiency in next-generation LIBs.

The increasing demand for high-performance LIBs calls for advanced anode materials with improved capacity, stability, and rate capability. High-entropy perovskite metal fluorides (HEPMFs) show great potential due to their unique pseudocapacitive properties, providing high specific capacity and excellent rate performance [65]. However, traditional solvothermal synthesis methods hinder large-scale production, restricting their practical use. High-entropy perovskite metal fluoride (HEPMF anode materials were synthesized using different methods, demonstrating promising electrochemical properties for LIBs. $K_{0.65}Li_{0.07}Mg_{0.19}Mn_{0.17}Co_{0.16}Ni_{0.17}Cu_{0.24}F_{2.70}$ (KLMF) was prepared via a direct liquid-phase method using spent $LiNi_{1/3}Co_{1/3}Mn_{1/3}O_2$/graphite (NCM111/G) [66] batteries as the primary precursors, offering a sustainable recycling approach for electrode materials. Meanwhile, $K_{0.9}(Mg_{0.2}Mn_{0.2}Co_{0.2}Ni_{0.2}Cu_{0.2})F_{2.9}$ was synthesized using a one-pot solution-based method, expanding the synthetic strategies for high-entropy perovskite fluorides. The schematic representation of the perovskite-based LIB application is shown in Figure 14.3.

The electrochemical evaluation revealed excellent rate capability and cycling stability for both materials. KLMF delivered a remarkable specific capacity of 1043.1 mAh/g at 0.1 A/g, maintaining 477.95 mAh/g at 6.4 A/g after 500 cycles at 2 A/g. Similarly, the $K_{0.9}(Mg_{0.2}Mn_{0.2}Co_{0.2}Ni_{0.2}Cu_{0.2})F_{2.9}$ anode demonstrated high rate performance, achieving a specific capacity ranging from 698 mAh/g at 0.1 A/g to 311 mAh/g at 3.2 A/g. These impressive electrochemical properties were attributed to the high density of vacancy defects, which enhanced ion transport by increasing the specific surface area and diffusion pathways. Such materials exhibited stable performance in full-cell configurations, demonstrating their viability for real-world LIB applications [61]. Their high-capacity retention and superior rate capabilities position HEPMFs as strong candidates for next-generation energy storage systems. The combination of scalable synthesis methods, structural stability, and enhanced ion transport properties highlights the potential of HEPMF-based anodes for high-performance LIBs, paving the way for their practical implementation in advanced battery technologies.

14.3.2 Sodium-Ion Batteries

Sodium-ion batteries (SIBs) have gained attention as an alternative to LIBs due to the abundance and low cost of sodium resources. However, the development of efficient anode materials remains a challenge. Perovskite-based materials have emerged as promising candidates for SIB anodes due to their unique crystal structures, high sodium storage capacity, and stable electrochemical performance [68]. This chapter explores the electrochemical properties and applications of perovskite materials such as $Na_{0.5}Bi_{0.5}TiO_3$ (NBTO) and $NaNiF_3$, and their potential applications as anodes for SIBs.

The perovskite $Na_{0.5}Bi_{0.5}TiO_3$ (NBTO) [69] was studied for its sodium storage potential. Electrochemical analysis confirmed its alloying/de-alloying reaction mechanism, enabling fast sodium ion kinetics and stable cycling. NBTO exhibited a high specific capacity of ~470 mAh g^{-1} at 100 mA g^{-1} and retained 230 mAh g^{-1} at 250 mA g^{-1}. After 50 cycles at 100 mA g^{-1}, it maintained ~215 mAh g^{-1}, making it competitive with various metal oxide anodes for SIBs. Additionally, nanostructured $NaNiF_3$ [70] perovskite was investigated as a SIB anode (Figure 14.4). Optimized synthesis using trisodium citrate dehydration and microwave heating yielded a favorable morphology. The $NaNiF_3$ electrode exhibited an initial discharge capacity of ~376 mAh g^{-1}, demonstrating promise for high-performance sodium storage.

The results show that perovskite materials, especially $Na_{0.5}Bi_{0.5}TiO_3$ and $NaNiF_3$, have excellent electrochemical potential as SIB anodes [71]. Their high capacity, stability, and efficient sodium storage make them strong candidates for next-generation SIBs. Further material optimization and electrode design could improve their performance for practical energy storage applications.

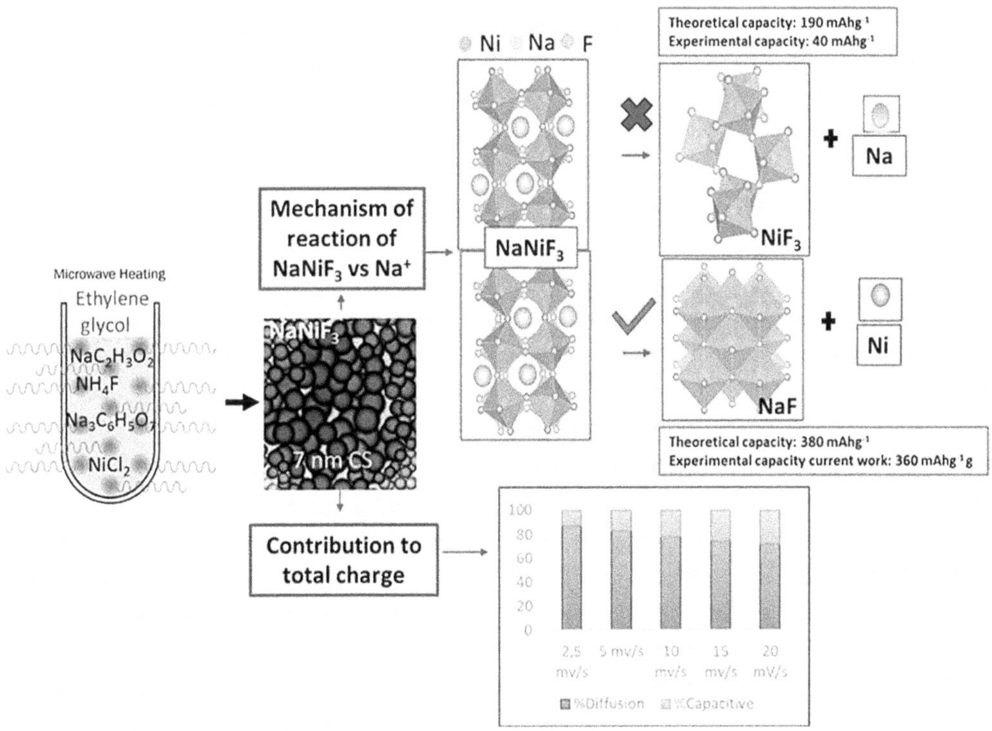

FIGURE 14.4 Scheme of perovskite in an Na-ion battery application [70].

14.4 ENHANCEMENT OF BATTERY PERFORMANCE

Energy Density: The energy density is a critical factor in battery performance, determining the amount of energy stored per unit mass or volume. Perovskite-based materials contribute to enhanced energy density due to their high electrochemical potential and superior charge storage capabilities. The ability to incorporate multi-valent cations in perovskite structures allows for higher capacity retention and increased volumetric energy density, making them attractive for electric vehicles and portable electronics.

Efficiency: Perovskites improve battery efficiency by reducing internal resistance and enhancing the ion transport pathways. The high ionic mobility in perovskite electrolytes ensures rapid charge transfer, reducing energy losses and improving overall cycle efficiency. Additionally, perovskite materials demonstrate lower polarization, leading to better charge utilization and minimal energy dissipation. Efficiency in battery systems refers to the ratio of the energy output to the energy input during charging and discharging cycles. Perovskite materials contribute to improved efficiency through enhanced ionic conductivity and favorable electrochemical properties:

Solid-State Electrolytes: Perovskite-based solid electrolytes, such as Li3xLa (2/3−x)TiO_3 (LLTO), exhibit high lithium-ion conductivity (~10^{-3} S/cm). This high conductivity reduces internal resistance, leading to higher coulombic efficiency in solid-state batteries.

Electrode Materials: The structural tunability of perovskites allows for the optimization of electrode materials, resulting in improved charge/discharge rates and overall energy efficiency.

Stability: Stability is a crucial factor determining the lifespan and safety of battery systems. Perovskite materials have shown both advantages and challenges in this regard.

Thermal Stability: Perovskite-based solid electrolytes offer superior thermal stability compared to conventional liquid electrolytes, enhancing the safety profile of batteries.

Chemical Stability: Some perovskite materials are prone to degradation upon exposure to moisture and oxygen, which can adversely affect battery performance. However, ongoing research focuses on developing more stable perovskite compositions and protective coatings to mitigate these issues.

14.5 INNOVATIONS AND BREAKTHROUGHS

Perovskite materials have gathered significant attention in energy storage and conversion technologies due to their exceptional optoelectronic properties and structural versatility. Recent innovations have focused on enhancing their performance and stability through various strategies, including structural engineering, surface modification, integration of hybrid perovskites, and approaches to extend battery lifespan.

14.5.1 Structural Engineering of Perovskite Materials

Structural engineering involves deliberate manipulation of the perovskite crystal lattice to optimize its properties for specific applications. This can be achieved through techniques such as doping, dimensionality control, and strain engineering.

Doping: Introducing foreign ions into the perovskite lattice can tailor electronic properties and improve stability. For instance, substituting lead with tin or germanium has been explored to reduce toxicity and enhance environmental stability.

Dimensionality Control: Adjusting the dimensionality from three-dimensional (3D) to two-dimensional (2D) or quasi-2D structures can enhance moisture resistance and thermal stability.

2D perovskites, with their natural quantum well structures, have shown improved exciton binding energies and environmental stability.

Strain Engineering: Applying mechanical strain to the perovskite lattice can modulate its band gap and carrier mobility, leading to enhanced photovoltaic performance. This approach has been utilized to fine-tune the optoelectronic properties without altering the material composition.

14.5.2 Surface Modification Techniques

Surface defects in perovskite materials often act as non-radiative recombination centers, adversely affecting device performance and stability. Surface modification techniques aim to passivate these defects and improve interfacial properties.

Passivation Layers: Applying thin passivation layers using materials such as phenethylammonium iodide (PEAI) or other organic molecules can effectively reduce surface trap states. This approach has been shown to enhance power conversion efficiency and operational stability in perovskite solar cells.

Self-Assembled Monolayers (SAMs): SAMs can be used to modify the surface energy and electronic properties of perovskite films, leading to improved charge extraction and reduced hysteresis in devices.

Interface Engineering: Optimizing the interfaces between perovskite layers and charge transport layers can minimize energy barriers and suppress interfacial recombination. Techniques such as inserting buffer layers or using graded compositions have been employed to achieve this.

14.5.3 Integration of Hybrid Perovskites for Improved Electrochemical Performance

Hybrid organic–inorganic perovskites (HOIPs) combine the advantageous properties of both organic and inorganic components, offering tunable band gaps, high absorption coefficients, and flexibility in processing.

Tandem Structures: Integrating HOIPs with other materials in tandem configurations can broaden the absorption spectrum and enhance overall device efficiency. For example, combining perovskites with silicon or other semiconductors has led to record-breaking solar cell efficiencies.

Composite Electrodes: Incorporating perovskite materials into composite electrodes with carbon-based materials or polymers can improve conductivity and mechanical stability, which are critical for battery applications.

Layered Architectures: Designing layered structures with alternating organic and inorganic components can enhance ion transport and structural integrity, leading to better electrochemical performance in devices such as supercapacitors and batteries.

14.5.4 Strategies for Extending Battery Lifespan

Enhancing the operational lifespan of perovskite-based batteries involves addressing degradation mechanisms and improving material stability.

Encapsulation: Protecting perovskite materials from environmental factors such as moisture and oxygen through encapsulation techniques can significantly extend device longevity. Materials like graphene or polymer coatings have been used for this purpose.

Additive Engineering: Incorporating additives into the perovskite precursor solution can improve film formation and passivate defects, leading to enhanced stability and longer cycle life in batteries.
Thermal Management: Implementing strategies to manage heat generation and dissipation within the device can prevent thermal degradation of perovskite materials, thereby extending battery lifespan.

14.6 CHALLENGES IN PEROVSKITE-BASED BATTERIES

However, several challenges impede their commercialization and widespread adoption. This section delves into key issues such as material degradation and environmental stability, toxicity concerns, defect formation, and cost and scalability issues, supported by legitimate references.

14.6.1 Material Degradation and Environmental Stability

Perovskite materials are notably sensitive to environmental factors like moisture, oxygen, light, and heat, leading to rapid degradation. Exposure to moisture and oxygen can cause the decomposition of perovskite structures, resulting in diminished battery performance. Light and thermal stress further exacerbate this degradation, affecting the long-term stability of perovskite-based devices. Encapsulation techniques have been employed to mitigate these effects; however, they often add complexity and cost to the manufacturing process. Compositional and interfacial engineering strategies are being explored to enhance stability, but achieving long-term durability remains a significant hurdle [72].

14.6.2 Toxicity Concerns

The incorporation of lead in many perovskite compositions raises substantial environmental and health concerns due to its toxicity. Lead can leach into the environment during the production, usage, or disposal of perovskite-based batteries, posing risks to ecosystems and human health. To address this, research is focused on developing lead-free perovskite materials by substituting lead with less toxic elements such as tin or germanium. However, these alternatives often suffer from lower stability and efficiency, presenting a trade-off between reducing toxicity and maintaining performance [73].

14.6.3 Formation of Defects

Defects in perovskite materials, including vacancies, interstitials, and grain boundaries, can significantly impact their electronic properties and, consequently, battery performance. These defects can act as non-radiative recombination centers, reducing charge carrier lifetimes and increasing energy losses. Moreover, defects can facilitate ion migration, leading to further degradation of the material. Strategies such as defect passivation through chemical treatments and optimizing fabrication processes are being investigated to minimize defect densities and enhance material quality [72].

14.6.4 Cost and Scalability Issues

While perovskite materials offer the advantage of low-cost raw materials, challenges in manufacturing processes hinder their scalability. Achieving uniform and high-quality perovskite films over large areas is difficult, often requiring precise control over processing conditions. Additionally, the need for encapsulation to protect against environmental degradation adds to production costs.

Innovations in material synthesis, processing techniques, and supply chain management are essential to drive down production costs and improve the competitiveness of perovskite-based batteries in the renewable energy market [72]. Ongoing research aims to develop more stable, non-toxic, and defect-free perovskite materials, alongside scalable and cost-effective manufacturing processes, to realize the full potential of perovskite-based energy storage systems.

14.7 FUTURE PERSPECTIVES AND RESEARCH TRENDS

14.7.1 Outline of Ongoing Research Efforts

Current research in perovskite-based energy solutions is multi-dimensional, focusing on enhancing material properties, device performance, and scalability:

Material Optimization: Efforts are directed toward improving the intrinsic properties of perovskite materials, such as ionic conductivity and structural stability. This includes compositional engineering to develop lead-free perovskites, aiming to mitigate toxicity concerns while maintaining high performance [74].

***Stability Enhancement*:** Researchers are investigating methods to improve the environmental stability of perovskite materials. Strategies include surface passive action, incorporation of hydrophobic layers, and development of robust encapsulation techniques to protect against moisture and oxygen degradation.

***Scalability and Manufacturing*:** Scaling up perovskite-based devices from laboratory-scale prototypes to commercial products is a significant focus. This involves developing scalable fabrication techniques, such as roll-to-roll printing and vapor deposition methods, to produce uniform, high-quality perovskite films over large areas [75].

***Integration into Tandem Devices*:** Combining perovskite materials with traditional photovoltaic technologies, such as silicon, to form tandem solar cells is an area of active research. This approach aims to surpass the efficiency limits of single-junction cells by utilizing the complementary absorption spectra of different materials.

14.7.2 Exploration of Critical Challenges and Possible Solutions

Despite the promising advancements, several challenges persist in the development of perovskite-based energy solutions:

***Environmental Stability*:** Perovskite materials are prone to degradation under environmental stressors like moisture, oxygen, and heat. To address this, researchers are exploring the use of encapsulation techniques and the development of more stable perovskite compositions.

***Toxicity Concerns*:** The presence of lead in many perovskite formulations raises environmental and health concerns. Research is ongoing to develop lead-free perovskite materials by substituting lead with less toxic elements, such as tin or germanium, though these alternatives currently face challenges related to stability and performance [74].

***Defect Formation*:** Defects in perovskite crystals can act as non-radiative recombination centers, adversely affecting device performance. Strategies to mitigate this issue include defect passivation through chemical treatments and optimizing fabrication processes to reduce defect densities [76].

***Scalability and Manufacturing Challenges*:** Transitioning from laboratory-scale fabrication to large-scale manufacturing presents challenges in maintaining material quality and device performance. Developing scalable and cost-effective manufacturing processes is crucial for commercial viability [75].

14.7.3 Vision for Sustainable, High-Performance Energy Solutions Using Perovskites

The future of perovskite-based energy solutions is promising, with the potential to revolutionize the renewable energy landscape:

***High-Efficiency Devices*:** Continued research into material optimization and device engineering is expected to yield perovskite-based devices with efficiencies surpassing current technologies. For instance, perovskite-silicon tandem cells have already demonstrated efficiencies exceeding 28%, indicating a pathway toward more efficient energy conversion.

***Sustainable Materials*:** The development of lead-free perovskite materials will address toxicity concerns, contributing to more environmentally friendly energy solutions. Additionally, the use of abundant and non-toxic elements will enhance the sustainability of perovskite-based devices [74].

***Scalable Manufacturing*:** Advancements in scalable fabrication techniques will facilitate the mass production of perovskite-based devices, making them more accessible and affordable. This includes the development of roll-to-roll printing and other high-throughput manufacturing processes [75].

***Integration into Diverse Applications*:** Beyond traditional photovoltaic applications, perovskite materials are being explored for use in various energy storage and conversion devices, including batteries and capacitors, due to their tunable electronic properties and high ionic conductivity [4].

14.8 CONCLUSIONS

Perovskite materials are emerging as promising candidates for next-generation battery technologies due to their exceptional electrochemical properties, tunable electronic structures, and material versatility. Their applications in LIBs, SIBs, and photo-rechargeable batteries highlight their potential to revolutionize energy storage with advantages such as high ionic conductivity, lightweight structures, and scalable manufacturing compatibility. However, challenges such as environmental instability, material degradation, and moisture sensitivity must be addressed through advancements in encapsulation and compositional engineering. Additionally, concerns over lead toxicity necessitate the development of eco-friendly alternatives, while defect formation and scalability issues require innovative fabrication and defect-passivation strategies. Future research should focus on refining perovskite compositions, optimizing device architectures, and integrating them with solid-state and hybrid battery technologies to enhance energy density, efficiency, and longevity. Collaboration among academia, industry, and policymakers is crucial for transitioning perovskite-based batteries from laboratory research to large-scale commercialization. While challenges remain, continuous advancements in materials science and engineering indicate a promising future where perovskites could significantly contribute to sustainable and high-performance battery solutions.

REFERENCES

[1] Yang, L. Kong, X. Li, F. Hao, H. Cheng, Z. Liu, H. Li, J.F. and Zhang, S. 2019. Perovskite lead-free dielectrics for energy storage applications. *Prog. Mater. Sci.* 102:72–108.

[2] Kostopoulou, A. Brintakis, K. Nasikas, N.K. and Stratakis, E. 2019. Perovskite nanocrystals for energy conversion and storage. *Nanophotonics*. 8(10):1607–1640.

[3] Xu, X. Su, C. and Shao, Z. 2021. Fundamental understanding and application of Ba0. 5Sr0. 5Co0. 8Fe0. 2O3− δ perovskite in energy storage and conversion: Past, present, and future. *Energy & Fuels*. 35(17):13585–13609.

[4] Veerapandiyan, V. Benes, F. Gindel, T. and Deluca, M. 2020. Strategies to improve the energy storage properties of perovskite lead-free relaxor ferroelectrics: A review. *Materials*. 13(24):5742.

[5] Liang, J. Zhu, G. Lu, Z. Zhao, P. Wang, C. Ma, Y. Xu, Z. Wang, Y. Hu, Y. Ma, L. and Chen, T. 2018. Integrated perovskite solar capacitors with high energy conversion efficiency and fast photo-charging rate. *J. Mater. Chem. A*. 6(5):2047–2052.

[6] Jafarizadeh, H. Yamini, E. Zolfaghari, S.M. Esmaeilion, F. Assad, M.E.H. and Soltani, M. 2024. Navigating challenges in large-scale renewable energy storage: Barriers, solutions, and innovations. *Energy Rep*. 12:2179–2192.

[7] Shah, A.U.I. and Meyer, E.L. 2025. Perovskite-based solar cells in photovoltaics for commercial scalability: Current progress, challenges, mitigations and future prospectus. *Solar Energy*. 286:113172.

[8] Dastgeer, G. Nisar, S. Zulfiqar, M.W. Eom, J. Imran, M. and Akbar, K. 2024. A review on recent progress and challenges in high-efficiency perovskite solar cells. *Nano Energy.* 132 (110401): 1–19.

[9] Alao, K.T. Alao, T.O. Adebanjo, A.U. and Alara, O.R. 2025. Potential uses of perovskite-based photovoltaics for hydrogen production: A pathway to sustainable energy solutions. *Next Energy*. 7:100215.

[10] Gaurav, A. Das, A. Paul, A. Jain, A. Boruah, B.D. and Abdi-Jalebi, M. 2024. Could halide perovskites revolutionalise batteries and supercapacitors: A leap in energy storage. *J. Energy Storage*. 88:111468.

[11] Han, B. Zhao, J. Luo, Z. Cai, F. Yuan, Z. and Zeng, H. 2023. Energy storage research of metal halide perovskites for rechargeable batteries. *Nano Energy*. 155:108646.

[12] Monama, G.R. Ramohlola, K.E. Iwuoha, E.I. and Modibane, K.D. 2022. Progress on perovskite materials for energy application. *Results Chem*. 4:100321.

[13] Monama, G.R. Modibane, K.D. Ramohlola, K.E. Molapo, K.M. Hato, M.J. Makhafola, M.D. Mashao, G. Mdluli, S.B. and Iwuoha, E.I. 2019. Copper (II) phthalocyanine/metal organic framework electrocatalyst for hydrogen evolution reaction application. *Int. J. Hydrogen Energy*. 44(34):18891–18902.

[14] Dehghani-Sanij, A.R. Tharumalingam, E. Dusseault, M.B. and Fraser, R. 2019. Study of energy storage systems and environmental challenges of batteries. *Renew. Sust. Energy Rev*. 104:192–208.

[15] Hara, M. Nakano, H. Dokko, K. Okuda, S. Kaeriyama, A. and Kanamura, K. 2009. Fabrication of all solid-state lithium-ion batteries with three-dimensionally ordered composite electrode consisting of Li0. 35La0. 55TiO3 and LiMn2O4. *J. Power Sources*. 189(1):485–489.

[16] Zhang, L. Mei, L. Wang, K. Lv, Y. Zhang, S. Lian, Y. Liu, X. Ma, Z. Xiao, G. Liu, Q. and Zhai, S. 2023. Advances in the application of perovskite materials. *Nano-Micro Letters*. 15(1):177.

[17] Jia, Z. Cheng, C. Chen, X. Liu, L. Ding, R. Ye, J. Wang, J. Fu, L. Cheng, Y. and Wu, Y. 2023. Applications of all-inorganic perovskites for energy storage. *Mater. Adv.* 4(1):79–104.

[18] Zhang, L. Miao, J. Li, J. and Li, Q. 2020. Halide perovskite materials for energy storage applications. *Adv. Funct. Mater*.30(40):2003653.

[19] Kumar, R. Kumar, A. Shukla, P.S. Varma, G.D. Venkataraman, D. and Bag, M. 2022. Photorechargeable hybrid halide perovskite supercapacitors. *ACS Appl. Mater. Interfaces*.14(31):35592–35599.

[20] Gaurav, A. Das, A. Paul, A. Jain, A. Boruah, B.D. and Abdi-Jalebi, M. 2024. Could halide perovskites revolutionalise batteries and supercapacitors: A leap in energy storage. *J. Energy Storage*. 88:111468.

[21] Chen, Q. De Marco, N. Yang, Y.M. Song, T.B. Chen, C.C. Zhao, H. Hong, Z. Zhou, H. and Yang, Y. 2015. Under the spotlight: The organic–inorganic hybrid halide perovskite for optoelectronic applications. *Nano Today*. 10(3):355–396.

[22] Kojima, A. Teshima, K. Shirai, Y. and Miyasaka, T. 2009. Organometal halide perovskites as visible-light sensitizers for photovoltaic cells. *J. Am. Chem. Soc.* 131(17):6050–6051.

[23] Teo, S.H. Ng, C.H. Ng, Y.H. Islam, A. Hayase, S. and Taufiq-Yap, Y.H. 2022. Resolve deep-rooted challenges of halide perovskite for sustainable energy development and environmental remediation. *Nano Energy*. 99:107401.

[24] Kore, B.P. Jamshidi, M. and Gardner, J.M. 2024. The impact of moisture on the stability and degradation of perovskites in solar cells. *Mater. Adv.* 5:2200–2217.

[25] Znidi, F. Morsy, M. and Uddin, M.N. 2024. Navigating challenges and solutions for metal-halide and carbon-based electrodes in perovskite solar cells (NCS-MCEPSC): An environmental approach. *Heliyon*. 10(12):1-21.

[26] Ravi, V.K. Mondal, B. Nawale, V.V. and Nag, A. 2020. Don't let the lead out: New material chemistry approaches for sustainable lead halide perovskite solar cells. *ACS Omega*. 5(46):29631–29641.

[27] Zhang, L. Mei, L. Wang, K. Lv, Y. Zhang, S. Lian, Y. Liu, X. Ma, Z. Xiao, G., Liu, Q. and Zhai, S. 2023. Advances in the application of perovskite materials. *Nano-Micro Letters*. 15(1):177.

[28] López-Fernández, I. Valli, D. Wang, C.Y. Samanta, S. Okamoto, T. Huang, Y.T. Sun, K., Liu, Y. Chirvony, V.S. Patra, A. and Zito, J. 2024. Lead-free halide perovskite materials and optoelectronic devices: Progress and prospective. *Adv. Funct. Mater.* 34(6):2307896.

[29] Huang, Y.T. Kavanagh, S.R. Scanlon, D.O. Walsh, A. and Hoye, R.L. 2021. Perovskite-inspired materials for photovoltaics and beyond—from design to devices. *Nanotechnology*. 32(13):132004.

[30] Diouf, B. Muley, A. and Pode, R. 2023. Issues, challenges, and future perspectives of perovskites for energy conversion applications. *Energies*. 16(18):6498.

[31] Kumar, S. Sharma, V. Kumari, N. Kaur, G.A. Saha, A. Thakur, S. and Shandilya, M. 2024. Recent advances in perovskite materials: Exploring multifaceted properties for energy harvesting applications. *Ionics*. 30(9):5159–5188.

[32] Pandiyan, A. Veeramuthu, L. Yan, Z.L. Lin, Y.C. Tsai, C.H. Chang, S.T. Chiang, W.H. Xu, S. Zhou, T. and Kuo, C.C. 2023. A comprehensive review on perovskite and its functional composites in smart textiles: Progress, challenges, opportunities, and future directions. *Prog. Mater. Sci.* 140:101206.

[33] Yang, C. Hu, W. Liu, J. Han, C. Gao, Q. Mei, A. Zhou, Y. Guo, F. and Han, H. 2024. Achievements, challenges, and future prospects for industrialization of perovskite solar cells. *Light: Sci. Appl.* 13(1):227.

[34] Jain, A. Wang, Y.G. and Shi, L.N. 2022. Recent developments in BaTiO3 based lead-free materials for energy storage applications. *J. Alloys Compd.* 928:167066.

[35] Amat, E. Mosconi, E. Ronca, C. and Quarti, P. Umari.2014. Cation-induced band-gap tuning in organohalide perovskites: Interplay of spin-orbit coupling and octahedra tilting. *Nano Lett.* 14(6):3608–3616.

[36] Goyal, S. McKechnie, D. Pashov, W. Tumas, M. and Schilfgaarde. 2018. Origin of pronounced non-linear band gap behavior in lead–tin hybrid perovskite alloys. *Chem. Mater.* 30(11):3920–3928.

[37] Fan, Z. Sun, K. and Wang, J. 2015. Perovskites for photovoltaics: A combined review of organic-inorganic halide perovskites and ferroelectric oxide perovskites. *J. Mater. Chem. A* 3(37):18809–18828.

[38] Buin, R. Comin, J.X. Xu, A.H. and Ip, E.H. 2015. Sargent, Halide dependent electronic structure of organolead perovskite materials. *Chem. Mater.* 27(12):4405–4412.

[39] Stranks, S.D. Eperon, G.E. Grancini, G. Menelaou, C. and Alcocer, M.J.P. 2013. Electron-hole diffusion lengths exceeding 1 micrometer in an organometal trihalide perovskite absorber. *Science*. 342(6156):341–344.

[40] Ye, Q. Zhao, Y. Mu, S. Ma, F. and Gao, F. 2019. Cesium lead inorganic solar cell with efficiency beyond 18% via reduced charge recombination. *Adv. Mater.* 31(49):1905143.

[41] Ren, M. Cao, S. Zhao, J. Zou, B. and Zeng R. 2021. Advances and challenges in two-dimensional organic-inorganic hybrid perovskites toward high-performance light-emitting diodes. *Nano-Micro Lett.* 13:163.

[42] Han, Y. Yue, S. and Cui, B.B. 2021. Low-dimensional metal halide perovskite crystal materials: Structure strategies and luminescence applications. *Adv. Sci.* 8(15):2004805.

[43] Huo, C. Cai, B. Yuan, Z. Ma, B. and Zeng, H. 2017. Two-dimensional metal halide perovskites: Theory, synthesis, and optoelectronics. *Small Methods* 1(3):1600018.

[44] Inaguma, Y. Liquan, C. Itoh, M. Nakamura, T. Uchida, T. Ikuta, H. and Wakihara, M. High ionic conductivity in lithium lanthanum titanate. *Solid State Commun.* 86(1993):689–693.

[45] Bohnke, O. Mechanism of ionic conduction and electrochemical intercalation of lithium into the perovskite lanthanum lithium titanate. *Solid State Ion.* 91(1996):21–31.

[46] Ji, S. Wang, X. Li, K. Huan, Y. Ma, G. Su, Y. and Wei, T. 2022. 3D vertically aligned microchannel three-layer all ceramic Lithium-ion battery for high-rate and long-cycle electrochemical energy storage. *Small*. 18:2107442.

[47] Aguesse, F. López del Amo, J.M. Roddatis, V. Aguadero, A. and Kilner, J.A. 2014. Enhancement of the grain boundary conductivity in ceramic Li 0.34 La 0.55 TiO 3 electrolytes in a moisture-free processing environment. *Adv. Mater. Interfaces*. 1:1300143.

[48] Inada, R. Kimura, K. Kusakabe, K. Tojo, T. and Sakurai, Y. 2014. Synthesis and lithium-ion conductivity for perovskite-type Li3/8Sr7/16Ta3/4Zr1/4O3 solid electrolyte by powder-bed sintering. *Solid State Ion.* 261:95–99.

[49] Luo, J. Zhong, S. Huang, Z. Huang, B. and Wang, C.-A. 2014. HighLi+-conductive perovskite Li3/8Sr7/16Ta3/4Zr1/4O3 electrolyte prepared by hot-pressing for all-solid-state Li-ion batteries. *Solid State Ion*. 338:1–4.

[50] Zhao, Y. Liu, Z. Xu, J. Zhang, T. Zhang, F. and Zhang, X. 2019. Synthesis and characterization of a new perovskite-type solid-state electrolyte of Na1/3La1/3Sr1/3ZrO3 for all-solid-state sodium-ion batteries. *J. Alloys Compd*. 783:219–225.

[51] Mulmi, S. and Thangadurai, V. 2019. Solid-state electrolytes: Structural approach. *Solid Electrolytes for Advanced Applications: Garnets and Competitors*. 3–24.

[52] Li, Y. et al. 2018. A P2-type layered superionic conductor Ga-doped $Na_2Zn_2TeO_6$ for all-solid-state sodium-ion batteries. *Chem. Eur. J*. 24:1057–1061.

[53] Nalbandyan, V., Petrenko, A. A. and Evstigneeva, M. 2013. Heterovalent substitutions in $Na_2M_2TeO_6$ family: Crystal structure, fast sodium ion conduction and phase transition of $Na_2LiFeTeO_6$. *Solid State Ion*. 233:7–11.

[54] Li, Y., Han, J.-T., Wang, C.-A., Xie, H. and Goodenough, J. B. 2012. Optimizing Li+ conductivity in a garnet framework. *J. Mater. Chem*. 22:15357–15361.

[55] Arbi, K. Hoelzel, M. Kuhn, A. García-Alvarado, F. and Sanz, J. 2013. Structural factors that enhance lithium mobility in fast-ion Li1+xTi2−xAlx(PO4)3 (0 ≤x≤ 0.4) conductors investigated by neutron diffraction in the temperature range 100–500 K. *Inorg. Chem*. 52:9290–9296.

[56] Maus, O. et al. 2023. On the discrepancy between local and average structure in the Fast Na+ ionic conductor $Na_{2.9}Sb_{0}.9W_{0^{.}1}S_4$. *J. Am. Chem. Soc*. 145:7147–7158.

[57] Tang, W. et al. 2016. Stabilizing superionic-conducting structures via mixed-anion solid solutions of monocarba-closo-borate salts. *ACS Energy Lett*. 1:659–664.

[58] Zhang, Z. et al.2020. Enhancing ionic conductivity of solid electrolyte by lithium substitution in halogenated Li-Argyrodite. *J. Power Sources*. 450:227601.

[59] Zhang, L. Zhang, X. Tian, G. et al. 2020. Lithium lanthanum titanate perovskite as an anode for lithium ion batteries. *Nat Commun*.11:3490.

[60] Yan, S. Yim, C.H. Pankov, V. Bauer, M. Baranova, E. Weck, A. Merati, A. and Abu-Lebdeh Y. 2021. Perovskite solid-state electrolytes for lithium metal batteries. *Batteries*. 7(4):75.

[61] Chintha, S., Atif, S., Chaupatnaik, A., Golubnichiy, A., Abakumov, A. M. and Barpanda, P. 2024. Na 0.5 Bi 0.5 TiO 3 perovskite anode for lithium-ion batteries. *Sustainable Energy Fuels*. 8(21):5058–5064.

[62] Tewari, N. et al. 2022. Recent advancements in batteries and photo-batteries using metal halide perovskites. *APL Mater*. 10:040905.

[63] Lei, Z Miao, J. Li, J. and Li, Q. 2020. Halide perovskite materials for energy storage applications. *Adv. Funct. Mater*. 30(40):2003653.

[64] Beladona, S. U. M. Purwanto, F. Jumiati, J. Simanjuntak, E. R. Simarmata, S. N. Pasaribu, M. H. and Iqbal, R. M. 2024. A review of perovskite-based lithium-ion battery materials. *Vietnam J. Sci*. 62(5):813–835.

[65] Su, M. Zhu, S. Yu, R. Li, X. Lu, S. Huang, Y. and Zhan, H. 2025. Synthesis of high-entropy perovskite metal fluoride anode materials for lithium-ion batteries via a one-pot solution method. *J. Alloys Compd*. 1010:177458.

[66] Su, M. Yu, R. Li, X. Lu, S. Zhu, S. Huang, Y. and Zhan, H. 2024. Superior performance high-entropy perovskite metal fluoride electrodes derived from spent LiNi1/3Co1/3Mn1/3O2/graphite lithium-ion batteries. *Mater. Lett*. 370:136784.

[67] Mathieson, A. Feldmann, S. and De Volder, M. 2022. Solid-state lithium-ion batteries as a method for doping halide perovskites with an in situ optical readout of dopant concentration. *JACS Au*. 2(6):1313–1317.

[68] Weng, G. M. Kong, J. Wang, H. Karpovich, C. Lipton, J. Antonio, F. and Taylor, A. D. 2020. A highly efficient perovskite photovoltaic-aqueous Li/Na-ion battery system. *Energy Storage Mater*. 24:557–564.

[69] Bharathi, K. K. Moorthy, B. Dara, H. K. Durai, L. and Kim, D. K. 2019. Electrochemical properties of Na 0.5 Bi 0.5 TiO 3 perovskite as an anode material for sodium ion batteries. *J. Mater. Sci*. 54:13236–13246.

[70] Ch, L. T. L. Medina, A. Jaramillo, F. Calderón, J. A. Lavela, P. and Tirado, J. L. 2023. New insights on the reaction mechanism and charge contribution of $NaNiF_3$ perovskite as an anode for sodium-ion batteries. *Electrochimica Acta*. 453:142341.

[71] Yan, T. Ding, R. Huang, Y. Ying, D. Tan, C. Huang, Y. and Liu, E. 2021. A novel sodium-ion supercabattery based on vacancy defective Ni–Co–Mn ternary perovskite fluoride electrode materials. *J. Mater. Chem. A*. 9(25):14276–14284.

[72] Duan, L. and Uddin, A. 2022. Defects and stability of perovskite solar cells: A critical analysis. *Mater. Chem. Front*. 6(4):400–417.

[73] Khan, A.D. Mustajab, M. Moeen, S. Imran, M. Ikram, M., Khan, Q. and Khan, M. 2024. Advancements in the stability, protection and lead-free strategies of perovskite solar cells: A critical review. *Environ. Sci.: Adv*. 3(7):1004–1029.

[74] Noman, M. Khan, Z. and Jan, S.T. 2024. A comprehensive review on the advancements and challenges in perovskite solar cell technology. *RSC Adv.* 14(8):5085–5131.

[75] Yan, J. Savenije, T.J. Mazzarella, L. and Isabella, O. 2022. Progress and challenges on scaling up of perovskite solar cell technology. *Sustainable Energy Fuels*. 6(2):243–266.

[76] Hossain, K. Nayak, S. and Kabra, D. 2024. Challenges and opportunities in high efficiency scalable and stable perovskite solar cells. *Appl. Phys. Lett*. 125:170501.

15 Advanced Artificial Intelligence and Machine Learning Strategies for Optimizing Lightweight Materials Processing for Electric Vehicles

Hariharasakthisudhan Ponnarengan, Sathickbasha Katharbasha, and Varun Kumar Arulvizhi

15.1 INTRODUCTION

The pursuit of lightweight materials has emerged as a cornerstone in modern engineering and automotive industries, driven by the need to optimize performance, reduce energy consumption, and minimize environmental impact. Lightweight materials, such as aluminum alloys, magnesium alloys, titanium composites, and advanced polymer matrix composites, offer an exceptional strength-to-weight ratio, corrosion resistance, and durability [1]. These characteristics make them indispensable in applications ranging from aerospace components to structural elements in the automotive sector [2]. By significantly reducing the overall mass of systems, lightweight materials enable enhanced fuel efficiency, higher payload capacities, and improved dynamic performance, positioning them as critical enablers of next-generation engineering solutions.

In the context of electric vehicles (EVs), the importance of lightweight materials becomes even more pronounced. As the automotive industry undergoes a paradigm shift toward electrification, the reduction of vehicular mass is pivotal in addressing the limitations of battery energy density and driving range [3]. Lightweight materials not only contribute to extending the range of EVs but also enable the optimization of critical subsystems, such as battery enclosures, suspension systems, and crash-resistant structures [4]. Furthermore, their inherent recyclability aligns with the sustainability goals of the EV sector, ensuring the development of environmentally responsible mobility solutions including battery management. The barriers for the usage of EVs are represented in Figure 15.1. The convergence of lightweight materials and EV applications underscores the imperative for advanced material processing techniques that cater to these specialized requirements [5].

The advent of artificial intelligence (AI) and machine learning (ML) has brought about a transformative shift in materials science and processing [6]. AI-driven predictive modeling and ML-based optimization algorithms have revolutionized the way materials are characterized and processed [7]. These technologies enable the rapid prediction of material properties under varying processing conditions, thus minimizing experimental iterations and resource wastage [8]. In materials characterization, AI algorithms excel in analyzing microstructural features through techniques like field emission scanning electron microscopy (FESEM) and energy-dispersive X-ray analysis (EDAX),

DOI: 10.1201/9781003596097-15

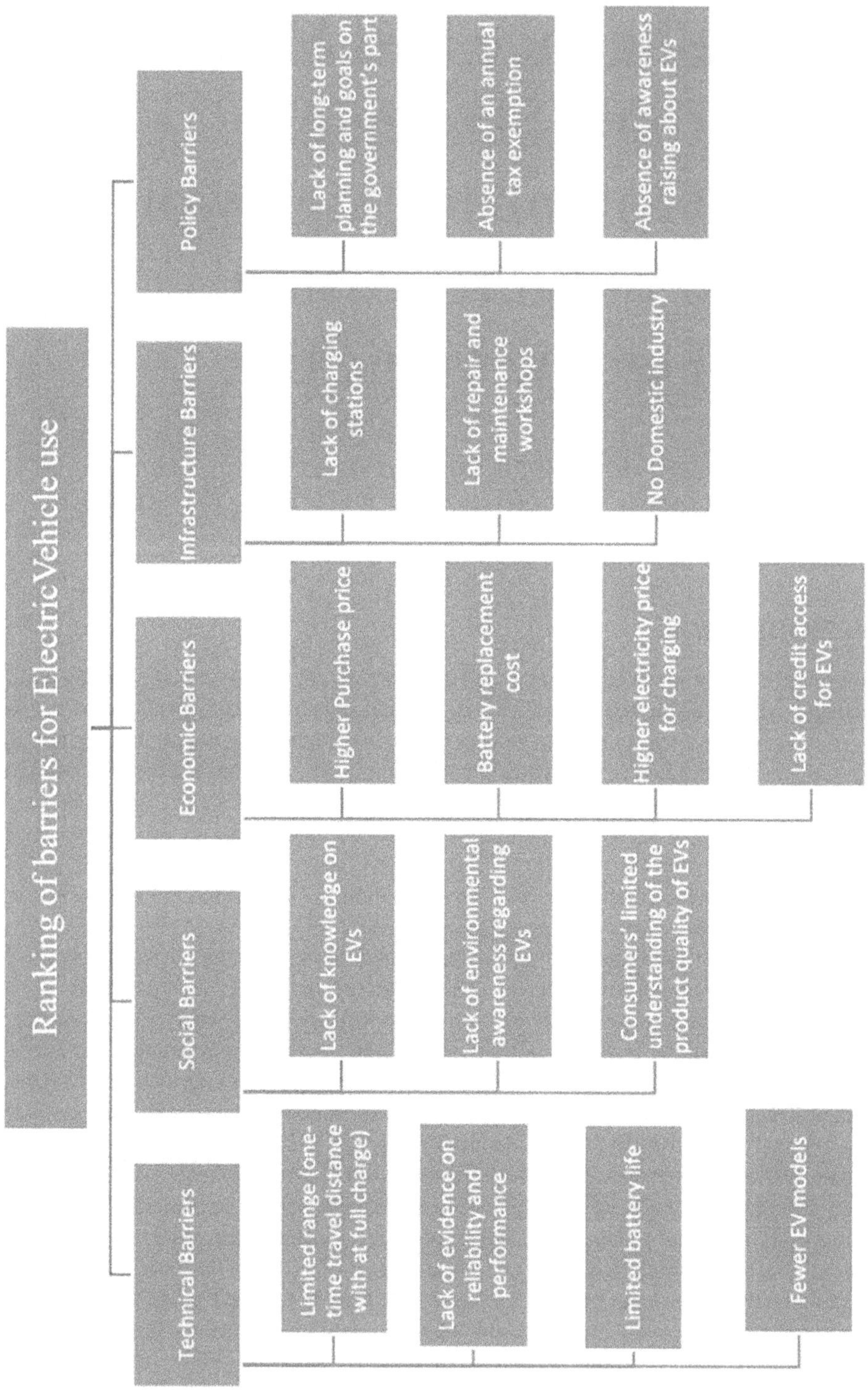

FIGURE 15.1 Barriers to the development of electric vehicle usage [5].

offering unparalleled accuracy and insights into material behavior [9]. Similarly, in processing, AI and ML facilitate the precise control of parameters in casting [10], welding [11], machining [12], and additive manufacturing (AM) [13], ensuring superior quality and consistency in end products. This synergy of AI and ML with materials science is paving the way for smarter, data-driven approaches to meet the complex demands of lightweight material development [14].

This chapter provides a comprehensive exploration of AI and ML applications in lightweight material processing, particularly for EVs. It begins with an overview of AI and ML fundamentals, discussing key techniques such as supervised learning, unsupervised learning, and reinforcement learning that optimize material characterization and processing. The discussion then transitions into lightweight materials, their significance, and the challenges associated with their manufacturing. Advanced AI-driven manufacturing techniques, including casting, welding, AM, and machining, are examined in detail, showcasing how predictive modeling and real-time monitoring enhance efficiency and precision. Case studies highlight successful AI-ML implementations in material optimization, defect detection, and predictive maintenance. The chapter also addresses challenges such as data limitations, computational constraints, and industry adoption barriers. Finally, future directions explore emerging AI methodologies like generative adversarial networks (GANs) and graph convolutional networks (GCNs), emphasizing their potential to revolutionize materials science and sustainable manufacturing.

15.2 FUNDAMENTALS OF AI AND ML IN MATERIALS SCIENCE

AI and ML have redefined the landscape of material processing, enabling unprecedented levels of precision, efficiency, and insight. These technologies leverage advanced algorithms to predict material behavior [15], optimize processing conditions [16], and extract complex patterns from experimental and simulated data [17]. By addressing the challenges of traditional trial-and-error methodologies, AI and ML provide transformative solutions that enhance the quality, performance, and sustainability of materials, particularly in applications requiring lightweight alloys and composites [18].

Supervised learning, a cornerstone of ML, involves training models on labeled datasets to predict specific material properties, such as tensile strength, thermal conductivity, and wear resistance, under varying conditions [19]. This technique is instrumental in optimizing process parameters, such as temperature and pressure, across casting, welding, and machining applications. Conversely, unsupervised learning focuses on discovering hidden patterns and correlations in unlabeled datasets, making it ideal for clustering materials based on microstructural characteristics or identifying anomalies during quality control processes [20]. Reinforcement learning (RL) offers a dynamic approach by enabling systems to learn optimal strategies through iterative interaction with their environments. RL is particularly effective in adaptive control scenarios, such as real-time optimization of AM processes or robotic-assisted material handling [21].

Key terminologies in AI-ML for materials science include artificial neural networks (ANNs) for deep learning [22], support vector machines (SVMs) [23] for classification and regression, and genetic algorithms (GAs) [24] for solving multi-objective optimization problems. Techniques like feature engineering, hyperparameter tuning, and high-dimensional data visualization are critical for achieving robust model performance [25]. Furthermore, methodologies such as ensemble learning and transfer learning enhance predictive accuracy and adaptability in diverse material processing contexts. Together, these tools and approaches are revolutionizing materials science by offering data-driven solutions to complex challenges in characterization, design, and manufacturing. AI and ML are transforming materials science by enabling precise prediction and optimization of lightweight materials tailored to meet the demanding requirements of EV applications. These tools provide robust solutions for understanding and controlling the complex relationships between material composition, processing conditions, and resulting properties. AI models, such as deep neural networks (DNNs) [26] and Gaussian process regression [27], are employed to predict key properties like tensile strength, fatigue life, and thermal conductivity. For example, in magnesium-aluminum-zinc alloys commonly used in EV chassis, DNNs can forecast mechanical properties by analyzing datasets that include processing variables like cooling rates, heat treatments, and alloying element concentrations [28]. Similarly, ML algorithms have been applied to predict thermal expansion coefficients of polymer composites used in EV battery packs, enabling the design of components with minimal deformation under thermal cycling [29].

ML significantly accelerates microstructural characterization by automating the analysis of large datasets generated by advanced imaging techniques. Convolutional neural networks (CNNs) are applied to FESEM images to identify and quantify grain boundaries [30], voids, and dislocations with high accuracy. For instance, in nickel-based alloys, CNN-based segmentation has been used to detect and quantify nanoscale precipitates [31], which are critical for enhancing yield strength. Unsupervised learning methods, such as principal component analysis (PCA) and t-distributed stochastic neighbor embedding (t-SNE), have been employed to reduce dimensionality in EDAX data, revealing correlations between elemental distributions and localized material performance [32].

AI excels in identifying failure mechanisms in lightweight materials by analyzing historical data and operational conditions. RL models have been used to predict and prevent fatigue failures in materials [33]. These models learn from operational stress patterns, identifying regions prone to crack initiation and providing actionable insights to improve design and processing. Similarly, stress corrosion cracking in pipeline materials has been mitigated through ML-based optimization [34].

Predictive modeling tools, such as Bayesian optimization and GAs, are utilized for multi-objective optimization of properties. For example, in AM of polymers, Bayesian optimization identifies optimal print speeds, layer heights, and curing profiles to achieve maximum tensile strength while minimizing defects [35]. In casting processes for aluminum alloys, GAs optimize cooling rates and mold designs to minimize defects and enhance properties [36]. These AI- and ML-driven advancements are not only improving the efficiency of material development but also ensuring that EV components meet stringent performance, safety, and sustainability requirements. By leveraging these technologies, researchers and manufacturers are bridging the gap between material capabilities and the evolving demands of EV engineering.

The integration of AI and ML in materials science has significantly enhanced the understanding and optimization of material properties, enabling precise control over processing conditions and improving overall efficiency. As these technologies continue to evolve, their application in lightweight materials becomes increasingly relevant, especially in sectors like automotive and aerospace, where weight reduction is critical for performance and sustainability. Understanding the properties, benefits, and challenges associated with lightweight materials is essential before delving into AI-driven processing techniques. The next section explores the fundamental characteristics of lightweight materials, their role in EVs, and the complexities involved in their manufacturing and processing.

15.3 LIGHTWEIGHT MATERIALS: AN OVERVIEW

Lightweight materials, pivotal in advanced engineering and automotive sectors, encompass a diverse range of categories including aluminum alloys, magnesium alloys, polymer-based composites, and metal matrix composites (MMCs). Aluminum alloys, valued for their exceptional strength-to-weight ratio, corrosion resistance, and recyclability, are widely used in structural applications [37]. Magnesium alloys, being the lightest structural metals, offer high specific strength but pose challenges in terms of creep resistance and corrosion susceptibility [38]. Polymers and polymer-based composites, such as carbon fiber-reinforced polymers (CFRPs), provide unparalleled design flexibility and superior mechanical performance, making them ideal for intricate geometries in EV components. Emerging lightweight materials and their structure are presented in Figure 15.2 [39].

The properties of lightweight materials critical to EV applications extend beyond mere weight reduction. A high strength-to-weight ratio is essential to ensure structural integrity while minimizing mass, directly enhancing vehicle range and energy efficiency [40]. Thermal stability and conductivity are equally crucial, particularly for components like battery housings and cooling systems, where heat management is critical [41]. Additional properties, such as impact resistance, fatigue durability, and recyclability, align with the performance and sustainability requirements of modern EVs [42].

Processing lightweight materials for EV applications presents significant challenges. The inherent high reactivity of magnesium alloys during casting or welding demands precise control of processing environments to prevent oxidation and defects [43]. Polymer composites require advanced molding and curing techniques to achieve uniform dispersion of reinforcements, such as carbon fibers, and to ensure consistent mechanical properties [44]. Moreover, joining dissimilar materials, a common requirement in multi-material EV structures, introduces complexities due to differences in thermal expansion coefficients and bonding mechanisms [45]. These challenges necessitate the adoption of advanced processing methods, underpinned by data-driven optimization through AI and ML, to ensure precision, efficiency, and sustainability in the manufacturing of lightweight materials for EV components.

Lightweight materials have gained significant traction in modern engineering applications due to their exceptional strength-to-weight ratio and sustainability benefits. However, their processing poses unique challenges, including issues related to formability, joining, and defect control, which

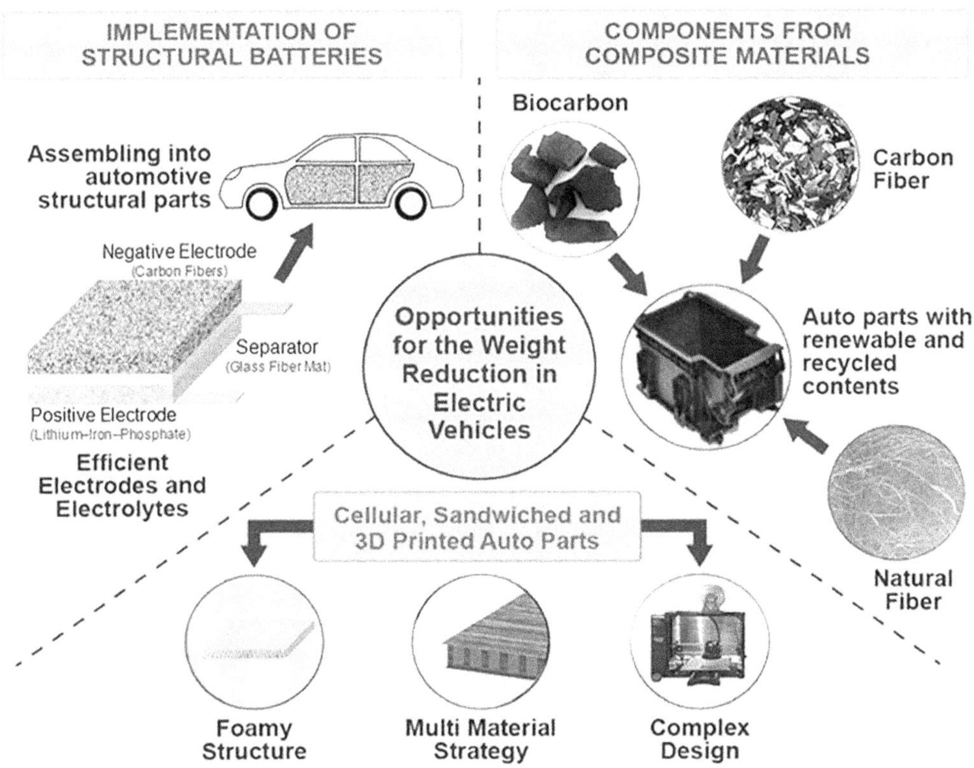

FIGURE 15.2 Emerging lightweight materials used in electric vehicles [39].

require precise optimization of manufacturing parameters. Traditional methods often struggle to meet these demands efficiently, leading to increased material waste and production costs. This is where AI and ML play a transformative role, enabling real-time process monitoring, defect prediction, and adaptive control. The following section explores how AI-driven techniques are revolutionizing advanced manufacturing processes such as casting, welding, AM, and machining to enhance material performance and production efficiency.

15.4 AI AND ML IN PROCESSING TECHNIQUES

AI and ML are transforming casting processes, particularly for lightweight alloys like aluminum and magnesium, which are pivotal in manufacturing EV components such as battery enclosures, suspension mounts, and structural reinforcements. AI-driven models, such as ANNs, are used to optimize critical casting parameters [46] like mold temperature, alloy composition, and cooling rates. For example, in high-pressure die casting of magnesium alloys, AI models predict solidification patterns to minimize defects like porosity and shrinkage, which directly impact mechanical strength and fatigue resistance. Advanced ML algorithms, including decision trees and SVMs, analyze historical data on defects to predict issues like gas entrapment, hot tearing, or cold shuts [47]. These predictions enable real-time process adjustments, such as modifying gate design or altering pouring speeds, to reduce defects and improve part integrity. A practical example is the use of AI in optimizing squeeze casting of aluminum-silicon alloys for EV chassis components, ensuring uniform microstructures with minimal inclusion formation, which is essential for crash resistance.

Welding and joining processes for lightweight materials in EVs face challenges due to the differing thermal and mechanical properties of dissimilar materials like aluminum, magnesium, and high-strength steels. RL algorithms have enabled adaptive control systems that dynamically adjust parameters such as welding speed, heat input, and electrode position. In friction stir welding (FSW) of aluminum alloys for EV battery trays [48], RL systems maintain optimal joint strength by adapting to variations in material thickness and composition. ML models, including random forests and deep learning frameworks, predict weld quality by analyzing high-speed imaging of the weld pool and thermal gradients. These models help identify defects such as cracks, voids, and lack of fusion before they compromise part performance. For EV applications, laser welding augmented with ML [49] ensures precise joining of dissimilar materials, creating lightweight yet strong battery housings capable of withstanding thermal and mechanical stresses.

AI has revolutionized AM of lightweight materials by enabling the optimization of complex geometries and ensuring consistency in mechanical properties. Deep learning models analyze layer-by-layer data to optimize print parameters such as nozzle temperature, layer height, and deposition speed, ensuring minimal surface roughness. Various ML models used in AM are presented in Figure 15.3 [50]. Real-time defect detection and correction, powered by CNNs and computer vision, allows for the identification of defects like voids, uneven deposition, or microcracks during the printing process. For example, in 3D printing of CFRPs [51] used for EV chassis parts, AI systems dynamically adjust deposition patterns to achieve optimal fiber orientation, resulting in enhanced strength and stiffness. AM is also used for producing lattice structures in EV crash zones, where AI-driven topology optimization ensures maximum energy absorption while minimizing material usage.

Machining lightweight materials like aluminum and magnesium presents challenges such as tool wear, burr formation, and thermal deformation, all of which are mitigated using predictive analytics powered by ML. Advanced ML algorithms predict tool wear by analyzing real-time data on cutting forces, vibrations, and temperature, enabling proactive tool replacement and reducing downtime. For example, during the machining of magnesium alloys for EV suspension arms, ML systems optimize feed rates and spindle speeds to minimize surface defects and extend tool life. In forming

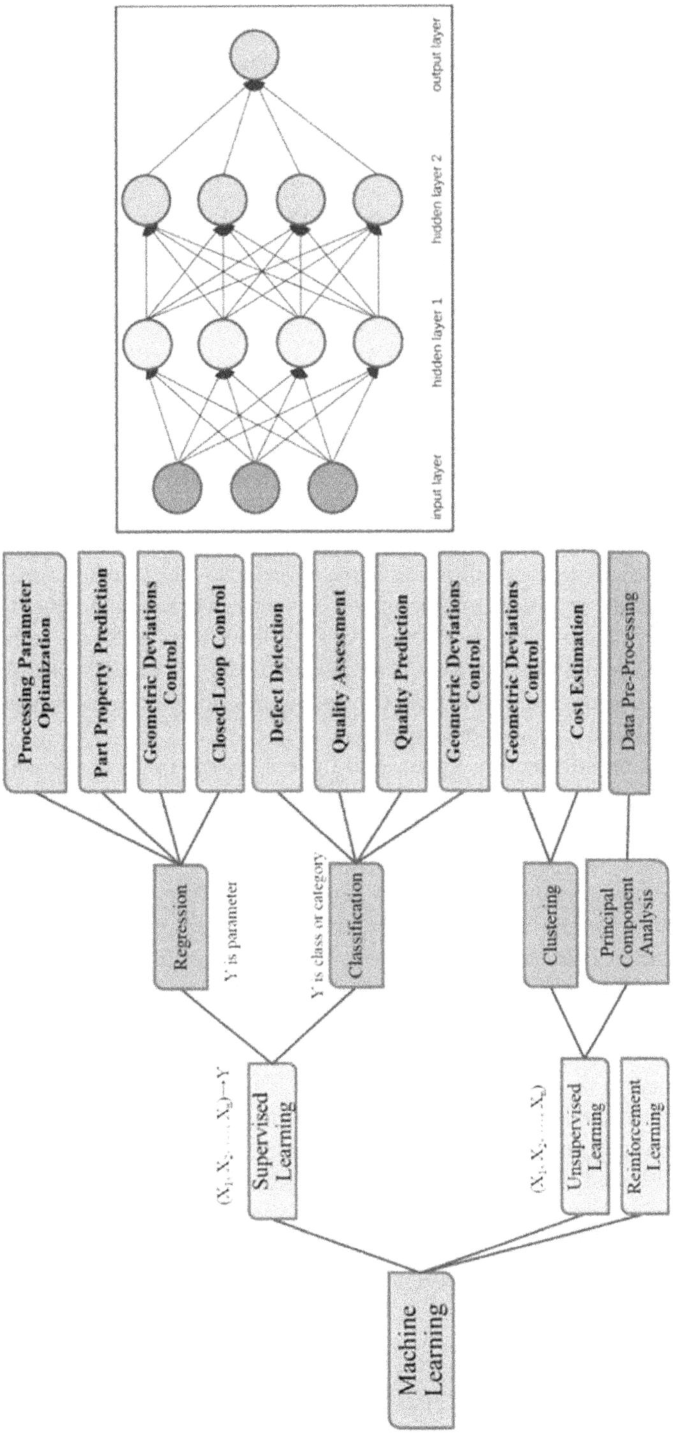

FIGURE 15.3 Applications of machine learning taxonomy in additive manufacturing [50].

processes, AI models integrate finite element simulations with ML techniques to predict outcomes like thinning, wrinkling, or springback. For instance, forming simulations for high-strength aluminum sheets used in EV body panels benefit from AI-enabled optimization, ensuring uniform deformation and reduced weight without compromising structural integrity.

The integration of AI and ML in advanced manufacturing techniques has demonstrated significant improvements in process optimization, defect minimization, and overall production efficiency. By leveraging predictive modeling, real-time monitoring, and adaptive control, these technologies enable precise control over processing parameters, ensuring the consistent quality of lightweight materials. However, beyond theoretical advancements, real-world applications provide critical insights into the practical benefits and challenges of AI-ML-driven material processing. The next section presents case studies that highlight successful implementations of AI and ML in optimizing material properties, detecting defects, and enhancing manufacturing reliability, demonstrating their transformative impact on the material processing industry.

15.5 CASE STUDIES

AI and ML technologies are revolutionizing the development and optimization of multi-material composites and lightweight alloys, addressing the complex demands of EVs in terms of performance, efficiency, and sustainability. Multi-material composites, such as aluminum-magnesium laminates reinforced with high-performance fibers like carbon, aramid, or glass, offer exceptional strength-to-weight ratios, thermal stability, and impact resistance. These materials are indispensable for EV body structures, battery enclosures, crash zones, and suspension components, where weight reduction and structural integrity are paramount. AI algorithms, such as RL, have been extensively applied to optimize composite stacking sequences and predict failure mechanisms. For example, RL models adjust ply orientations in CFRPs to maximize stiffness and reduce delamination risks under impact loading, critical for EV crashworthiness [52].

ML enables precise optimization of lightweight material properties by analyzing large datasets of experimental and simulation results. Advanced algorithms, such as gradient boosting and random forests, predict the influence of alloying elements, thermal treatments, and processing conditions on key properties like tensile strength, ductility, fatigue resistance, and thermal conductivity. For instance, in magnesium-zinc-aluminum (AZ91) alloys, ML has been used to identify the optimal grain refinement techniques and aging processes, enhancing creep resistance and thermal stability for EV drivetrain components. Similarly, support vector regression (SVR) models optimize the quenching and aging cycles for high-strength aluminum alloys, ensuring uniform microstructure and superior fatigue performance in chassis applications [53].

Real-world applications underscore the transformative potential of AI-ML in material processing. Predictive maintenance powered by Internet of Things (IoT)-enabled data collection is a game changer in magnesium alloy die casting, a key process for lightweight EV components [54]. Sensors monitor real-time parameters like die temperatures, injection pressures, cooling rates, and cycle times. AI models analyze these data to predict potential issues such as porosity, hot tearing, and tooling wear. For instance, Bayesian networks identify patterns leading to defects, while neural networks forecast wear patterns on dies, enabling proactive interventions. In addition, reinforcement learning systems dynamically adjust process parameters to prevent defect formation during production, ensuring consistent quality.

One practical example is the production of magnesium alloy battery housings, where AI-ML-driven optimization reduces weight while maintaining crash safety and thermal management. In CFRP manufacturing for EV body parts, CNNs detect fiber misalignment during automated layup processes, enabling real-time corrective actions. Furthermore, AI-driven topology optimization enhances multi-material composite designs, creating intricate lattice structures for EV crash zones that absorb impact energy effectively while minimizing material usage [55].

By integrating AI and ML into multi-material composite development and lightweight alloy processing, EV manufacturers are achieving unprecedented levels of material performance, production efficiency, and sustainability. These technologies are setting a new benchmark for innovation, aligning the automotive industry with future demands for energy-efficient, high-performance vehicles.

The case studies discussed illustrate the tangible benefits of AI and ML in lightweight material processing, showcasing their ability to optimize manufacturing parameters, enhance product quality, and reduce defects. While these advancements highlight the transformative potential of AI-driven approaches, several challenges must still be addressed for widespread industrial adoption. Issues such as data availability, computational demands, and integration complexities pose significant barriers to fully realizing AI's potential in material processing. The next section delves into these challenges, examining the limitations that hinder seamless AI-ML implementation and exploring potential strategies to overcome these obstacles for a more efficient and sustainable manufacturing ecosystem.

15.6 CHALLENGES AND LIMITATIONS

The adoption of AI and ML technologies in lightweight material processing for EV applications faces multifaceted challenges, with data availability and quality being one of the foremost hurdles, as presented in Figure 15.4. In the context of advanced materials, the data required for model training

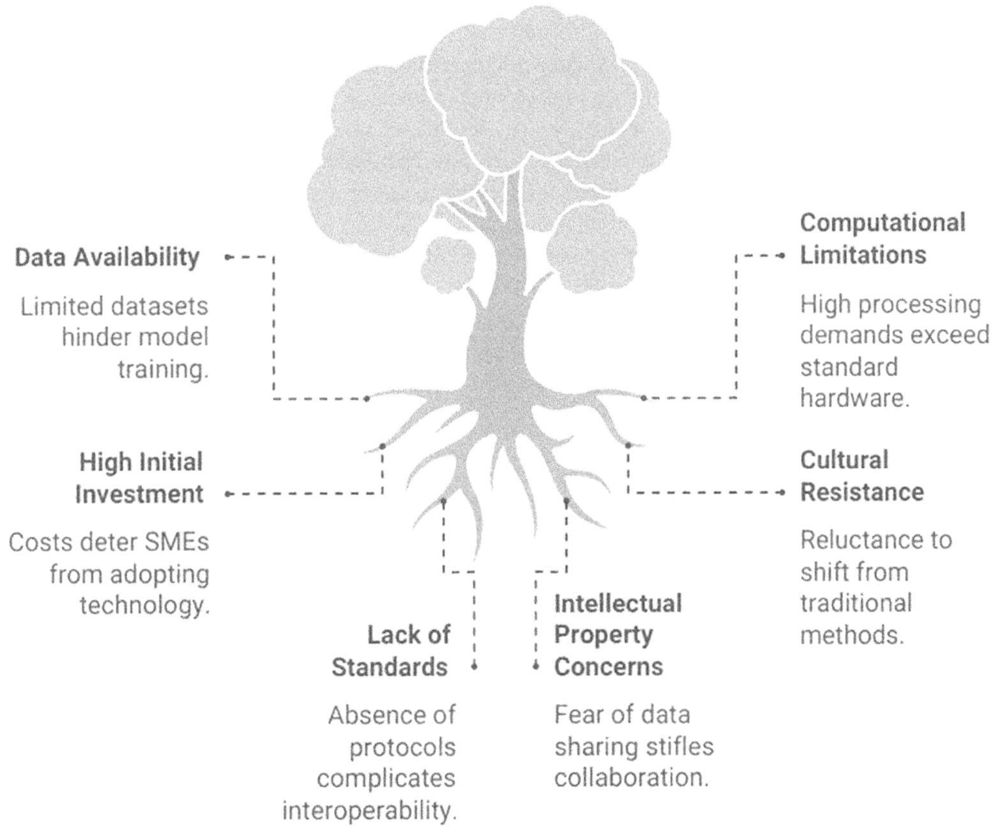

FIGURE 15.4 Barriers in adopting AI and ML in lightweight materials processing.

and validation often involve highly specialized experiments, such as in situ mechanical testing under extreme conditions or high-speed thermal imaging during casting. For example, capturing accurate datasets for magnesium alloy die casting necessitates real-time data on melt flow, temperature gradients, and solidification patterns, which are difficult to obtain without cutting-edge instrumentation. Furthermore, data heterogeneity—stemming from variations in raw material batches, environmental conditions, and operator expertise—complicates the development of generalized models. These challenges are exacerbated in hybrid composite systems, where multi-scale interactions between constituents (e.g., fiber–matrix bonding) demand highly granular datasets that are both expensive and labor-intensive to generate.

Computational challenges also present significant barriers, particularly for real-time AI-ML applications. Processes such as laser-based AM or high-speed extrusion generate massive data streams that must be processed in milliseconds to enable adaptive control. Predicting outcomes like thermal distortion, layer adhesion, or grain refinement during these processes involves solving complex, high-dimensional models. For instance, RL systems used in AM require iterative training cycles to adjust print parameters dynamically, which is computationally intensive and often unfeasible on standard hardware. Similarly, optimizing the microstructural uniformity of lightweight alloys during forming processes demands parallel computations across thermal, mechanical, and chemical domains, overwhelming conventional computational infrastructures. Moreover, latency in data transfer from IoT-enabled devices to centralized AI systems can impede real-time decision-making, undermining process efficiency.

Barriers to industry adoption extend beyond technical constraints. The initial capital investment required for IoT sensors, edge computing devices, and AI-ML software platforms is prohibitively high for many small-to-medium-sized enterprises (SMEs), which constitute a significant portion of the manufacturing sector. Additionally, cultural resistance to transitioning from traditional, empirically driven methodologies to AI-powered workflows slows adoption. This resistance is compounded by a lack of regulatory frameworks and standards governing the implementation of AI-ML in material processing. For example, the absence of uniform protocols for data handling and model evaluation in EV manufacturing makes interoperability across supply chains challenging. Intellectual property concerns further hinder collaboration between companies and AI solution providers, particularly when proprietary data must be shared for model development. The path to overcoming these challenges requires targeted investments in data standardization, accessible computational infrastructure, and workforce upskilling programs, ensuring that the transformative potential of AI-ML technologies can be fully realized in lightweight material processing for EVs and beyond. Lightweight materials continuously evolved with their superior properties and performance in the automotive and other industries. The next section deals with the recent advancements in lightweight materials and their impact on EVs.

15.7 RECENT ADVANCEMENTS IN LIGHTWEIGHT MATERIALS AND THEIR IMPACT ON EVS

Recent advancements in lightweight materials have profoundly impacted the EV industry, enabling manufacturers to design vehicles that are not only more energy-efficient but also structurally robust and environmentally sustainable. One of the most significant breakthroughs is the development of high-strength aluminum alloys, such as the 7xxx and 6xxx series, which offer excellent mechanical properties and corrosion resistance while being significantly lighter than traditional steel [56]. These alloys are widely used in EV chassis, body panels, and battery housings to enhance durability without compromising weight savings [57].

Another crucial innovation is the emergence of magnesium-lithium (Mg-Li) alloys, which are among the lightest structural metals available [58]. These alloys offer a remarkable reduction in weight being up to 30% lighter than aluminum, making them ideal for applications requiring extreme weight minimization, such as battery enclosures and suspension components. However, challenges related to corrosion resistance and formability are being addressed through advanced alloying techniques and coatings [59]. Graphene-reinforced polymer composites have also gained traction in the EV sector due to their exceptional mechanical strength, electrical conductivity, and thermal stability [60]. These composites are being explored for use in lightweight body panels, structural reinforcements, and battery thermal management systems, where high strength and efficient heat dissipation are crucial. CFRPs and natural fiber composites [61], such as flax- or hemp-based reinforcements, are also seeing increased application in EV interiors and exteriors, offering sustainability benefits while maintaining mechanical performance. Table 15.1 presents the weight reduction achieved by different advanced lightweight materials.

The development of advanced MMCs, particularly aluminum or magnesium-based MMCs reinforced with ceramic particles (such as SiC or Al_2O_3), has further enhanced the potential of lightweight materials. These composites exhibit high wear resistance, superior strength-to-weight ratios, and improved thermal properties, making them suitable for structural and drivetrain components in EVs [62,63].

Beyond material innovations, significant advancements have also been made in processing techniques to optimize the manufacturability of these materials. FSW is now widely used for joining aluminum components in EVs, offering defect-free, high-strength welds while minimizing heat distortion [64]. Additionally, AM or 3D printing is playing a crucial role in fabricating lightweight composite and metal components with complex geometries, allowing for material optimization and reduction of waste [65,66]. AI-driven process optimizations in AM ensure that parameters such as layer thickness, printing speed, and cooling rates are precisely controlled, enhancing the quality and performance of printed parts.

These advancements in lightweight materials not only improve the structural integrity and crashworthiness of EVs but also directly impact their energy efficiency. By reducing the overall vehicle weight, these materials enable longer driving ranges, improved acceleration, and enhanced battery life—all of which are critical for increasing EV adoption. Moreover, the integration of recyclable and bio-based composites aligns with circular economy principles, reducing the carbon footprint of EV manufacturing and making vehicles more sustainable throughout their lifecycle. With continuous research and development in nano-engineered materials, hybrid composites, and self-healing

TABLE 15.1
Data on Weight Reduction Percentages Achieved by Different Materials

Material Type	Weight Reduction (%)	Key Applications in EVs	Source
Magnesium alloys	30–70%	Structural components, gearbox housings, steering columns	[67]
Carbon fiber composites	50–70%	Body panels, structural components, interior parts	
Aluminum and aluminum matrix composites	30–60%	Chassis components, body panels, battery enclosures	
Titanium	40–55%	Fasteners, exhaust systems, suspension components	
Glass fiber composites	25–35%	Interior components, non-structural panels	
Advanced high-strength steel (AHSS)	15–25%	Structural reinforcements, safety cages, crumple zones	
High-strength steels, composites, and hybrid materials	Up to 40%	Chassis components	[68]
Plastic wheels and components	30%	Wheels, doors, exteriors, interiors	[69]

materials, the future of lightweight materials in EVs looks promising. These innovations are expected to push the boundaries of what is possible in sustainable vehicle design, paving the way for next-generation EVs that are lighter, stronger, and more energy-efficient than ever before.

15.8 FUTURE DIRECTIONS

The evolution of AI and ML techniques is driving groundbreaking innovations in lightweight material development, particularly for the high-performance requirements of EVs. Emerging methodologies, such as GANs, GCNs, and attention-based transformers, are enabling unprecedented advances in material design and process optimization. GANs are being employed to create synthetic datasets, simulate material microstructures, and propose novel compositions, such as magnesium-lithium alloys or graphene-reinforced composites, which combine extreme lightness with exceptional strength. These capabilities allow researchers to explore unconventional alloying strategies or fiber–matrix combinations, drastically reducing the development time for materials suited to EV applications.

GCNs excel in modeling atomic and molecular interactions, providing insights into phase stability, grain growth kinetics, and defect propagation under specific processing conditions. For instance, GCNs have been applied to predict the anisotropic behavior of magnesium alloys during extrusion, enabling the tailoring of material properties for EV structural beams. Attention-based transformers, adapted from natural language processing, are advancing multi-scale modeling by effectively integrating data from experimental, computational, and sensor-based sources. These models streamline the understanding of complex phenomena, such as the interplay between thermal stresses and microstructural evolution during AM of hybrid composites for EV chassis components.

The opportunities for integrating these techniques into EV manufacturing processes are vast. AI-ML models can be paired with real-time process monitoring systems to create adaptive manufacturing environments. For example, in friction stir welding of dissimilar materials for EV battery enclosures, reinforcement learning algorithms adjust weld parameters dynamically, ensuring defect-free joins despite material and thermal mismatches. In advanced AM, AI-guided topology optimization allows for the creation of ultra-lightweight, high-strength lattice structures that maximize crash energy absorption while minimizing material usage. These designs are particularly relevant for EV crash zones and suspension components.

However, significant research gaps remain in lightweight material processing for EV applications. The lack of high-quality, multi-scale datasets capturing interdependencies between composition, microstructure, and performance across various manufacturing techniques is a critical bottleneck. Furthermore, current computational resources often fall short in enabling real-time AI-ML implementations for high-speed manufacturing processes, such as roll-to-roll composite fabrication or laser cladding of lightweight alloys. Addressing these challenges requires a coordinated effort to standardize data collection across labs and industries, invest in high-throughput experimental methods such as combinatorial synthesis, and develop hybrid computational architectures combining cloud computing with edge processing for localized AI-ML applications.

By addressing these research gaps and leveraging emerging AI-ML techniques, the industry can accelerate the development of innovative lightweight materials, pushing the boundaries of performance, efficiency, and sustainability in EV manufacturing. These advancements not only meet current demands but also lay the foundation for the next generation of electric mobility solutions.

15.9 CONCLUSION

The integration of AI and ML into lightweight material processing represents a paradigm shift in materials science and advanced manufacturing, enabling unprecedented levels of innovation, efficiency, and precision. Emerging techniques, such as transformer-based models for multi-scale material property prediction and Bayesian optimization for multi-objective parameter tuning, have revolutionized the development and processing of materials like aluminum-magnesium alloys, high-performance composites, and graphene-reinforced polymers. These technologies empower researchers and manufacturers to design materials with exceptional strength-to-weight ratios, thermal stability, and durability, while significantly reducing energy consumption and material wastage. For instance, in magnesium alloy extrusion for EV chassis components, AI optimizes grain refinement techniques to achieve superior crash resistance and fatigue life, ensuring both performance and safety.

In the context of EV manufacturing, the application of AI and ML extends beyond material development into process optimization and scalability. Predictive models tailored for high-speed manufacturing environments, such as AM and roll-to-roll processing, allow real-time adjustments to parameters like deposition rates and heat inputs, ensuring defect-free production. Advanced AI-driven digital twins simulate entire manufacturing workflows, predicting the impact of process conditions on material performance before physical trials, enabling faster iterations and reduced downtime. For example, reinforcement learning systems in AM enable the dynamic adjustment of laser power and scanning speeds, ensuring uniform microstructure and mechanical properties for intricate EV parts like lattice crash zones and battery cooling plates.

Sustainability emerges as a critical advantage of AI and ML in lightweight material processing. These technologies facilitate waste minimization through advanced process control and enable circular manufacturing practices, such as efficient recycling of aluminum and polymer composites into secondary applications without significant loss of performance. AI-based lifecycle analysis models evaluate the environmental impact of materials and processes, guiding manufacturers toward greener alternatives that align with global sustainability goals. For example, AI is being employed to identify optimal end-of-life strategies for composite materials, such as pyrolysis for fiber recovery or remelting for alloys.

Despite these advancements, the journey toward full-scale AI-ML adoption in material processing is still evolving. Future efforts must address the development of high-fidelity, multi-scale datasets, standardization of AI protocols for manufacturing, and the deployment of scalable hybrid computational architectures that combine cloud and edge computing. These innovations will bridge existing gaps, enabling a seamless integration of AI-ML into the complex workflows of EV manufacturing. Ultimately, this convergence will not only elevate the capabilities of materials science but also redefine the landscape of sustainable, high-performance EV production, ensuring a smarter and greener future for mobility.

REFERENCES

[1] Zhang W, Xu J. Advanced lightweight materials for Automobiles: A review. *Materials & Design*. 2022 Sep 1;221:110994.

[2] Taub AI, Luo AA. Advanced lightweight materials and manufacturing processes for automotive applications. *Mrs Bulletin*. 2015 Dec;40(12):1045–54.

[3] Delogu M, Zanchi L, Dattilo CA, Pierini M. Innovative composites and hybrid materials for electric vehicles lightweight design in a sustainability perspective. *Materials Today Communications*. 2017 Dec 1;13:192–209.

[4] Husain I, Ozpineci B, Islam MS, Gurpinar E, Su GJ, Yu W, Chowdhury S, Xue L, Rahman D, Sahu R. Electric drive technology trends, challenges, and opportunities for future electric vehicles. *Proceedings of the IEEE*. 2021 Jan 8;109(6):1039–59.

[5] Adhikari M, Ghimire LP, Kim Y, Aryal P, Khadka SB. Identification and analysis of barriers against electric vehicle use. *Sustainability*. 2020;12(12):4850. https://doi.org/10.3390/su12124850

[6] Wei J, Chu X, Sun XY, Xu K, Deng HX, Chen J, Wei Z, Lei M. Machine learning in materials science. *InfoMat*. 2019 Sep;1(3):338–58.

[7] Morgan D, Jacobs R. Opportunities and challenges for machine learning in materials science. *Annual Review of Materials Research*. 2020 Jul 1;50(1):71–103.
[8] Costa AP, Seabra MR, de Sá JM, Santos AD. Manufacturing process encoding through natural language processing for prediction of material properties. *Computational Materials Science*. 2024 Mar 25;237:112896.
[9] Senthilnathan A, Saseendran V, Acar P, Yamamoto N, Sundararaghavan V. Comparison and validation of stochastic microstructure characterization and reconstruction: Machine learning vs. deep learning methodologies. *Acta Materialia*. 2024 Oct 1;278: 120220.
[10] Park S, Youm S. Establish a machine learning based model for optimal casting conditions management of small and medium sized die casting manufacturers. *Scientific Reports*. 2023 Oct 11;13(1):17163.
[11] Kahnamouei JT, Moallem M. Advancements in control systems and integration of artificial intelligence in welding robots: A review. *Ocean Engineering*. 2024 Nov 15;312:119294.
[12] Ullrich K, von Elling M, Gutzeit K, Dix M, Weigold M, Aurich JC, Wertheim R, Jawahir IS, Ghadbeigi H. AI-based optimisation of total machining performance: A review. *CIRP Journal of Manufacturing Science and Technology*. 2024 Jun 1;50:40–54.
[13] Kumar S, Gopi T, Harikeerthana N, Gupta MK, Gaur V, Krolczyk GM, Wu C. Machine learning techniques in additive manufacturing: A state of the art review on design, processes and production control. *Journal of Intelligent Manufacturing*. 2023 Jan;34(1):21–55.
[14] Wang H, Li B, Gong J, Xuan FZ. Machine learning-based fatigue life prediction of metal materials: Perspectives of physics-informed and data-driven hybrid methods. *Engineering Fracture Mechanics*. 2023 May 19;284:109242.
[15] Krzywanski J, Sosnowski M, Grabowska K, Zylka A, Lasek L, Kijo-Kleczkowska A. Advanced computational methods for modeling, prediction and optimization—A review. *Materials*. 2024 Jul 16;17(14):3521.
[16] Jain NK, Jain VK, Deb K. Optimization of process parameters of mechanical type advanced machining processes using genetic algorithms. *International Journal of Machine Tools and Manufacture*. 2007 May 1;47(6):900–19.
[17] Choudhary K, DeCost B, Chen C, Jain A, Tavazza F, Cohn R, Park CW, Choudhary A, Agrawal A, Billinge SJ, Holm E. Recent advances and applications of deep learning methods in materials science. *npj Computational Materials*. 2022 Apr 5;8(1):59.
[18] Tang X, Li X, Ding Y, Song M, Bu Y. The pace of artificial intelligence innovations: Speed, talent, and trial-and-error. *Journal of Informetrics*. 2020 Nov 1;14(4):101094.
[19] Chibani S, Coudert FX. Machine learning approaches for the prediction of materials properties. *APL Materials*. 2020 Aug 1;8(8):1–8.
[20] Ceriotti M. Unsupervised machine learning in atomistic simulations, between predictions and understanding. *Journal of Chemical Physics*. 2019 Apr 21;150(15):1–15.
[21] Mohamed AM, Careri F, Khan RH, Attallah MM, Stella L. A novel porosity prediction framework based on reinforcement learning for process parameter optimization in additive manufacturing. *Scripta Materialia*. 2025 Jan 15;255:116377.
[22] Haque ME, Sudhakar KV. ANN back-propagation prediction model for fracture toughness in microalloy steel. *International Journal of Fatigue*. 2002 Sep 1;24(9):1003–10.
[23] Gopalakrishnan K, Kim S. Support vector machines approach to HMA stiffness prediction. *Journal of Engineering Mechanics*. 2011 Feb;137(2):138–46.
[24] Li S, Li S, Liu D, Zou R, Yang Z. Hardness prediction of high entropy alloys with machine learning and material descriptors selection by improved genetic algorithm. *Computational Materials Science*. 2022 Apr 1;205 :111185.
[25] Lee M, Park T, Min K. Matini-Net: Versatile material informatics research framework for feature engineering and deep neural network design. *Journal of Chemical Information and Modeling*. 2024 Nov 21;64(21):9424–9432.
[26] Feng S, Zhou H, Dong H. Using deep neural network with small dataset to predict material defects. *Materials & Design*. 2019 Jan 15;162:300–10.
[27] Deringer VL, Bartók AP, Bernstein N, Wilkins DM, Ceriotti M, Csányi G. Gaussian process regression for materials and molecules. *Chemical Reviews*. 2021 Aug 16;121(16):10073–141.

[28] Mohanty I, Banerjee R, Santara A, Kundu S, Mitra P. Prediction of properties over the length of the coil during thermo-mechanical processing using DNN. *Ironmaking & Steelmaking*. 2021 Sep 14;48(8):953–61.
[29] Liu Y, Zhao T, Ju W, Shi S. Materials discovery and design using machine learning, *Journal of Materiomics*. 2017;3(3):159–77. https://doi.org/10.1016/j.jmat.2017.08.002.
[30] Abhilash PM, Ahmed A. Convolutional neural network–based classification for improving the surface quality of metal additive manufactured components. *The International Journal of Advanced Manufacturing Technology*. 2023 Jun;126(9):3873–85.
[31] Senanayake NM, Carter JL. Computer vision approaches for segmentation of nanoscale precipitates in nickel-based superalloy IN718. *Integrating Materials and Manufacturing Innovation*. 2020 Dec;9(4):446–58.
[32] Ryu J, Kim H, Kim RM, Kim S, Jo J, Lee S, Nam KT, Joo YC, Yi GC, Lee J, Kim M. Dimensionality reduction and unsupervised clustering for EELS-SI. *Ultramicroscopy*. 2021 Dec 1;231:113314.
[33] Jin Y, Misra S. Controlling mixed-mode fatigue crack growth using deep reinforcement learning. *Applied Soft Computing*. 2022 Sep 1;127:109382.
[34] Alamri AH. Application of machine learning to stress corrosion cracking risk assessment. *Egyptian Journal of Petroleum*. 2022 Dec 1;31(4):11–21.
[35] Sattari K, Wu Y, Chen Z, Mahjoubnia A, Su C, Lin J. Physics-constrained multi-objective Bayesian optimization to accelerate 3d printing of thermoplastics. *Additive Manufacturing*. 2024 Apr 25;86:104204.
[36] Li XP, Zhao GQ, Guan YJ, Ma MX. Optimal design of heating channels for rapid heating cycle injection mold based on response surface and genetic algorithm. *Materials & Design*. 2009 Dec 1;30(10):4317–23.
[37] Li Y, Hu A, Fu Y, Liu S, Shen W, Hu H, Nie X. Al alloys and casting processes for induction motor applications in battery-powered electric vehicles: A review. *Metals*. 2022 Jan 24;12(2):216.
[38] Kulekci MK. Magnesium and its alloys applications in automotive industry. *The International Journal of Advanced Manufacturing Technology*. 2008 Nov;39:851–65.
[39] Amar KM, Singaravelu V, Neelima T, Poritosh R, Michael RS, Lawrence TD, Manjusri M. Sustainable composites for lightweight and flame retardant parts for electric vehicles to boost climate benefits: A perspective. *Composites Part C: Open Access*. 2023;12:100380, https://doi.org/10.1016/j.jcomc.2023.100380.
[40] Matli PR, Krishnan AV, Manakari V, Parande G, Chua BW, Wong SC, Lim CY, Gupta M. A new method to lightweight and improve strength to weight ratio of magnesium by creating a controlled defect. *Journal of Materials Research and Technology*. 2020 May 1;9(3):3664–75.
[41] Duan X, Naterer GF. Heat transfer in phase change materials for thermal management of electric vehicle battery modules. *International Journal of Heat and Mass Transfer*. 2010 Nov 1;53(23–24):5176–82.
[42] Hossain MS, Kumar L, Islam MM, Selvaraj J. A comprehensive review on the integration of electric vehicles for sustainable development. *Journal of Advanced Transportation*. 2022;2022(1):3868388.
[43] Czerwinski F. Welding and joining of magnesium alloys. *Magnesium Alloys-Design, Processing and Properties*. 2011 Jan 14:469–90.
[44] Jose JP, Joseph K. Advances in polymer composites: Macro-and microcomposites–state of the art, new challenges, and opportunities. *Polymer Composites*. 2012 Mar 21;1:1–6.
[45] Sadeghian A, Iqbal N. A review on dissimilar laser welding of steel-copper, steel-aluminum, aluminum-copper, and steel-nickel for electric vehicle battery manufacturing. *Optics & Laser Technology*. 2022 Feb 1;146:107595.
[46] Zheng J, Wang Q, Zhao P, Wu C. Optimization of high-pressure die-casting process parameters using artificial neural network. *The International Journal of Advanced Manufacturing Technology*. 2009 Oct;44:667–74.
[47] Chen S, Kaufmann T. Development of data-driven machine learning models for the prediction of casting surface defects. *Metals*. 2021 Dec 21;12(1):1.
[48] Patel V, De Backer J, Hindsefelt H, Igestrand M, Azimi S, Andersson J, Säll J. High speed friction stir welding of AA6063-T6 alloy in lightweight battery trays for EV industry: Influence of tool rotation speeds. *Materials Letters*. 2022 Jul 1;318:132135.

[49] Kim IS, Lee MG, Jeon Y. Review on machine learning based welding quality improvement. *International Journal of Precision Engineering and Manufacturing-Smart Technology*. 2023 Jul 5;1(2):219–26.

[50] Kumar S, Gopi T, Harikeerthana N, et al. Machine learning techniques in additive manufacturing: A state of the art review on design, processes and production control. *Journal of Intelligent Manufacture*. 2023;34:21–55. https://doi.org/10.1007/s10845-022-02029-5

[51] Sanei SH, Popescu D. 3D-printed carbon fiber reinforced polymer composites: A systematic review. *Journal of Composites Science*. 2020 Jul 24;4(3):98.

[52] Chu Y, Sun L, Li L. Lightweight scheme selection for automotive safety structures using a quantifiable multi-objective approach. *Journal of Cleaner Production*. 2019 Dec 20;241:118316.

[53] Valizadeh A, Sahara R, Souissi M. Alloys innovation through machine learning: A statistical literature review. *Science and Technology of Advanced Materials: Methods*. 2024 Dec 31;4(1):2326305.

[54] Compare M, Baraldi P, Zio E. Challenges to IoT-enabled predictive maintenance for industry 4.0. *IEEE Internet of Things Journal*. 2019 Dec 2;7(5):4585–97.

[55] Alfouneh M, Hoang VN, Luo Z, Luo Q. Topology optimization for multi-layer multi-material composite structures. *Engineering Optimization*. 2023 May 4;55(5):773–90.

[56] Czerwinski F. Aluminum alloys for electrical engineering: A review. *Journal of Materials Science*. 2024;59(32):14847–92.

[57] Lombardi A, Byczynski G, Vidanalage BG, Fatima A, Kar N. Development of a novel high-strength aluminum-cerium based rotor alloy for electric vehicle induction motor applications. *SAE International Journal of Advances and Current Practices in Mobility*. 2023;5(2023-01-0878):2365–72.

[58] Gao S, Zhao H, Li G, Sun G, Zhou L, Zhao Y. Strengthening mechanism and corrosion behavior of friction stir welded LAZ933 magnesium-lithium alloy. *Journal of Magnesium and Alloys*. 2023;11(100846):1–12.

[59] Han G, Lu Y, Jia H, Ding Z, Wu L, Shi Y, Pan F. Magnesium-based energy materials: Progress, challenges, and perspectives. *Journal of Magnesium and Alloys*. 2023;11(11):3896–25.

[60] Cui J, Zeng F, Wei D, Wang Y. Unraveling the effects of geometrical parameters on dynamic impact responses of graphene reinforced polymer nanocomposites using coarse-grained molecular dynamics simulations. *Physical Chemistry Chemical Physics*. 2024;26(28):19266–81.

[61] Luo Y, Shi Z, Qiao S, Tong A, Liao X, Zhang T, Xu W. Advances in nanomaterials as exceptional fillers to reinforce carbon fiber-reinforced polymer composites and their emerging applications. *Polymer Composites*. 2025;46(1):54–80.

[62] Muthampatty SS, Subramanian S, Shanmugasundaram R, Samudram MK. Modular multilevel converter-based hybrid energy storage system for electric vehicles: Design, simulation, and performance evaluation. *Energy Sources, Part A: Recovery, Utilization, and Environmental Effects*. 2024;46(1):16777–93.

[63] Zhong Z, Jiang X, Sun H, Wu Z, Yang L, Matamoros-Veloza A. Recent research on the optimization of interfacial structure and interfacial interaction mechanisms of metal matrix composites: A review. *Advanced Engineering Materials*. 2024;26(23):2401392.

[64] Dai H, Liang G, Liu L, Pan S. Research on friction stir welding of electric control system accessories of new energy vehicle charger. *Transactions of the Indian Institute of Metals*. 2024;77(2):405–16.

[65] Sarzyński B, Śnieżek L, Grzelak K. Metal additive manufacturing (MAM) applications in production of vehicle parts and components—A review. *Metals*. 2024;14(2):195.

[66] De A, Ramasubramian B, Ramakrishna S, Chellappan V. Advances in additive manufacturing techniques for electrochemical energy storage. *Advanced Materials Technologies*. 2024;9(4):2301439.

[67] Singh R. Lightweight metals and alloys in electric vehicle manufacturing: Enhancing performance and efficiency. *International Journal of Science and Research Archive*. 2024;13(2):1735–42.

[68] Barbu L. Novel lightweight materials to reduce weight of chassis components of Electric Vehicles by 40%. https://eurecat.org/en/novel-lightweight-materials-to-reduce-weight-of-chassis-components-of-electric-vehicles-by-40/

[69] Erk A, Charoensirisomboon P, Sun M. World's lightweight future. www.basf.com/tw/en/media/BASF-Information/Innovation/Worlds-lightweight-future

16 Adaptive Vibration Isolation in Electric Vehicles Using Magnetorheological Elastomers

Characterization and Applications

Praveen Shenoy Kaup, Anarghya Ananda Murthy, Nagaraja Kamath, Shrinivasa Mayya D., and Gangadharan K.V.

16.1 INTRODUCTION

Manufacturers struggle to control ride comfort, noise, vibration, and harshness (NVH) levels as electric car use increases worldwide. Primarily composed of high-frequency rotating motors, electric vehicle (EV) drivetrains produce different vibration patterns that contrast with those of internal combustion engines. Lighter, more robust parts used in EV architectures can help to improve vibrations. In the long term, electric cars run against particular challenges in vibration isolation.

16.1.1 Sources and Origins of Vibration in Electric Vehicle (EV) and Hybrid Electric Vehicle (HEV) Powertrains

An electric motor generates numerous vibrations; the most common source in EVs is the torque ripple from the electromagnetic forces produced inside the motor. Commonly used in EVs for their efficiency, permanent magnet synchronous motors (PMSMs) and switched reluctance motors (SRMs) generate ripple due to their design characteristics, including cogging torque and harmonics from air-gap flux [1,2]. The research conducted by Chuan et al. [3] examines the application of proportional integral resonant (PIR) controllers to mitigate torque ripple in PMSMs. This technology reduces speed ripple by 80%, demonstrating that electronic control can substantially assist in vibration mitigation. It recognizes the drivetrain natural frequency as a critical element in the design of these controls, as vibrations intensify when torque ripple frequencies coincide with this natural frequency. HEVs [4], which integrate ICEs with electric motors, pose supplementary torsional vibration issues due to the intricate dynamics of the transition between power sources. Further, Chen et al. [5] devised a model reference adaptive control (MRAC) system for HEV powertrains, which exhibited significant effectiveness in mitigating vibrations during transitions, particularly at crucial phases such as clutch engagement. Their findings demonstrate a substantial decrease in motor speed variability (up to 97%) and enhanced engine stability.

DOI: 10.1201/9781003596097-16

16.1.2 Approaches to Modeling and Simulation for Vibration Analysis

Different studies in the numerical simulation for vibration isolation highlight the need to reproduce the dynamics of EV and HEV powertrains to properly predict and control torsional vibrations.

Cao et al. [6] employed a PMSM model to simulate torsional vibrations in an EV driveline utilizing MATLAB/Simulink. A fuzzy-tuned proportional–integral–derivative (PID) controller was designed to react to load fluctuations, thereby reducing vibration and improving torque delivery to the wheels. These models offer significant insights for integrating magnetorheological elastomer (MREs), enabling researchers to forecast the interaction between MRE-based isolators and the vibrational properties of the drivetrain. Chen and Hu [7] present a more thorough model that produced a real-time engine torque model for estimating torsional vibrations in hybrid drivetrains. Their harmonic injection technique reduces vibration at specific frequencies, improving drivetrain stability. Also, Rocca and Russo [8] numerically investigated an MRE-based torsional vibration damper to address the start-up transient gear rattles during and after rapid clutch engagements. The effectiveness of the mechanism were evaluated by using the gear rattle index (GRI). These models highlight the potential of integrating advanced modeling with material advances such as MREs to address complex vibrational dynamics in EVs.

16.1.3 Active Vibration Mitigation in Electric Powertrains

With the various sources of vibrations being identified and numerical analysis being carried out, research has also been conducted experimentally to reduce the vibrations at the source. Various active control solutions have been devised to alleviate vibrations. Sun et al. [9] examined the application of active damping techniques in HEVs to regulate transient engine torque via harmonic current injection so vibration amplitudes can be efficiently diminished by addressing harmonic frequencies inside the drivetrain. Furthermore, research on active vibration control measured in all-wheel-drive EVs yielded encouraging outcomes. Ivanov et al. [10] showed that an active vibration controller (AVC), combined with wheel slip controllers and direct yaw moment controllers, improves vehicle stability and traction under diverse operating situations. Krishna et al. [11] conducted a study that examines adaptive filters, namely field-oriented control (FOC) systems, aimed at mitigating NVH in EVs. Their findings indicated the need for robust control systems to reduce the frequent power fluctuations in electric traction, directly affecting ride quality and passenger comfort. Adaptive models utilizing real-time input have also demonstrated their utility, particularly for vibration-sensitive systems such as HEVs, where power sources regularly alternate between engine and electric motor modes.

16.1.4 Experimental Validation and Practical Applications

Validation of the approaches' effectiveness through experimental setups and pragmatic implementations strengthens them. Ivanov et al. [10] carried out research on all-wheel-drive EVs equipped with onboard motors illustrating the practical implementation of vibration management via an AVC, validated in the European Union's FP7 E-VECTOORC project. This experimental investigation showed that MREs with active controllers could efficiently reduce vibrations without compromising vehicle handling. Reducing the torque interruption and enhancing its shifting comfort helped to build an magnetorheological (MR)-based dual-clutch system to maximize EV transmissions. This method enhanced the gear-shifting performance using exact controllability and fast response times. The smooth gearshifts lowered power losses during transitions, improving the vehicle's general efficiency. The maximum driving torque was obtained by using mathematical models developed from the relationships between the input current and the output torque of the internal and external clutches. Experimental validation revealed a less than 2 Nm difference, confirming the model's accuracy [12–14]. MR-based clutches are also used in two-wheeler EVs, where accurate torque transmission is necessary. The torque transmission capability of two variations of clutches is evaluated by conducting rigorous experiments to characterize the torque response, hysteresis effects, and thermal performance of magnetorheological fluid MRF) clutches. The design with slits in the clutch plates offers enhanced torque transmissibility due to improved fluid flow and better thermal dissipation as shown by Lokhande and Patil [15]. Also, a driving-adapt strategy is used to improve and utilize the maximum effect of implementing MR-based systems [25]. These techniques dynamically change the behavior of the gearbox to fit different driving circumstances and vehicle dynamics, optimizing powertrain efficiency and performance. The adaptive control system can maximize gear-changing schedules, torque distribution, and clutch engagement/disengagement timings through constant vehicle speed, acceleration, and road grade. This results in improved fuel economy, lower emissions, and a smoother and more comfortable driving experience. Moreover, EVs run on varied terrains, which results in ineffective performance. A novel dual-motor powertrain architecture incorporating an MR-based planetary transmission was used to address this issue by Deng et al. [16]. This configuration offers enhanced flexibility and controllability for EV drivetrains by enabling seamless transitions between driving modes and operating conditions. The MRF-based planetary transmission provides a wide range of gear ratios and torque distribution capabilities, allowing for optimized performance across various driving scenarios. The vehicle can achieve improved acceleration, efficiency, and regenerative braking performance by integrating two electric motors with this advanced transmission system.

Experimental validation is equally crucial for MRE applications. Research on real-time testing of MREs in diverse magnetic fields has demonstrated that these materials may adaptively respond to varied vibrational frequencies, rendering them suitable for EV vibration isolation. Incorporating MRE-based dampers into vehicle suspension and drivetrain systems effectively mitigates high-frequency torque ripple and low-frequency torsional vibrations.

16.1.5 Torsional Vibration in Dual-Motor Drive Systems of Electric Vehicles

Cao, Zhanyong, and He et al. [17] extensively studied the torsional vibration characteristics and suppression techniques for EVs using a dual-motor coupling drive system. A four-inertia modeling method derived from a simplified lumped mass model was proposed to capture these drive system dynamics precisely; this model simplified the complexity of torsional vibration while maintaining essential vibrational characteristics. The study pointed out that rapid torque changes in dual-motor configurations lead to significant torsional vibrations, negatively impacting vehicle comfort and performance. Simulation studies demonstrate that it accurately reflects the torsional behavior of EVs under various driving conditions, including abrupt torque changes and standard driving cycles. To solve torsional vibrations, Yue et al. [18] suggested an active control approach combining fuzzy logic, PID control, and Kalman filtering. This approach dynamically changes motor torque depending on real-time feedback from vehicle acceleration, reducing torsional vibrations and increasing structural resilience along several structural criteria. Simulation results using this control strategy highlight its effectiveness in reducing vibration amplitudes and enhancing driving comfort.

16.1.6 Torsional Vibration in Hybrid Electric Vehicles with Compound Planetary Gear Sets

Tang et al. [19] explored torsional vibration in HEVs with compound planetary power-split transmissions. This study concentrated on HEVs with intricate powertrain topologies, in which a compound planetary gear system connects an ICE and two electric motors (E1 and E2). The main objectives were to find the ideal system parameters to minimize torsional vibrations and identify their characteristics under varied operating situations. Using multi-body dynamic simulations in ADAMS software, they developed a torsional vibration model that considers important parts, including the torsional damper, half-shafts, and wheels. The study shows that the vehicle's body and wheels experienced a lot of low-frequency torsional vibrations. Higher frequencies, on the other hand, were associated with parts like the planetary gears and differential. The investigation discovered that the engine's rotational inertia and the half-shaft's torsional rigidity and damping had a major impact on the vibration amplitudes. Optimal damping and stiffness values, which minimize vibration, were identified as 15 N·m·s/rad and 618 N·m/rad for the torsional damper and 2760 N·m/rad for the half-shaft.

The study employed forced vibration analysis by exposing the model to sinusoidal torque inputs at different frequencies. Torsional damping is essential for reducing resonance-induced vibrations, as demonstrated by the identification of resonance conditions made possible by this method. The results of Tang et al. [20] confirmed that in order to attain the best NVH characteristics in HEV powertrains, mechanical parameters must be precisely adjusted. In order to enhance NVH performance, both studies emphasize the significance of active and passive vibration reduction techniques in EVs and HEVs. Tang et al. [20] highlighted the mechanical tuning components, such as torsional dampers and half-shafts in HEVs. In contrast, Yue et al. [18] concentrated on an active control technique using real-time feedback in dual-motor EVs. These approaches reflect two complementary paths toward vibration management: using real-time adaptive control systems and optimizing mechanical properties to reduce inherent vibrational tendencies.

Regarding applications, the findings are particularly relevant for vehicles with dual-motor setups, where rapid torque adjustments are standard, and an active control approach is essential to maintain stability. These combined insights are invaluable for developing advanced torsional vibration control strategies in EVs and HEVs, with the potential for further research into hybrid solutions that integrate active control and optimized mechanical design to achieve comprehensive vibration suppression. You et al. [21] explored the design of an automotive mechanical automatic transmission system based on torsional damping. The system utilizes physical damping elements to reduce torsional vibrations, enhancing gear-shifting smoothness and driver comfort without requiring complex control systems.

It is evident from the various vibration isolator methods that a complete isolator must be created for the isolation process to be independent of the system's external input. Smart materials like MREs can be used to design and develop such isolators. These isolators work effortlessly, irrespective of the input given to the system, such that they could be implemented in an active vibration isolation system.

16.1.7 Magnetorheological Elastomers as Semi-Active Vibration Isolators

MREs comprise elastomeric matrix-integrated magnetic particles, such as powdered carbonyl iron. When exposed to external magnetic fields, increased vibration damping and shock attenuation are made possible by MREs' changing elastomeric characteristics. The field-dependent control of damping and stiffness offers efficient isolations essential for electric car applications since driving conditions affect the vibrations caused by the motor and the road. Because MRE-based isolation systems allow for stiffness modifications based on operating conditions, they offer a tunable solution to these needs. The physical characteristics that alter when subjected to magnetic fields must be understood to assess the MRE isolator's performance. Typically, the two crucial parameters under study include:

Stiffness Modulation: MREs provide stiffness variations under the influence of magnetic fields, allowing for dynamic control by the field's intensity. Variations in the isolator stiffness directly impact the system's natural frequency, enabling EV vibration isolators to adjust to variations in vibration frequency. The advantage over passive isolation includes the adaptability of the isolator to external variations in the external frequency.

Damping Characteristics: Studies show that the damping behavior of MREs varies under different input displacements and magnetic fields, even though they exhibit varying degrees of damping. However, only stiffness adjustments can improve isolation effectiveness for particular frequency ranges. Because of the change in the inherent frequency of the system, the amplitude is reduced, which is the leading cause of the stiffness-induced isolation. While studies show that damping is a defined pattern in lateral vibrations, it is not a defined pattern in torsional vibrations.

16.2 METHODOLOGY

Two-part room temperature vulcanizing (RTV) silicone rubber is one of the key ingredients of the isotropic MRE. For this isolator, MoldSil 102LL (Part-A) was utilized. Performance polymers CAT 9 (Part B) is included with this. Carbonyl iron particles (CIP) from BASF, which have an average diameter of 5 µm, are also a component of the isotropic MRE. The percentage is computed using data from an earlier study by Poojary and Gangadharan [22], where the CIP and silicone rubber particles maintain a volume fraction of 73:27. Throughout the experiment, this ratio remains constant. This method entails converting each component element into the proper mass while accounting for the volume of the mold and the density of each substance. Furthermore, the mixture contains silicone oil, making up 10% of the volume of the silicone rubber.

The necessary amounts of silicone polymer (Part A) and CIP are thoroughly combined for 30 minutes to create the MRE sample. This guarantees that the iron particles are dispersed uniformly throughout the matrix of polymers. The mixture is then thoroughly mixed for two minutes with the catalyzer, which is Part B and includes 5% of the silicone matrix by Bodelot et al. [23]. In the following step, the mixture is poured into a mold with a diameter of 50 mm and a thickness of 10 mm. The degasification procedure, which follows, is carried out in a desiccator and eliminates any potential air bubbles using a vacuum pump. This ensures that the CIP's magnetic permeability stays consistent throughout the sample. The mixture was left to cure at room temperature and constant pressure for 24 hours by Li et al. [24] and Liao et al. [25]. Each step of the process is illustrated in Figure 16.1.

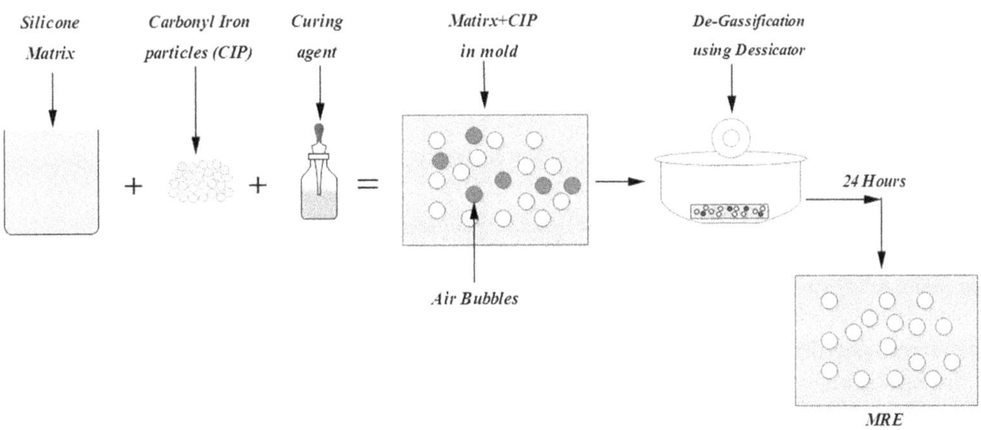

FIGURE 16.1 Preparation of an MRE isolator.

Experimental studies were conducted on a single-degree-of-freedom (SDoF) torsional system to assess the viability of magnetorheological elastomers for EV applications. Two aspects were looked into in the isolation processes:

- *Transmissibility Ratios:* The MRE-based system's transmissibility ratio decreased as the magnetic field intensity increased. Decreased transmissibility indicates improved isolation effectiveness, as the MRE efficiently separates more vibrational energy.
- *Frequency Shift:* In high magnetic fields, the MREs demonstrated a notable change in natural frequency, facilitating the evasion of resonant frequencies at different vehicle speeds. This field-responsive frequency shift is particularly beneficial for EVs, given that their motor speed varies significantly across various operating conditions.

As shown in Figure 16.1, a custom-made SDoF system is designed to comprehend the isolation of torsional vibrations. Figure 16.2 presents a schematic representation of the torsional system that operates with an SDoF. Two aluminum shafts have a negligible mass when putting together the SDoF system. The input shaft is connected to the electrodynamic shaker through a stinger to facilitate the engagement. The torsional system obtains harmonic disturbances of known amplitude from the shaft, which acts as the vibrating basis from which these disturbances are controlled. An AGILENT 33220A function generator is used to supply the shaker with its harmonic input, which is then amplified by an amplifier. The input accelerations are measured by connecting two YMC accelerometers to the serial configuration on a massless disc. The acceleration values are numerically combined to determine the system's torsion vibration amplitudes. The MRE isolator is positioned between the two aluminum shafts to provide the torsional system with semi-active isolation. Installing known inertia discs on the opposite end of the second shaft is necessary. In the same manner as the input, two accelerometers are connected to the inertial component in a serial configuration to measure the amplitudes of the output acceleration. The system is subjected to torsional vibrations to support the shafts and ensure they remain in line. This is accomplished by installing bearings in the appropriate positions. LabVIEW 2014 performs the processing once the signals have been received and transmitted through a data acquisition system [National Instruments (NI) 9234] at a sampling rate of 25.6 kS/s.

16.3 RESULTS AND DISCUSSION

16.3.1 Isolation Effect and Transmissibility

Figure 16.3 shows the field-dependent acceleration transmissibility plot for the rotor of inertia of 0.0002 kg.m^2 and angular displacement of 0.002 rad. The ordinate represents the ratio of the output acceleration to the input acceleration, while the abscissa represents the frequency of the torsional system. It is observed that the torsional natural frequency of the system in the absence of an external magnetic field is 25 Hz. With the increase in the magnetic field, there is a rightward shift in the frequency and it reaches a maximum of 28 Hz at 0.28 T. With an overall change of 3 Hz, the graph reveals the field-induced increment in the torsional natural frequency of the MRE.

As discussed, the field-stiffening action of the MRE explains the field-dependent rise in the torsional natural frequency (Figure 16.4). The filler iron particles generate dipoles under the influence of the magnetic field, which raises the compressive force between the filler and the matrix, hence improving the stiffness of the MRE. The torsional frequency of the MRE increases with the increasing stiffness.

Figure 16.5 displays the fluctuations in the isolation effect over changing magnetic fields. The fraction of the isolation effect rises as T, the magnetic field, rises. This implies that efficacy

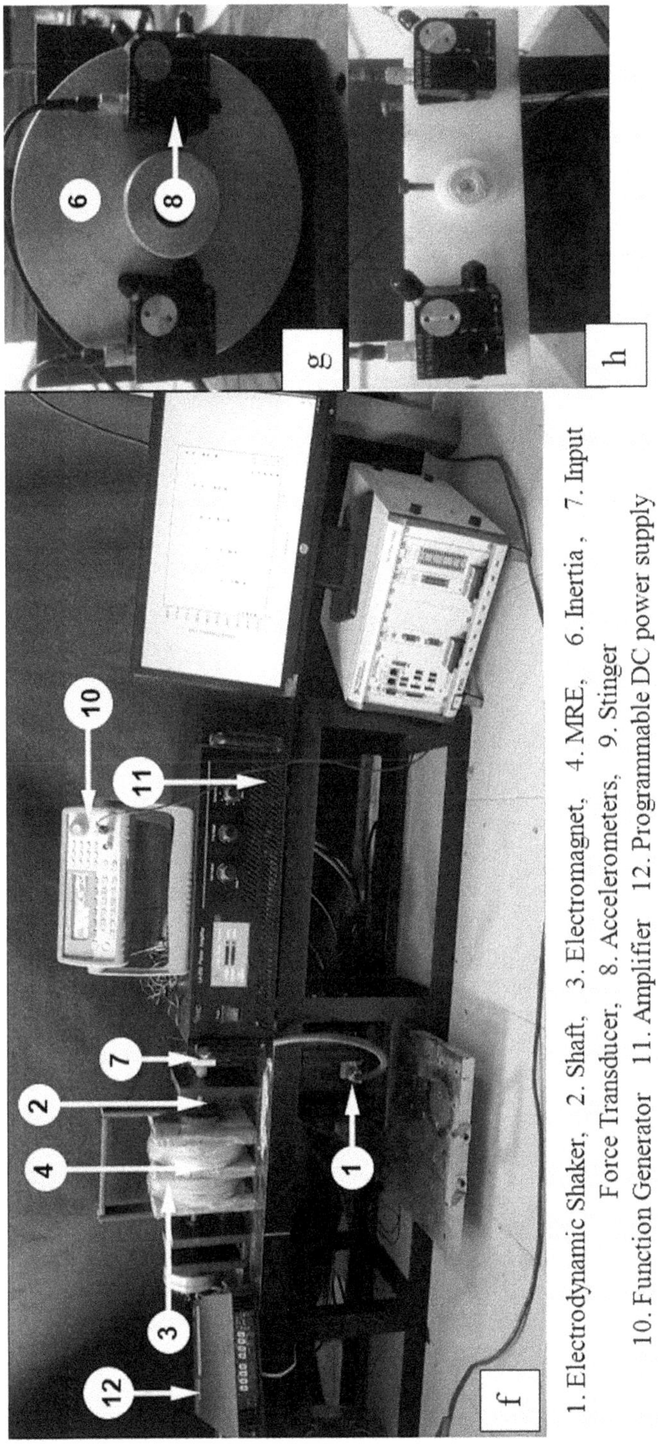

1. Electrodynamic Shaker, 2. Shaft, 3. Electromagnet, 4. MRE, 6. Inertia, 7. Input Force Transducer, 8. Accelerometers, 9. Stinger
10. Function Generator 11. Amplifier 12. Programmable DC power supply

FIGURE 16.2 Experimental setup for an SDoF torsional vibration isolation system.

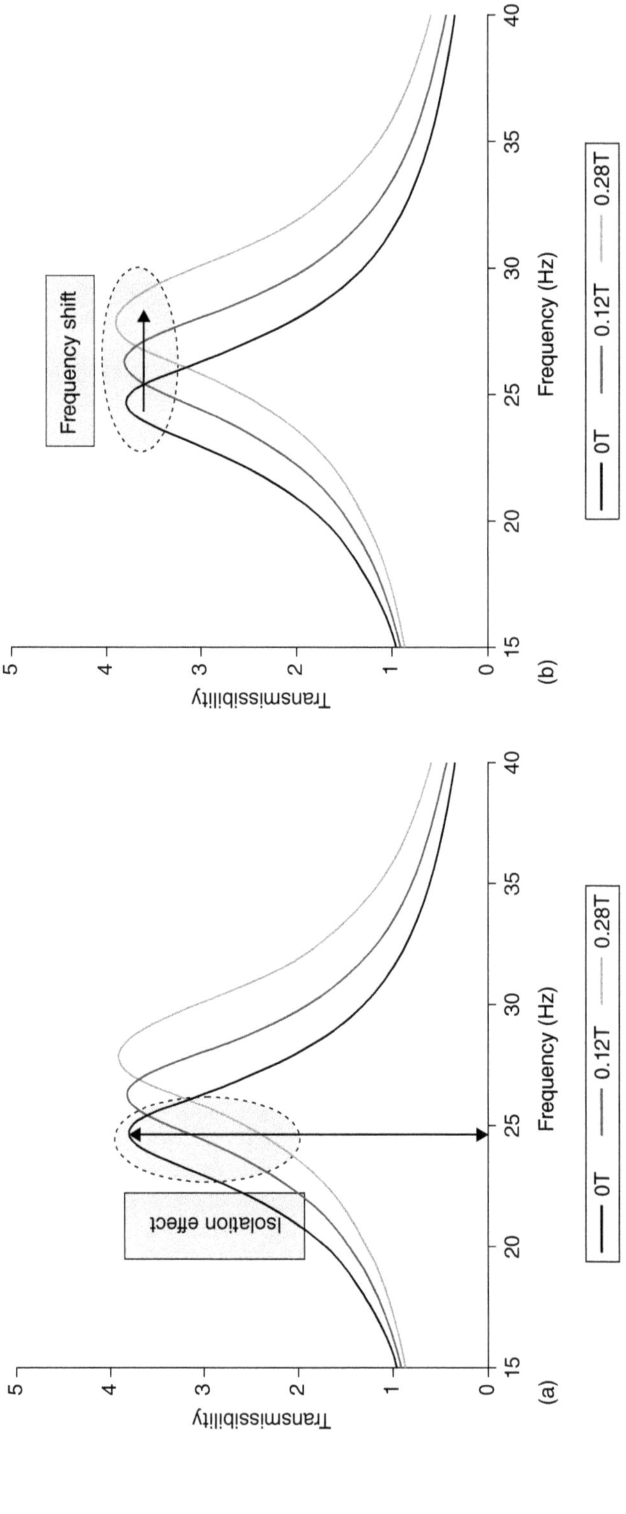

FIGURE 16.3 Isolation effect and frequency shift for an SDoF torsional isolation system.

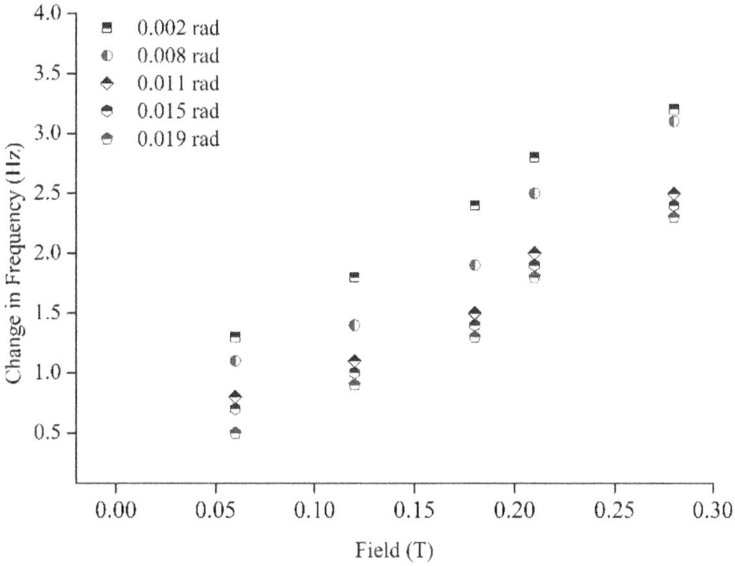

FIGURE 16.4 Natural frequency variations for varying input displacements.

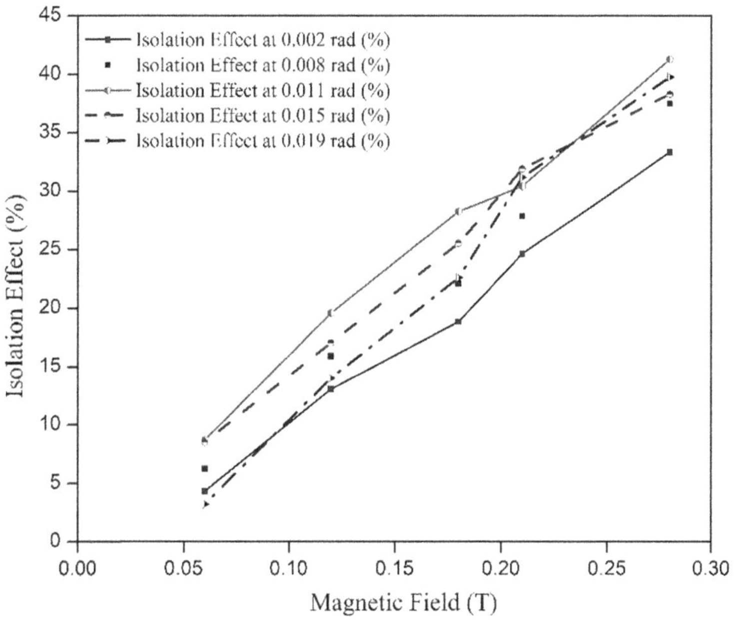

FIGURE 16.5 Field-induced isolation effect.

of isolation increases with increasing magnetic fields. The isolation effect exhibits an obvious apparent incremental trend. For 0.002 rad, for instance, the isolation effect increases from 0% to 6.57% within 0.00 T to 0.06 T. A clear change is seen by a growth of 17.36% from 0.06 T to 0.12 T. The isolation effect reaches a value of 25.78% with a magnetic field strength of 0.18 T, therefore showing a further improvement. Further, at 0.21 T, the variations increase to 32.89%. It reaches 37.36% at 0.28 T.

Between the first field values (from 0.00 T to 0.12 T), the isolation effect accelerates more rapidly, demonstrating increased sensitivity to changes in the field at lower levels. The enhancement in isolation effect persists even if it is less when the field climbs over 0.18 T, implying a possible tendency wherein developments in isolation may start to stabilize at higher fields.

16.3.2 Transmissibility Ratio vs. Magnetic Field for Torsional Vibrations

As observed at all vibration frequencies, the transmissibility ratio lowers with higher magnetic field strength. This suggests that applying a magnetic field in electric car torsional isolation systems can minimize the transmission of vibrations from the drivetrain to the vehicle structure, therefore lowering unwanted torsional oscillations. For electric cars, this is especially important since rapid throttle changes or regenerative braking cause high-frequency torsional vibrations. A stronger magnetic field could help to lower these vibrations, therefore facilitating more smooth power transmission. Furthermore, a minimum change in transmissibility is seen at lower magnetic fields. As the field approaches 0.21 T and beyond, the decrease gains ever more relevance. This suggests that a large magnetic field effectively separates the drivetrain from high-frequency torsional vibrations, which is necessary to ensure dependable and comfortable running of an electric car.

16.3.3 Isolation Effects for Torsional Vibration Control

Torsional vibration transmission to the EV's chassis can be helped to be reduced by raising the magnetic field strength over 0.21 T. This helps especially to reduce the high-frequency torsional oscillations typical of dynamic driving circumstances. Increasing the magnetic field strength greatly increases the isolation effect, therefore enabling better control of torsional vibrations. Good isolation will reduce vibration transfer from the drivetrain to the car, therefore enhancing passenger comfort and protection of drivetrain components. Higher frequencies—that is, 0.011 rad and above—show a more marked isolation effect in relation to magnetic fields than lower frequencies. For electric cars especially, this is crucial since quick changes in torque demand often produce high-frequency torsional vibrations. Reducing these vibrations helps drivetrain components last longer and provides more effective power transfer. The isolation effect across many frequencies converges at about 0.28 T. This suggests that using a magnetic field of about 0.28 T could provide homogeneous torsional vibration isolation benefits in both high-speed and low-speed driving scenarios.

Maximum torsional vibration isolation across several operating conditions is best achieved from a magnetic field in the 0.21 T–0.28 T range. Especially during torque changes, this increases riding comfort and extends drivetrain lifetime.

For various magnetic flux densities, Figure 16.6 displays the isolation effect (%) as a function of angular displacements. The isolation effect first increases with angular displacements for every magnetic flux density; with greater angular displacements, it either begins to level out or somewhat decrease. This trend implies that isolation efficiency increases up to a given angular displacement and thereafter usually decreases. Fascinatingly, higher magnetic flux densities, such as 0.28 T and 0.21 T, always show a greater isolation effect across all angular displacements than lower magnetic flux densities, suggesting that stronger magnetic fields improve the isolation performance over a spectrum of angular displacements. The system is more sensitive to angular displacement changes at higher magnetic fields as the steeper slopes of higher fields demonstrate.

Figure 16.7 shows the rate of change in isolation effect against frequency for varied magnetic flux densities. Certain values exhibit an initial increase in isolation effect with angular displacements; with larger angular displacements, the rate of change turns negative, therefore indicating a reduction

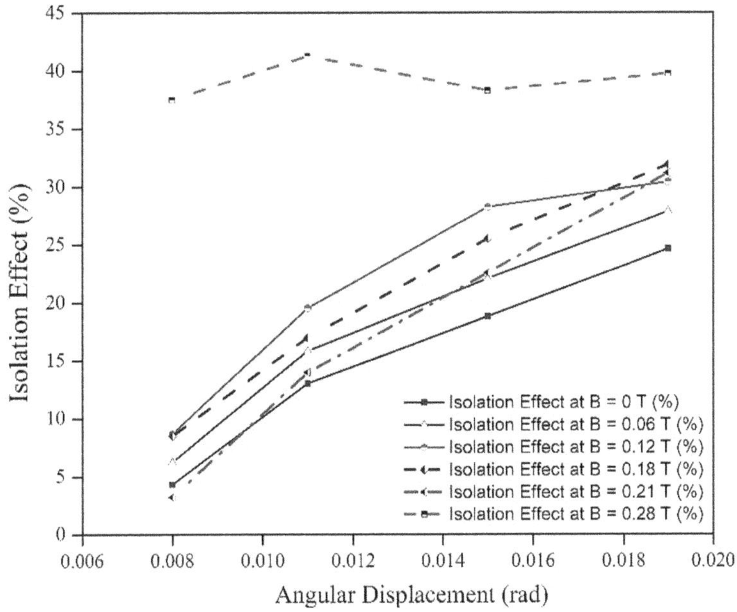

FIGURE 16.6 Isolation effect as a function of angular displacement.

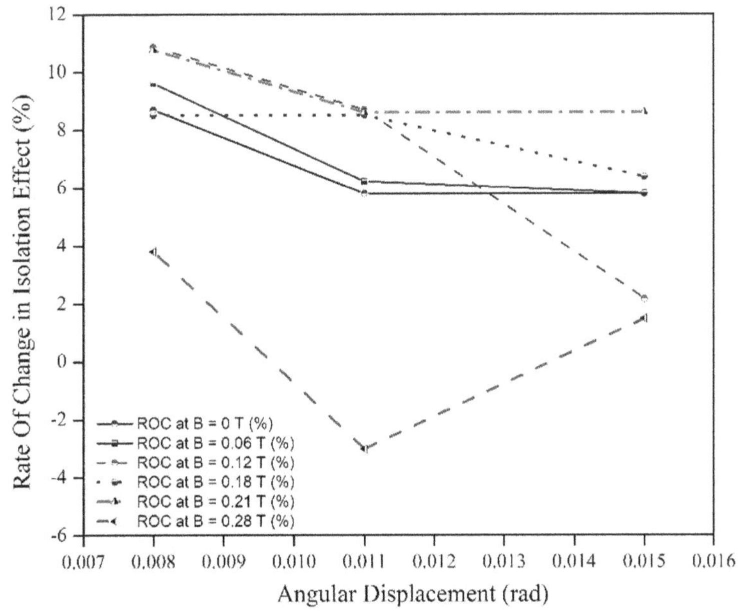

FIGURE 16.7 Rate of change of isolation effect.

in isolation efficacy. Large oscillations in the data for 0.28 T indicate that the isolation effect at this magnetic field is sensitive to rotational displacements and shows both major positive and negative changes. The rate of change stays rather steady for magnetic flux densities of 0.18 and 0.12 T; smaller slopes indicate less susceptibility to angular displacement changes.

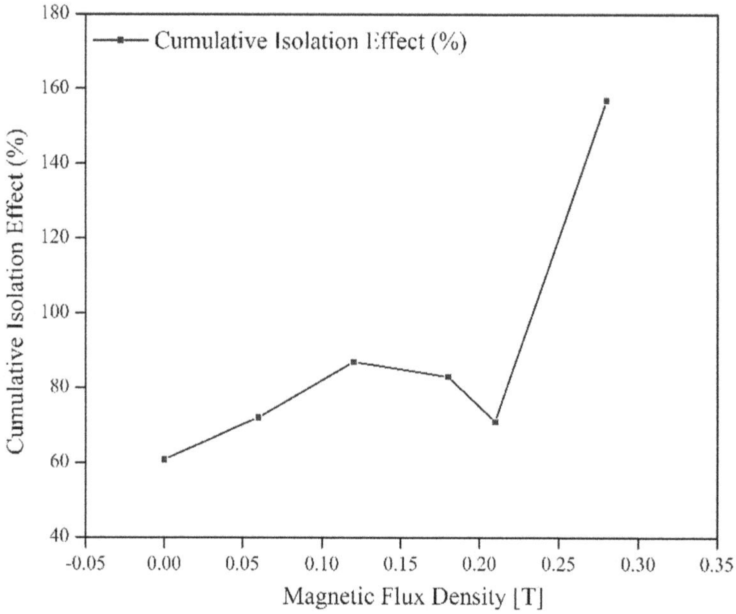

FIGURE 16.8 Cumulative isolation effect for varying magnetic fields.

Figure 16.8 displays the isolation effects over all the accessible frequencies for every magnetic flux density. This reveals the field values with the optimum overall isolation performance and implies the isolation capabilities at various magnetic fields.

16.4 CONCLUSION

This chapter elucidates the possible application of a smart torsional vibration isolator using a magnetorheological elastomer for EVs and HEVs. An SDoF torsional vibration system was developed, and the properties of the MRE isolator were studied across various loading conditions. The results indicate that the stiffness of the MRE-based torsional isolator escalates with the intensity of the applied magnetic field. This characteristic is advantageous for EVs, where fluctuating load conditions necessitate adjustable stiffness. Furthermore, an approximate 35% increase in stiffness was observed, indicating substantial control over the system's inherent frequency. Further, the increase in angular displacement reduces the MRE system's natural frequency, attributable to an associated decline in effective stiffness. This adaptive feature would enable EV systems to modify their vibration response according to real-time torque requirements, hence improving ride comfort and component longevity. The MRE-based system attains reduced transmissibility ratios in magnetic fields, resulting in diminished vibrational energy transfer. At a field strength of 0.06 T, the isolation effect improved by as much as 15%, indicating that these isolators could markedly diminish detrimental vibrations in sensitive EV components. Further, the transmissibility ratio, a measure of vibration isolation efficacy, fluctuates according to the applied field and displacement. In EVs, the isolator's efficacy can be tuned to adapt to operational variations, facilitating dynamic vibration control during high-torque or regenerative braking conditions.

REFERENCES

[1] Wang, X., Wang, D., Peng, C., Wang, B., & Wang, X. (2023). Torsional vibration analysis and suppression of interior permanent magnet synchronous motor with staggered segmented rotor for

electric vehicles. *IEEE Transactions on Transportation Electrification, 11*(10320389), 1–12. https://doi.org/10.1109/TTE.2023.3333906

[2] Sahu, A. K., Emadi, A., & Bilgin, B. (2023). Noise and vibration in switched reluctance motors: A review on structural materials, vibration dampers, acoustic impedance, and Noise masking methods. *IEEE Access, 11*, 27702–27718. https://doi.org/10.1109/ACCESS.2023.3257124

[3] Chuan, H., Fazeli, S. M., Wu, Z., & Burke, R. (2020). Mitigating the torque ripple in electric traction using proportional integral resonant controller. *IEEE Transactions on Vehicular Technology, 69*(10), 10820–10831. https://doi.org/10.1109/TVT.2020.3013414

[4] Yue, H., He, H., & Han, M. (2023a). Study on torsional vibration characteristics and suppression of electric vehicles with dual-motor drive system. *Journal of the Franklin Institute, 360*(1), 380–402. https://doi.org/10.1016/J.JFRANKLIN.2022.11.008

[5] Chen, X., Peng, D., Wu, W., Liu, H., & Zheng, X. (2022). Active control of torsional vibration during mode switching of hybrid powertrain based on adaptive model reference. *Machines, 10*(8), 1–18. https://doi.org/10.3390/machines10080647

[6] Cao Zhanyong and He, F. and L. H. and X. Z. (2018a). Research on torsional vibration suppression of electric vehicle driveline. In J. and C. M. and Y. Z. and N. Q. Li Kang and Zhang (Ed.), *Advances in Green Energy Systems and Smart Grid* (pp. 61–69). Springer Singapore.

[7] Chen, S., & Hu, M. (2023). Active torsional vibration suppression of hybrid electric vehicle drive system based on optimal harmonic current instruction calculation method. *Mechanical Systems and Signal Processing, 205*, 110837. https://doi.org/10.1016/J.YMSSP.2023.110837

[8] Rocca, E., & Russo, R. (2022). Gear Rattle during a start-up transient in a driveline equipped with a tuneable torsional Vibration Damper with Magneto-Rheological Elastomers. *Proceedings of the Institution of Mechanical Engineers, Part K: Journal of Multi-Body Dynamics, 236*(1), 140–150. https://doi.org/10.1177/14644193211073016/ASSET/IMAGES/LARGE/10.1177_14644193211073016-FIG10.JPEG

[9] Sun, S. (2020). *Research on Intelligent Suppression for Torsional Vibration of Electric Vehicle Drivetrain Certificate of Original Authorship* (Vol. 15(4), e180521177964: pp. 1–22). Bentham Science Publishers.

[10] Ivanov, V., Savitski, D., Orus, J., Fortun, J. M. R., Sorniotti, A., & Gruber, P. (2015). All-wheel-drive electric vehicle with on-board motors: Experimental validation of the motion control systems. In *IECON 2015–41st Annual Conference of the IEEE Industrial Electronics Society* (pp. 1729–1734). https://doi.org/10.1109/IECON.2015.7392351

[11] Krishna, K., Mahesha, G. T., Hegde, S., & Satish Shenoy, B. (2023). A review on vibrations in electric and hybrid electric vehicles. *Journal of The Institution of Engineers (India): Series C, 104*(2), 423–438. https://doi.org/10.1007/s40032-023-00930-3

[12] Zhang, H., Deng, L., Zhao, J., Li, W., & Du, H. (2024). Comparative analysis and optimization of nonlinear hysteresis models for a magnetorheological fluid dual-clutch of an electric vehicle transmission. *Smart Materials and Structures, 33*(9), 095045. https://doi.org/10.1088/1361-665X/AD6ECD

[13] Zhang, H., Du, H., Sun, S., Li, W., & Wang, Y. (2019). Design and analysis of a novel magnetorheological fluid dual clutch for electric vehicle transmission. In *SAE Technical Papers, 2019-January*(January). https://doi.org/10.4271/2019-01-5014

[14] Zhang, H., Du, H., Sun, S., Zhao, J., Ning, D., Li, W., & Wang, Y. (2021). A novel magneto-rheological fluid dual-clutch design for two-speed transmission of electric vehicles. *Smart Materials and Structures, 30*(7), 075035. https://doi.org/10.1088/1361-665X/AC0674

[15] Lokhande, S. B., & Patil, S. R. (2021). Experimental characterization and evaluation of magnetorheological clutch for an electric two-wheeler application. *Measurement, 175*, 109150. https://doi.org/10.1016/J.MEASUREMENT.2021.109150

[16] Deng, L., Zhao, J., Ning, D., Wong, P., Zhao, J., Li, W., & Du, H. (2024). Design of a dual-motor powertrain with magnetorheological planetary transmission for electric vehicles. *SAE Technical Papers*. https://doi.org/10.4271/2024-01-2636

[17] Cao, Z., He, F., Li, H. and Xu, Z. (2018b). Research on torsional vibration suppression of electric vehicle driveline. In K. Li, J. Zhang, M. Chen, Z. Yang, & Q Niu (Eds.), *Advances in Green Energy Systems and Smart Grid* (pp. 61–69). Springer Singapore.

[18] Yue, H., He, H., & Han, M. (2023b). Study on torsional vibration characteristics and suppression of electric vehicles with dual-motor drive system. *Journal of the Franklin Institute*, *360*(1), 380–402. https://doi.org/10.1016/J.JFRANKLIN.2022.11.008

[19] Tang, X., Zhang, J., Zou, L., Yu, H., & Zhang, D. (2014a). Study on the torsional vibration of a hybrid electric vehicle powertrain with compound planetary power-split electronic continuous variable transmission. *Proceedings of the Institution of Mechanical Engineers, Part C: Journal of Mechanical Engineering Science*, *228*(17), 3107–3115. https://doi.org/10.1177/0954406214526162

[20] Tang, X., Zhang, J., Zou, L., Yu, H., & Zhang, D. (2014b). Study on the torsional vibration of a hybrid electric vehicle powertrain with compound planetary power-split electronic continuous variable transmission. *Proceedings of the Institution of Mechanical Engineers, Part C: Journal of Mechanical Engineering Science*, *228*(17), 3107–3115. https://doi.org/10.1177/0954406214526162/ASSET/IMAGES/LARGE/10.1177_0954406214526162-FIG17.JPEG

[21] You, Z. (2022). Erosion wear analysis of coconut shell dust filled epoxy composites using computational fluid dynamics and Taguchi method. *Applied Mathematics and Nonlinear Sciences*, *8*(1), 989–998. https://doi.org/10.2478/amns.2022.2.0086

[22] Poojary, U. R., & Gangadharan, K. V. (2018). Integer and fractional order-based viscoelastic constitutive modeling to predict the frequency and magnetic field-induced properties of magnetorheological elastomer. *Journal of Vibration and Acoustics, Transactions of the ASME*, *140*(4). https://doi.org/10.1115/1.4039242

[23] Bodelot, L., Voropaieff, J. P., & Pössinger, T. (2018). Experimental investigation of the coupled magneto-mechanical response in magnetorheological elastomers. *Experimental Mechanics*, *58*(2), 207–221. https://doi.org/10.1007/S11340-017-0334-7/FIGURES/12

[24] Li, Y., Li, J., Li, W., & Du, H. (2014). A state-of-the-art review on magnetorheological elastomer devices. *Smart Materials and Structures*, *23*(12), 123001. https://doi.org/10.1088/0964-1726/23/12/123001

[25] Liao, P., Ning, D., Wang, T., & Du, H. (2022). A driving-adapt strategy for the electric vehicle with magneto-rheological fluid transmission considering the powertrain characteristics. *Sensors*, *22*, 9619. https://doi.org/10.3390/S22249619

17 Elevating Electric Vehicle Battery Management with Emerging Technologies

Kavitha K.J., Ananda M.K., Madhu Basavaraj, Sohit Reddy Kalluru, and Prasanna Kumar Reddy Gurijala

17.1 INTRODUCTION

As the global push toward sustainable transportation intensifies, electric vehicles (EVs) have emerged as a cornerstone of this shift, offering an eco-friendly alternative to conventional internal combustion engine vehicles. The heart of every EV lies in its energy storage system, typically a lithium-ion battery which is responsible for powering the vehicle. However, the effectiveness, safety, and longevity of these batteries are heavily reliant on an efficient battery management system (BMS) [1–5]. A BMS plays a crucial role in monitoring and controlling key battery functions such as charge levels, temperature, and voltage to ensure optimal performance [6–8]. Without a sophisticated BMS, the battery's efficiency could be compromised, leading to reduced driving range, overheating, and even safety hazards like thermal runaway, where battery cells overheat and potentially catch fire. Given the high energy densities in modern batteries and the need for fast charging, managing these risks is critical. Moreover, the energy storage technologies within EVs are evolving rapidly, transitioning from conventional lithium-ion batteries to emerging systems like solid-state batteries (SSBs), lithium-sulfur, and sodium-ion batteries [9–14]. Each of these technologies introduces new challenges in terms of thermal regulation, charge/discharge cycling, and energy density management, requiring the BMS to adapt accordingly. As EV adoption continues to grow, the need for advanced materials in BMSs is becoming more apparent. These materials can enhance thermal management, improve battery longevity, and ensure the safety and reliability of energy storage systems under various driving conditions. In this context, exploring emerging materials in BMSs not only addresses current technological limitations but also paves the way for the next generation of energy-efficient and safer EVs [14–20].

Despite the growing success of EVs, current BMSs face several challenges that limit battery performance, longevity, and safety. Some of the most critical challenges include:

Thermal Management: Batteries generate heat during charging and discharging, especially during fast charging cycles. Inefficient thermal regulation can lead to thermal runaway, causing battery degradation, reduced efficiency, and safety risks, such as overheating and potential fires. Current BMS technologies struggle to maintain optimal temperature control across all cells in a battery pack, especially in extreme climates [21–24].

Charge/Discharge Cycles: Repeated charge and discharge cycles degrade battery materials over time, reducing their capacity and performance. Managing the balance between maximizing battery life and optimizing energy usage presents a significant challenge for BMS technology. Accurate state of charge (SoC) and state of health (SoH) estimation are critical [25], but existing BMS solutions often fall short, leading to reduced battery efficiency and lifespan.

DOI: 10.1201/9781003596097-17

Safety Concerns: As batteries in EVs store large amounts of energy, any fault in the system can have catastrophic consequences. Current BMS designs face difficulties in real-time monitoring of the battery's internal health and predicting potential failures, especially under high stress or fast charging scenarios. Existing materials used in BMS sensors and components can limit the ability to detect dangerous conditions in time to take preventive action.

Energy Efficiency: Maximizing the energy available for vehicle operation while minimizing losses during power conversion is a key challenge for BMSs. Conventional materials in BMS architecture do not fully support high-efficiency power conversion and energy management, especially when dealing with next-generation battery chemistries, which require more precise control.

To address the above-mentioned limitations in current BMS technology, there is a need for emerging materials that can significantly enhance the system's performance, safety, and reliability. Emerging materials offer innovative solutions to improve the thermal management, energy efficiency, and safety features of BMSs in several ways:

Advanced Thermal Management: New materials such as phase-change materials (PCMs), graphene, and ceramics can provide superior heat dissipation [20], helping to regulate battery temperatures more effectively during operation and fast charging. These materials can enable the design of BMSs that prevent overheating, thereby extending battery life and reducing the risk of thermal runaway.

High-Precision Sensors: Nanomaterials and conductive polymers can improve the sensitivity and accuracy of sensors used in BMSs to monitor voltage, temperature, and current in real time. These materials allow for better prediction of SoC and SoH, leading to more efficient management of charge/discharge cycles and extending the overall battery lifespan.

Safety Enhancements: The use of solid electrolytes and non-flammable materials in emerging BMS architectures can drastically reduce the risk of internal short circuits and overheating. Additionally, integrating wide bandgap semiconductors like silicon carbide (SiC) and gallium nitride (GaN) in power electronics can increase the efficiency of energy transfer and reduce thermal stress on battery cells.

Efficiency Improvements: New materials that support more efficient energy transfer, such as high-conductivity alloys and composite materials, can improve the overall energy efficiency of EVs. These materials can reduce energy losses in power conversion, allowing the BMS to make better use of the stored energy and enabling longer driving ranges.

The integration of these emerging materials into BMS technology holds the potential to overcome current limitations, ensuring that EVs can meet future demands for safety, efficiency, and sustainability.

17.2 BATTERY TECHNOLOGIES IN ELECTRIC VEHICLES

17.2.1 Overview of Batteries

The rapid evolution of EV battery technologies is at the core of improving vehicle range, performance, and safety. Several battery chemistries are under development or currently deployed in EVs, each offering distinct advantages and challenges. These technologies also require advanced BMSs to optimize their performance and ensure safety. Below is an overview of key battery technologies and the role of BMSs in enhancing their efficiency. An overview of battery technologies used in EVs is provided below:

Lithium-Ion Batteries (Li-ion): These batteries are the most widely used technology in EVs due to their high energy density, relatively long cycle life, and lightweight nature. They consist of a graphite anode, a lithium-metal oxide cathode (e.g., $LiCoO_2$), and a liquid electrolyte. At the same time, the current challenges with Li-ion batteries are prone to overheating, especially during fast charging, and they can degrade over time due to repeated charge/discharge cycles [23]. Additionally, they present safety risks such as thermal runaway if not properly managed.

Solid-State Batteries (SSBs): These batteries are seen as the next frontier in battery technology. Instead of using liquid electrolytes, they employ solid electrolytes, which offer significant safety improvements by reducing the risk of leakage and fire. They also promise higher energy density, enabling longer driving ranges for EVs. Despite their potential, SSBs face significant barriers in terms of scalability and production costs. Additionally, the materials used in solid-state electrolytes need to overcome issues such as poor ionic conductivity at lower temperatures.

Lithium-Sulfur Batteries (Li-S): These batteries offer a much higher theoretical energy density compared to Li-ion batteries. They replace traditional lithium-metal oxide cathodes with sulfur, which is cheaper and more abundant. Also, such batteries suffer from rapid capacity fading and poor cycle life, primarily due to the dissolution of sulfur compounds in the electrolyte (the "shuttle effect"). Managing these degradation mechanisms is crucial for making Li-S batteries viable for EVs.

Sodium-Ion Batteries (Na-ion): These batteries are an emerging technology offering a lower-cost alternative to Li-ion batteries. They are based on abundant and cheaper sodium resources, making them more sustainable. However, the lower energy density and larger ionic size of sodium compared to lithium lead to reduced energy storage capacity. Additionally, Na-ion batteries require significant improvements in performance to match Li-ion batteries' lifespan and efficiency.

17.2.2 Role of BMSs in Enhancing Battery Life, Safety, and Performance

Regardless of the battery chemistry used, the BMS plays a pivotal role in maximizing performance, ensuring safety, and extending the lifespan of EV batteries. Key roles of BMSs in enhancing battery performance include:

Temperature Monitoring and Control: Thermal management is crucial to prevent overheating and ensure consistent battery performance. The BMS monitors and regulates temperature across all cells, preventing thermal imbalances that can lead to degradation or thermal runaway.

State-of-Charge (SoC) and State-of-Health (SoH) Estimation: BMS systems ensure accurate estimations of the battery's SoC and SoH, which are essential for optimizing charging strategies and predicting battery lifespan. This enables precise control over charging and discharging, preventing overcharging or deep discharging, both of which can reduce battery capacity.

Cell Balancing: To ensure uniform performance across the battery pack, the BMS performs cell balancing equalizing the charge across individual cells. This improves overall efficiency and prevents weaker cells from degrading faster, enhancing the longevity of the battery.

Fault Detection and Safety: The BMS acts as the primary safety system, detecting anomalies such as over-voltage, short circuits, or temperature spikes. In case of a fault, the BMS can disconnect the battery from the powertrain to prevent accidents, fires, or battery damage.

Fast Charging Management: Fast charging introduces high thermal stress and increased risk of degradation. A robust BMS can carefully manage current and voltage during fast charging cycles, ensuring the battery operates within safe thermal limits while delivering quicker charging times.

17.2.3 Limitations of Traditional Materials Used in BMSs

The performance of BMSs is also limited by the materials currently used in their design, especially when handling newer battery chemistries:

Graphite Anodes: Graphite has been the standard anode material in Li-ion batteries due to its stability and abundance. However, it has a relatively low energy capacity compared to emerging anode materials like silicon or lithium metal. BMSs struggle to optimize performance for next-generation materials like silicon due to graphite's inherent limitations in energy density and cycling stability.

Conventional Liquid Electrolytes: Liquid electrolytes are commonly used in Li-ion batteries, but they pose safety risks such as leakage, flammability, and poor performance at extreme temperatures. As new chemistries like SSBs emerge, conventional liquid electrolytes are no longer viable, and current BMS systems need to adapt to handle new solid-state materials.

Traditional Thermal Management Materials: Current thermal management materials used in BMS, such as standard heat sinks or cooling fluids, are inadequate for newer high-energy-density batteries [26]. They may not dissipate heat quickly enough in high-demand scenarios like fast charging or high-power driving, leading to reduced battery life and safety issues.

Sensors and Conductors: Traditional BMSs rely on basic sensor technologies and copper conductors, which have limitations in accuracy, conductivity, and durability. The need for higher precision in real-time monitoring and more efficient power conversion is driving the demand for advanced materials like nanomaterials, wide-bandgap semiconductors, and lightweight alloys.

The transition to advanced battery technologies in EVs, such as SSBs and lithium-sulfur batteries, requires corresponding advancements in BMSs. The limitations of traditional materials, like graphite and liquid electrolytes, pose significant challenges in managing new high-performance batteries. By integrating emerging materials into BMS architecture, such as advanced thermal management systems, high-precision sensors, and solid electrolytes, EV batteries can achieve higher efficiency, safety, and longer lifespans, paving the way for the next generation of sustainable transportation.

17.3 EMERGING MATERIALS IN BMS

Solid Electrolytes: Solid electrolytes are pivotal in SSBs as they replace flammable liquid electrolytes, greatly enhancing safety by reducing the risk of leaks and fires. Unlike liquid electrolytes, solid electrolytes are non-volatile and provide better electrochemical stability, making them less prone to dendrite growth, which is a common issue in liquid-based lithium-ion batteries. This stability extends the cycle life of batteries and enables the use of more energy-dense materials, improving the overall performance of SSBs. Solid electrolytes also allow for compact battery designs, improving energy density.

High-Thermal Conductivity Materials: Materials like graphene, carbon nanotubes (CNTs), and ceramics are crucial for thermal management in batteries. Their high thermal conductivity helps in dissipating heat generated during charging and discharging cycles, reducing the risk of thermal runaway. Graphene and CNTs, with their excellent electrical and thermal properties, can be integrated into battery components to ensure uniform heat distribution. Ceramics are used in thermal interface materials and separators to control temperature. These materials improve the safety and lifespan of batteries by maintaining optimal operating temperatures.

Advanced Cathode and Anode Materials: New materials such as cobalt-free cathodes (e.g., nickel-rich or manganese-based alternatives) and silicon-based anodes are gaining attention for their higher energy density and sustainability. Cobalt-free cathodes reduce reliance on expensive and scarce materials while silicon-based anodes, with their higher capacity than graphite, can store more lithium ions, enhancing the battery's energy density. BMSs play a critical role in managing these advanced materials by carefully monitoring their electrochemical behavior. Silicon anodes, for instance, expand and contract significantly during charge–discharge cycles, which the BMS must manage to avoid structural degradation.

Nanomaterials: Nanomaterials are increasingly being used to enhance the performance of BMS sensors. Their high surface area and sensitivity improve the detection of parameters such as voltage, temperature, and current with greater accuracy. Nanomaterials like quantum dots, carbon nanostructures, and metal oxides can be used in sensor systems to improve real-time monitoring. This ensures that the BMS can detect minute changes in battery conditions and respond quickly, thus enhancing the overall safety, efficiency, and lifespan of the battery.

Phase-Change Materials (PCMs): PCMs are being explored for thermal regulation in batteries, particularly during fast charging and discharging cycles. PCMs absorb excess heat and transition from solid to liquid, storing thermal energy. This helps maintain the battery temperature within a safe range, preventing overheating without the need for active cooling systems. When the battery cools, the PCM solidifies, releasing the stored heat. This passive thermal management is particularly beneficial for EVs, where space and energy efficiency are critical. To effectively know the use and efficiency of emerging materials in BMSs for EVs, we need to establish key categories that comprehensively capture the necessary information, as given in Table 17.1.

From Table 17.1, we can identify the following key points:

Graphene-Based Research Findings: Research advocates such as Johan Liu and peers have been instrumental in exploring graphene and its applications for thermal management, indicating a sustained push toward graphene integration into BMSs [4].

PCM and Hybrid Cooling Systems Developers: Researchers like Changcheng Liu and Que Huang have consistently contributed extensive reviews and experimental studies on PCM-based systems, characterizing their thermal benefits [3,18]. Their work has been crucial in driving PCM adoption in battery management.

Innovators in Composite Material Solutions: Research groups led by authors like Simon Schmidt have focused on integrating lightweight and efficient materials into modular designs, aligning with sustainability goals [5]. Such contributions are critical to the ongoing evolution of EV battery technologies.

Early research took place from 2012–2015, focused on exploring single materials like graphite coatings for thermal management [14]. This initial phase laid the groundwork for understanding basic thermal properties and efficiency needs. Significant strides were made during 2017–2020, with the integration of PCMs such as paraffin and expanded graphite to manage heat more effectively in lithium-ion batteries [3,6]. Research during this time emphasized the limitations of traditional air-cooling methods and the benefits of PCM-based systems [3,24]. Later research (2019–2022) began incorporating graphene due to its high thermal conductivity and ability to enhance thermal efficiency in BMSs [4,13,22]. Studies highlighted the improvement in temperature uniformity and cooling performance with graphene applications. The recent trend (2020–present) involves developing hybrid cooling systems combining PCMs with additional technologies like liquid cooling and metal fins, reflecting a move toward more complex and efficient solutions [2,11,12,16]. There's also greater exploration into varied compositions and hybrid solutions for optimized thermal performance. Overall, the landscape of research indicates a trajectory from basic thermal management materials toward more integrated and sustainable material solutions, facilitated by key research figures

TABLE 17.1
Overview of Recent Studies on Thermal Management Strategies in Battery Systems

Sl. No.	Year	Materials Focus	Study Type	Thermal Management Strategy	Key Results	Efficiency and Safety Outcomes	Applications or Context
[1]	2023	Expanded graphite/EG, PEG	Experimental	Passive PCM system without extra power	Effective reduction of battery temperature discrepancy	Temperature discrepancy reduction	Pure electric vehicles
[2]	2024	Paraffin (n-octadecane)	Experimental	PCM with fins	Fins with PCM manage hot spots most effectively	Temperature contained effectively	Lithium-ion batteries, hybrid vehicles
[3]	2020	PCMs, expanded graphite	Review	PCM-based BTMS trend	PCMs combined with filler materials significantly improve BTMS	N/A	General review
[4]	2020	Graphene	Experimental	Graphene film heat sink/ water cooling	76% thermal resistance reduction compared to copper-based systems	Temperature decreased by 11 °C	Propulsion batteries in electric vehicles
[5]	2022	Aluminum-foam PCM	Experimental	Modular design	Cooling power reduced by ~20%, faster disassembly with "debonding-on-demand"	High resource efficiency	Vehicle battery modules
[6]	2017	Paraffin/EG	Experimental	Various PCMs evaluated	Temperature reductions of 10–20% at different discharge rates	Improved heat dissipation	Lithium-ion battery modules
[7]	2020	Various PCMs	Simulation/ modeling	PCMs in extreme temperature analysis	Satisfactory temperature control within safe operating conditions	Enhanced heat transfer, safety	Lithium polymer pouch batteries
[8]	2023	PCMs	Review	Passive BTMS with PCMs	Highlights machine learning and optimization for improved PCM use	N/A	General electric vehicle applications
[9]	2022	Graphite nanoplatelets	Simulation	Thermal conduction with impregnated graphite	Delays heat propagation and enhances safety during thermal runaway	Improved heat dissipation	Battery modules
[10]	2019	Carbon graphite	Simulation/ experimental	Graphite sheets to enhance heat dissipation	Demonstrated feasibility for improved heat management	Enhanced thermal management	Pouch type cell applications
[11]	2023	PCMs with liquid cooling	Simulation	Hybrid PCM and liquid BTMS	Better performance than liquid cooling alone, especially with 10% EG in paraffin wax	Lower T_{max} and ΔT_{max}	General electric vehicle applications

[12]	2019	Expanded graphite/paraffin (EG/P)	Experimental/ simulation	Thermoelectric cooling and heat pipe	Effective surface temperature maintenance and energy saving	Improved uniformity	Ternary lithium-ion power battery
[13]	2022	Carbon, graphene	Experimental	Natural convection with carbon/graphene coatings	Demonstrated heat dissipation under varying current demands	Temperature uniformity	Lithium polymer cells for automotive
[14]	2023	PCMs	Review	PCM systems for heat storage	Focus on enhancement techniques such as improved conductivity	N/A	Electric vehicle battery systems
[15]	2024	CPCM with liquid cooling	Simulation	Novel hybrid cooling model	Improved performance under high-power conditions	Reduced maximum temperature and ΔT	General electric vehicle applications
[16]	2024	Leak-proof PCM systems	Simulation	PCM with liquid-cooling strategies	Significant reduction of Tmax and ΔTmax compared to liquid-cooling alone	Improved thermal safety	Large-scale battery modules
[17]	2021	PCMs with metal foams	Review	PCM methods with nanoparticles	Effective heat dissipation when using nanoparticles	N/A	Fast-charging electric vehicles
[18]	2021	Expanded graphite PCM	Simulation/ Experimental	Hybrid BTMS with air and PCM cooling	High PCM requirement for purely passive solutions	Lower battery pack temperatures	Electric van battery packs
[19]	2022	CPCM/liquid cooling	Simulation	Optimization of hybrid BTMS	Optimal parameters identified for minimizing temperature differences	Improved thermal control	Lithium-ion battery systems

TABLE 17.2
EV Battery Thermal Management Research

Year Range	Key Focus Areas
2012–2015	- Single materials (e.g., graphite coatings) - Basic thermal properties and efficiency
2017–2020	- Integration of PCMs (paraffin, expanded graphite) - Limitations of traditional air-cooling - Benefits of PCM-based systems
2019–2022	- Incorporation of graphene - High thermal conductivity - Enhanced thermal efficiency and cooling performance
2020–present	- Development of hybrid cooling systems (PCMs + liquid cooling, metal fins) - Exploration of varied compositions - Focus on optimizing thermal performance

focusing on graphene, PCMs, and hybrid cooling systems [27]. Table 17.2 depicts an overview of the key focus areas in the respective years.

17.4 INNOVATIONS IN BMS ARCHITECTURE

As EVs and energy storage applications demand higher efficiency, safety, and scalability, this research addresses the crucial need for advancements in battery materials and BMS architectures. By investigating the role of solid electrolytes, advanced anode and cathode materials, and the incorporation of nanomaterials and phase-change technologies, this chapter aims to shed light on how these innovations are shaping the future of BMSs.

Sensor Technology: Advances in sensor technology are crucial for improving the precision and reliability of BMSs. New materials like nanocomposites, graphene, and metal oxides are being integrated into sensors to enhance their ability to detect thermal, pressure, and chemical changes in batteries. For instance, graphene-based sensors can provide real-time monitoring of thermal and electrochemical conditions due to their high sensitivity and rapid response times. Similarly, pressure sensors using piezoelectric materials can detect minute changes in battery cell expansion, which is critical for identifying potential swelling or damage due to overheating or overcharging. Chemical composition sensors using nanomaterials can monitor electrolyte degradation or the build-up of harmful gases, enabling more precise safety controls within the BMS.

Power Electronics: Wide bandgap semiconductors, such as SiC and GaN, are revolutionizing power electronics in BMSs by offering higher efficiency in energy conversion and management. These materials enable faster switching frequencies and can operate at higher temperatures, which reduces energy losses and enhances overall system efficiency. SiC and GaN are particularly advantageous in EVs because they allow for smaller, lighter, and more efficient power inverters and converters. This improves the power density of the battery system, allowing for longer driving ranges and quicker charging times. Additionally, these materials improve thermal management due to their higher thermal conductivity, contributing to the overall safety and longevity of the battery system.

Battery Health Monitoring: Innovative electrode materials, such as lithium-titanate, silicon-carbon composites, and solid electrolytes, are enabling more accurate SoC and SoH predictions.

These materials provide better cycling stability, making it easier to track the electrochemical processes occurring within the battery over time. For instance, silicon-carbon composite anodes offer high-capacity retention, making SoC calculations more precise by minimizing capacity fading over many charge cycles. Furthermore, solid electrolytes, with their stability and reduced degradation rates, help extend the battery's life and make SoH monitoring more accurate by allowing for consistent performance data over time. The combination of these materials with machine learning algorithms in BMSs enables predictive maintenance, allowing for early detection of potential battery failures and more efficient energy management.

These innovations collectively may enhance the overall functionality, safety, and longevity of battery systems, particularly in EVs, where energy efficiency and reliability are critical.

17.5 SUSTAINABILITY AND RECYCLING

As the demand for EVs continues to rise, the environmental impact of battery production, use, and disposal becomes a pressing concern [24]. Addressing these challenges involves improving the sustainability of batteries through recycling and reuse strategies, as well as developing eco-friendly materials for use in both batteries and BMSs. Emerging materials offer promising solutions to enhance the recyclability, reuse potential, and overall environmental sustainability of EV batteries.

The roles of emerging materials in enhancing battery recycling and reuse strategies include the following:

- ***Facilitating Material Recovery:*** Emerging materials used in battery cells and BMS components can simplify the recycling process by making it easier to recover valuable metals such as lithium, cobalt, nickel, and manganese. Traditional recycling methods struggle with extracting these metals from complex battery structures. However, new material designs—such as solid electrolytes or less-reactive anode materials allow for more efficient separation and recovery of critical resources. For example:

  Solid-State Batteries: The use of solid electrolytes in SSBs can improve recycling processes since they eliminate the need for flammable liquid electrolytes, reducing the risk of hazardous waste and allowing for more straightforward recycling procedures.

  Non-Toxic Materials: Batteries using cobalt-free cathodes or replacing rare and toxic elements with more abundant materials, such as manganese or iron, contribute to reducing the environmental and health risks during recycling.

  Designing for Disassembly: Emerging materials are enabling new design approaches that make batteries easier to disassemble and recycle. By using modular components and recyclable adhesives or fasteners, batteries can be broken down more efficiently at the end of their life cycle, reducing the energy required for recycling and increasing the recovery rate of valuable materials.

  Bonding Agents: Some research is focused on developing dissolvable binders or self-healing materials that simplify the dismantling process, allowing for easier recovery of individual components such as electrodes and current collectors.

  Extending Battery Life for Reuse: Advanced materials used in BMSs can improve the ability to monitor battery health over time, allowing for early detection of degradation. This makes it possible to identify cells that can be reused in secondary applications, such as energy storage systems for homes or businesses, before they are fully recycled.

  Smart Sensors: The incorporation of nanomaterials and intelligent sensors in BMSs can monitor cell health more accurately, identifying which cells can still be useful after their primary life in EVs, thus prolonging their lifespan and delaying their entry into the recycling stream.

Refurbishment Potential: Batteries made with durable cathode and anode materials that degrade more slowly (e.g., lithium iron phosphate [LFP] batteries) can be repurposed and refurbished for second-life applications in less demanding energy storage systems.

17.6 ECO-FRIENDLY MATERIALS REDUCING THE ENVIRONMENTAL IMPACT OF BATTERY PRODUCTION AND DISPOSAL

***Cobalt-Free and Nickel-Free Cathodes*:** Traditional lithium-ion batteries often rely on cathodes that contain cobalt and nickel, both of which pose environmental and ethical concerns, particularly related to mining practices. Emerging cathode materials, such as LFP and manganese-rich compounds, eliminate the need for these problematic elements. These materials are not only more abundant but also more environmentally benign, reducing the negative impacts associated with battery production and disposal.

***LFP Batteries*:** LFP batteries are gaining popularity due to their non-toxic composition, excellent thermal stability, and long cycle life, making them an eco-friendly alternative to cobalt-based batteries.

***Recyclable Electrolytes*:** The development of recyclable and less hazardous electrolytes is a critical area for reducing the environmental impact of batteries. Traditional liquid electrolytes are often flammable and difficult to dispose of safely. Emerging alternatives such as solid electrolytes and water-based electrolytes offer a safer and more sustainable solution.

***Solid Electrolytes*:** These eliminate the need for flammable and toxic liquid electrolytes, simplifying recycling and reducing the generation of hazardous waste.

***Ionic Liquids*:** Some ionic liquid electrolytes are designed to be non-volatile, non-flammable, and recyclable, offering a safer alternative to traditional organic solvents.

***Bio-Based and Sustainable BMS Materials*:** Beyond the batteries themselves, components of the BMS can be developed from sustainable and bio-based materials to reduce the overall environmental impact of EV production.

***Bio-Polymers*:** BMS components such as housings, connectors, and insulation can be made from bio-based polymers derived from renewable sources like plant fibers, reducing reliance on petroleum-based plastics.

***Recycled Metals*:** Using recycled aluminum and copper in BMS components can help reduce the carbon footprint of producing new materials, as the extraction of raw metals is energy-intensive and often damaging to the environment.

17.7 REDUCING ENERGY INTENSITY IN MANUFACTURING

Emerging materials and manufacturing techniques, such as additive manufacturing and low-temperature synthesis processes, can significantly reduce the energy consumption associated with producing battery components. These advancements not only lower the carbon footprint of battery manufacturing but also improve the sustainability of the overall supply chain. For energy-efficient production, for example, silicon anodes or high-energy-density cathodes can be synthesized using methods that require lower temperatures and less energy than traditional materials, making the battery production process more eco-friendly.

The integration of emerging materials in EV batteries and BMSs offers significant potential to enhance the sustainability of both battery production and recycling. From reducing the reliance on toxic and scarce materials like cobalt and nickel to improving the recyclability and reuse of battery components, these innovations can help minimize the environmental impact of EVs. Furthermore, eco-friendly alternatives such as solid electrolytes, bio-polymers, and recyclable electrolytes are poised to revolutionize the entire battery life cycle, ensuring that the transition to EVs is both environmentally sustainable and resource-efficient.

17.8 CHALLENGES AND OPPORTUNITIES

Scalability, cost efficiency, and standardization are critical factors for the widespread adoption of EVs, particularly when it comes to their BMSs and materials used.

- ***Scalability*:** The global transition to EVs requires battery technologies that can be scaled up to meet growing demand. Scalability involves both manufacturing processes and the ability to produce batteries in large volumes while maintaining quality, performance, and safety. Key factors that impact scalability in EV battery production include the following:

  Material Availability: Advanced materials such as solid electrolytes, high-capacity cathodes, and nanomaterials need to be readily available in large quantities. Some materials, like lithium and cobalt, have limited reserves or are concentrated in specific regions, which can pose challenges in ensuring sufficient supply chains for scalable production.

  Manufacturing Technology: The production processes for advanced materials must be optimized for mass manufacturing. For example, integrating solid electrolytes or phase-change materials into battery cells requires technological advancements that enable large-scale, automated production. The ability to mass-produce these materials at an industrial scale is key to ensuring scalability.

  Infrastructure: To scale EV adoption, widespread charging infrastructure is necessary. Batteries must also be compatible with fast-charging networks, and this places additional requirements on the BMS to ensure safety and performance under rapid charging conditions.

- ***Cost Efficiency*:** Cost efficiency is a major barrier to the affordability of EVs. The BMS and battery materials account for a significant portion of the overall cost of EVs, and reducing these costs is essential for making EVs accessible to a broader market. Achieving cost efficiency involves:

  Material Cost: Advanced materials like silicon anodes, nickel-rich cathodes, and nanomaterials can be expensive due to their production complexity or rarity. Ongoing research is aimed at finding cheaper, yet effective, alternatives or optimizing existing material production processes to reduce costs. For example, SSBs, though promising in terms of performance, need to overcome high material and production costs to become economically viable.

  Manufacturing Costs: Advances in battery production technologies, such as automated assembly lines and improvements in battery chemistry, can significantly reduce manufacturing costs. Standardization of production processes, cell designs, and BMS architecture can drive economies of scale, lowering the per-unit cost of batteries and improving the overall cost efficiency of EVs.

  Lifecycle Costs: Cost efficiency also considers the entire lifecycle of the battery, including its longevity and recyclability. Improving the durability of EV batteries and making them easier to recycle can help spread the cost over a longer vehicle lifespan, while reducing the environmental and financial costs associated with battery disposal.

- ***Standardization*:** Standardization is vital for ensuring compatibility across different EV models and manufacturers. Without standardization, there would be fragmented ecosystems that hinder the mass adoption of EVs. The following aspects contribute to the importance of standardization in EV BMS and battery technology:

  BMS Protocols: A standardized BMS architecture allows manufacturers to create interoperable systems that can be used across various vehicle platforms. Standardization in communication protocols between the BMS and other vehicle components ensures that EV batteries can be easily integrated with different vehicle designs, charging systems, and energy grids.

  Charging Standards: Uniform charging standards, such as the Combined Charging System (CCS) or CHAdeMO, ensure that EVs can use a wide network of charging stations. As fast charging becomes more prevalent, the BMS must be standardized to handle different charging speeds and methods while maintaining safety and efficiency.

Regulatory Standards: Government regulations often set safety and performance standards for EV batteries. Meeting these standards ensures that EVs can be sold globally without the need for extensive modifications. Standardized safety features in the BMS, such as thermal management and fault detection, are critical for meeting regulatory requirements and building consumer trust.

For EVs to reach mass adoption, the BMS and related materials must be scalable, cost-efficient, and standardized. Scalability ensures that manufacturers can meet the rising demand for EVs, cost efficiency makes EVs affordable for consumers, and standardization ensures compatibility and safety across the industry. These factors are interconnected, and advancements in one area often contribute to improvements in the others. For example, standardized manufacturing processes can lead to lower costs and improved scalability, while innovations in cost-efficient materials can make it easier to produce batteries at scale.

17.9 FUTURE DIRECTIONS IN ELECTRIC VEHICLE BATTERY MANAGEMENT SYSTEMS

Hybrid Materials: Hybrid materials are emerging as a promising approach to enhancing battery performance by combining the strengths of different compounds. For example, integrating graphene with existing materials like lithium can improve conductivity, durability, and thermal management in EV batteries. Graphene's high electrical and thermal conductivity makes it ideal for applications where fast charging and efficient heat dissipation are crucial.

Graphene-Based Hybrids: When used with silicon anodes or nickel-rich cathodes, graphene helps stabilize the structure, preventing degradation over time and enhancing energy density. This can lead to longer-lasting batteries with greater capacity, solving current issues of battery lifespan and charging speed.

Metal Oxides and Composites: Other hybrid approaches include combining metal oxides or polymer composites with traditional battery materials to achieve improved stability and energy efficiency. These materials can enhance the safety of the BMS by providing more reliable temperature control and reducing the risk of overheating, which is critical for future EVs operating under high power demands [22].

Artificial Intelligence and Machine Learning Integration: Artificial intelligence (AI) and machine learning (ML) are increasingly being integrated into BMSs to optimize performance and ensure safety in real time. By combining advanced materials with AI-driven analytics, BMSs can become more predictive, adaptive, and self-optimizing.

Predictive Maintenance: AI can analyze data from the battery's sensors to predict potential failures or the need for maintenance before problems occur. This ensures that EVs can operate efficiently with minimal downtime, extending the lifespan of the battery and reducing costs.

Adaptive Energy Management: Machine learning models can dynamically adjust charging cycles, discharge rates, and thermal management based on real-time conditions. This is especially beneficial when using materials like silicon anodes, which require careful control of charging parameters to avoid wear and tear.

Enhanced Safety Features: AI can improve safety by learning to detect and respond to irregular patterns in voltage, temperature, or pressure, preventing hazardous conditions such as thermal runaway. When paired with hybrid materials like graphene or nanomaterials, the BMS can operate under more demanding conditions with greater precision.

Next-Generation EV Batteries: The ongoing development of next-generation batteries, such as SSBs and lithium-air batteries, is expected to transform the EV industry, with advanced materials playing a crucial role in these innovations.

- **Solid-State Batteries:** These batteries replace the flammable liquid electrolyte with a solid electrolyte, improving both safety and energy density. Materials like ceramic electrolytes and hybrid polymers are being developed to make SSBs commercially viable. The inclusion of nanomaterials or graphene in these batteries could further enhance conductivity and energy storage, potentially doubling the range of EVs.
- **Lithium-Air Batteries:** Lithium-air batteries are another cutting-edge technology that could revolutionize energy storage by offering significantly higher energy density than traditional lithium-ion batteries. However, the development of these batteries relies on finding suitable materials that can facilitate the oxygen reaction and provide long-term stability. Hybrid materials, including catalytic nanostructures or advanced composites, may unlock the potential of lithium-air batteries, allowing EVs to achieve much longer ranges without significantly increasing battery size or weight.

The future of EV batteries lies in the combination of hybrid materials, AI-driven BMSs, and next-generation battery technologies like SSBs and lithium-air batteries. By integrating these cutting-edge advancements, the performance, safety, and efficiency of EV batteries will significantly improve. These developments will enable longer driving ranges, faster charging times, and enhanced energy management systems, ultimately driving the EVs industry toward greater sustainability and mass adoption.

17.10 DISCUSSION

The advancement of EV battery technology is key to driving the global transition toward sustainable transportation. At the heart of this transformation lies the BMS, which ensures the safety, efficiency, and longevity of the battery, the most critical and expensive component of EVs. While current lithium-ion batteries and their associated management systems have been sufficient for the early stages of EV adoption, the limitations they present—such as thermal management issues, restricted energy density, and charging inefficiencies—necessitate the exploration of new materials to take EV performance to the next level. Emerging materials play a pivotal role in addressing these challenges. Solid electrolytes, for example, have emerged as a critical component for next-generation SSBs. Unlike traditional liquid electrolytes, solid electrolytes eliminate the risk of leakage, improve thermal stability, and enable higher energy densities. The use of these materials in BMS architecture could significantly enhance the safety profile of batteries, making EVs more reliable, particularly under the extreme conditions of fast charging or high discharge rates. As these materials become more commercially viable, SSBs will likely replace conventional lithium-ion batteries, ushering in a new era of safer, more efficient energy storage. High-thermal conductivity materials, such as graphene and CNTs, are another area of innovation. These materials are crucial for effective thermal management within batteries. As EVs demand faster charging and more extended driving ranges, the ability to dissipate heat becomes paramount. Integrating materials like graphene into battery components allows for better heat dissipation and more efficient cooling, thereby extending the battery's operational life and preventing damage from overheating. Advanced cathode and anode materials, such as cobalt-free cathodes and silicon-based anodes, also promise to revolutionize battery performance by offering greater energy storage capacity. However, these materials introduce new complexities, such as volume expansion in silicon anodes during charge cycles, which require the BMS to be more sophisticated in its management of charge–discharge behaviors. Nanomaterials further enhance the capabilities of BMSs by improving the accuracy of sensor technology. By providing real-time data on voltage, temperature, and current flow, nanomaterials can significantly improve the predictive and responsive capabilities of the BMS. This allows for more precise energy management and prevents the degradation that typically occurs in conventional batteries. Additionally, the use of PCMs for thermal regulation offers another method for enhancing the safety and efficiency of batteries, particularly during high-stress operations like fast charging. PCMs absorb and release heat during phase transitions, providing a passive thermal management solution that can prevent overheating. In terms of broader innovations in BMS architecture, advances in sensor technology, power electronics, and battery health monitoring all point toward a future where batteries are not only more efficient but also more intelligent. Wide bandgap semiconductors like SiC and GaN are driving improvements in power conversion efficiency, reducing energy losses, and making BMSs more effective at managing power flow within the battery. Similarly, advanced materials in electrode design are enabling more accurate SoC and SoH predictions, allowing for better long-term battery management and extending the usable life of EV batteries. Despite the promise these materials hold, challenges remain, particularly regarding scalability and cost efficiency. The mass production of advanced materials like graphene, solid electrolytes, and nanomaterials must be optimized to meet the growing demand for EVs. Moreover, while these materials can significantly improve battery performance, their higher production costs pose challenges for making EVs more affordable to a broader market. However, as research continues and manufacturing processes become more efficient, these cost barriers are likely to decrease over time.

17.11 CONCLUSION

In conclusion, emerging materials hold the key to overcoming the limitations of current BMS technology. By enhancing safety, thermal management, energy density, and efficiency, these materials will play a critical role in enabling the next generation of EVs. As SSBs, nanomaterials, and advanced cathodes and anodes become more widely adopted, EVs will become more practical, cost-effective, and sustainable. These innovations will drive the electrification of transportation, reducing greenhouse gas emissions and paving the way for a cleaner, more sustainable future. The integration of these new materials into BMSs is not just a technological upgrade but a necessity for realizing the full potential of electric mobility.

REFERENCES

[1] Xia, Z., Li, C., Yu, H., & Wang, Z. (2023). Experimental study of a passive thermal management system using expanded graphite/polyethylene glycol composite for lithium-ion batteries. *Energies*, *16*(23), 7786. https://doi.org/10.3390/en16237786

[2] Paciolla, P., & Papurello, D. (2024). Improved thermal management of Li-ion batteries with phase-change materials and metal fins. *Batteries*, *10*(6), 0190. https://doi.org/10.3390/batteries10060190

[3] Liu, C., Xu, D., Weng, J., Zhou, S., Li, W., Wan, Y., Jiang, S., Zhou, D., Wang, J., & Huang, Q. (2020). Phase change materials application in battery thermal management system: A review. *Materials*, *13*(20), 4622. https://doi.org/10.3390/ma13204622

[4] Liu, Y., Thiringer, T., Wang, N., Fu, Y., Lu, H., & Liu, J. (2020). Graphene-based thermal management system for battery cooling in electric vehicles. In *2020 IEEE 8th Electronics System-Integration Technology Conference (ESTC)* (pp. 1–4). University of South-Eastern Norway and Norwegian Centres of Expertise NCE Micro- and Nanotechnology, IEEE-*Xplore*. https://doi.org/10.1109/ESTC48849.2020.9229856

[5] Schmidt, S., Clausen, J., van der Auwera, R., Klapp, O., Schmerler, R., Löffler, D., Werner, M., & Block, L. (2022). Novel battery module design for increased resource efficiency. *World Electric Vehicle Journal*, *13*(10), 177. https://doi.org/10.3390/wevj13100177

[6] Wang, Z., Li, X., Zhang, G., Lv, Y., Wang, C., He, F., Yang, C., & Yang, C. (2017). Thermal management investigation for lithium-ion battery module with different phase change materials. *RSC Advances*, *7*(71), 42909–42918. https://doi.org/10.1039/C7RA08181B

[7] Talluri, T., Kim, T. H., & Shin, K. (2020). Analysis of a battery pack with a phase change material for the extreme temperature conditions of an electric vehicle. *Energies*, *13*(3), 507. https://doi.org/10.3390/en13030507

[8] Dolla, D. A., & Fetene, M. (2023). Investigations of phase change materials in battery thermal management systems for electric vehicles: A review. *Materials Research Express*, *11*(3), Article 032005. https://doi.org/10.1088/2053-1591/ad1949

[9] Palmieri, B., Cilento, F., Siviello, C., Bertocchi, F., Giordano, M., & Martone, A. (2022). Mitigation of heat propagation in a battery pack by interstitial graphite nanoplatelet layer: Coupled electrochemical-heat transfer model. *Journal of Composites Science*, *6*(10), 296. https://doi.org/10.3390/jcs6100296

[10] Choi, W., Kang, M.-S., & Han, S. (2019). Analysis of thermal characteristics of a battery module using a novel graphite material. In *2019 IEEE Transportation Electrification Conference and Expo, Asia-Pacific (ITEC Asia-Pacific)* (pp. 1–4). https://doi.org/10.1109/ITEC-AP.2019.8903867

[11] Sun, J., Dan, D., Wei, M., Cai, S., Zhao, Y., & Wright, E. (2023). Pack-level modeling and thermal analysis of a battery thermal management system with phase change materials and liquid cooling. *Energies, 16*(15), 5815. https://doi.org/10.3390/en16155815

[12] Zhang, C., Chen, S., Gao, H., Xu, K.-J., Xia, Z., & Li, S. (2019). Study of thermal management system using composite phase change materials and thermoelectric cooling sheet for power battery pack. *Energies*, *12*(10), 1937. https://doi.org/10.3390/en12101937

[13] Sequino, L., Sebastianelli, G., & Vaglieco, B. (2022). Carbon and graphene coatings for the thermal management of sustainable LMP batteries for automotive applications. *Materials*, *15*(21), 7744. https://doi.org/10.3390/ma15217744

[14] Diaconu, B., Cruceru, M., Anghelescu, L., Racoceanu, C., Popescu, C., Ionescu, M., & Tudorache, A. (2023). Latent heat storage systems for thermal management of electric vehicle batteries: Thermal performance enhancement and modulation of the phase transition process dynamics: A literature review. *Energies*, *16*(6), 2745. https://doi.org/10.3390/en16062745

[15] Xu, C., Ma, C., Souri, M., Moztarzadeh, H., Nasr Esfahani, M., Jabbari, M., & Hosseinzadeh, E. (2024). Numerical investigation of thermal management of a large format pouch battery using combination of CPCM and liquid cooling. *Batteries*, *10*(4), 113. https://doi.org/10.3390/batteries10040113

[16] Zhang, W., Zhang, J., Zhang, G., Hu, Y., Shao, D., Jiang, L., & Wen, Y. (2024). A novel leak-proof thermal conduction slot battery thermal management system coupled with phase change materials and liquid-cooling strategies. *Energies*, *17*(4), 939. https://doi.org/10.3390/en17040939

[17] Thakur, A., Sathyamurthy, R., Saidur, R., Pandey, A., Sharshir, S., & Christopher, S. (2021). A critical review of recent development in Li-ion battery cooling using advanced phase change materials. *ECS Transactions, 105*(1), 411. https://doi.org/10.1149/10501.0411ecst

[18] Pra, F., Al Koussa, J., Ludwig, S., & De Servi, C. M. (2021). Experimental and numerical investigation of the thermal performance of a hybrid battery thermal management system for an electric van. *Batteries*, *7*(2), 27. https://doi.org/10.3390/batteries7020027

[19] Zeng, X., Men, Z., Deng, F., & Chen, C. (2022). Optimization of the heat dissipation performance of a lithium-ion battery thermal management system with CPCM/liquid cooling. *Materials*, *15*(11), 3835. https://doi.org/10.3390/ma15113835

[20] Jin, B., Fei, Q., Lin, R., & Zou, W. (2023). Heat dissipation performance research of power lithium battery based on high thermal conductivity graphite film. *Journal of Physics: Conference Series*, *2433*, Article 012036. https://doi.org/10.1088/1742-6596/2433/1/012036

[21] Kumar, S. D., Kumar, D. D., & Manojkumar, R. (2023). Design and thermal analysis of battery thermal management system for EV. *SAE Technical Paper Series*. https://doi.org/10.4271/2023-28-0087

[22] Gepp, M., Filimon, R., Koffel, S., Lorentz, V., & März, M. (2015). Advanced thermal management for temperature homogenization in high-power lithium-ion battery systems based on prismatic cells. In Nogueiras, A. (Ed.) *2015 IEEE 24th International Symposium on Industrial Electronics (ISIE)* (pp. 1230–1235). https://doi.org/10.1109/ISIE.2015.7281648.

[23] Yang, C., & Cao, L. (2019). The role of interfacial thermal resistance in Li-ion battery thermal management. In Narumanchi, S., (Ed.) *ASME 2019 International Technical Conference and Exhibition on Packaging and Integration of Electronic and Photonic Microsystems* (Paper No. IPACK2019-6594). American Society of Mechanical Engineers (ASME). https://doi.org/10.1115/ipack2019-6594.

[24] Xiong, F., Ouyang, C.-Z., Zeng, J. P., Zhu, L., & Li, G. (2020). Investigation on battery thermal management of power battery based on environmental impact of electric vehicle. *Journal of Physics: Conference Series*, *1600*, Article 012073. https://doi.org/10.1088/1742-6596/1600/1/012073

[25] Cheng, K., Divakar, B., Wu, H., Ding, K., & Ho, H. F. (2011). Battery-management system (BMS) and SOC development for electrical vehicles. *IEEE Transactions on Vehicular Technology*, *60*(1), 76–88. https://doi.org/10.1109/TVT.2010.2089647

[26] Rohini, A., Abishek, A. S., & Jeeva, S. (2024). Analysis on a battery thermal management system of a lithium-ion powered battery with heat sink for an electric vehicle. *Journal of Physics: Conference Series*, *2748*, Article 012017. https://doi.org/10.1088/1742-6596/2748/1/012017

[27] Tamizharasi, A. R. S., Gopika, M. V., Keerthimathi, K., & Kavitha, D. (2024). Impact of electrode materials on battery performance in hybrid systems for EV applications. In Vijayan, R., Meenatchi, S., & Brindha, K., (Eds.) *2024 Second International Conference on Emerging Trends in Information Technology and Engineering (ICETITE)* (pp. 1–8). IEEE Xplore. https://doi.org/10.1109/ic-ETITE58242.2024.10493538

18 Polymers and Composites in EVs and HEVs

Jitendra Kumar and Dinesh Kumar Rao

18.1 INTRODUCTION TO POLYMERS AND COMPOSITES IN EVS AND HEVS

Electric vehicles (EVs) and hybrid electric vehicles (HEVs) continue to be at the forefront of leveling up the automotive industry in terms of sustainability and low emissions [1]. EVs exclude the traditional internal combustion engines and make use of rechargeable batteries as their sole source of power to drive completely electric motors [1]. These are vehicles that emit no emissions at all from the tailpipe, which makes them an effective tool in helping to slash both airborne pollutants and greenhouse gases. Increased use of EVs is a result of enhancement in battery energy production, the increased average range of cars, and the availability of charging stations. HEVs, on the other hand, have an internal combustion engine with an electric motor and a sized battery, thus power comes from both the engine and the electric motor [2]. In an HEV, an electric motor works alongside the internal combustion engine to improve fuel economy and emissions output. In general, HEVs are equipped with an electric motor and can be driven only by it at low speeds with the engine taking over at higher speeds or when the battery is almost discharged. However, HEVs are different from EVs because they do not need to be charged from outside sources, but through the battery recharging mechanism, by regenerative braking and the engine [3,4]. Thus, both EVs and HEVs are progressively significant in eliminating the juncture to fossil fuels, increasing general fuel economy, and providing improved forms of green power in conveyance. These types of vehicles are promising in the future to meet global climate goals as regulatory pressures continue to rise and technological advancement continues [5].

Polymers and composites are now a critical component of the construction of EVs and HEVs, as they help to improve the vehicle's performance, energy efficiency, and lifespan. Today, the automotive industry is opting for cleaner sources of energy, and the need for weight-saving, longer-lasting, and low-energy-consuming materials ensures polymers and composites are in high demand [6]. Polymer materials as lighter and more versatile, extending their usage to such parts as instrument panels, seats, and cable insulation. Composites, containing polymers reinforced with carbon or glass fibers, are three times stronger and stiffer than the polymers themselves; fatigue resistance is also a critical factor for structural applications [2,7,8]. The global EV composites market was valued at approximately USD 2.3 billion to USD 2.5 billion in 2024. It is projected to reach USD 5.1–6.0 billion by 2029–30 or USD 10.75 billion by 2032, depending on the forecast period. The market is expected to grow at a compound annual growth rate (CAGR) of 17.1% to 17.6% from 2024 to 2029 or 2032, as shown in Figure 18.1.

In EVs and HEVs, the application of polymers and composites results in weight optimization and thus improved electrical efficiency and driving range. Let's take the case of carbon fiber-reinforced composites used in battery enclosures. Both rigidity and weight reduction are considered important in EVs [9]. Furthermore, the elements needed for the addition of these materials also provide high resistance to corrosion and wear, which is useful in parts that are exposed to different climate conditions such as the underbody and body shells. In addition to performance aspects, polymers used in the automotive sector and their accompanying composites are also conducive to more sustainable automobile construction [10]. This ability to develop components based on recyclable polymers

DOI: 10.1201/9781003596097-18

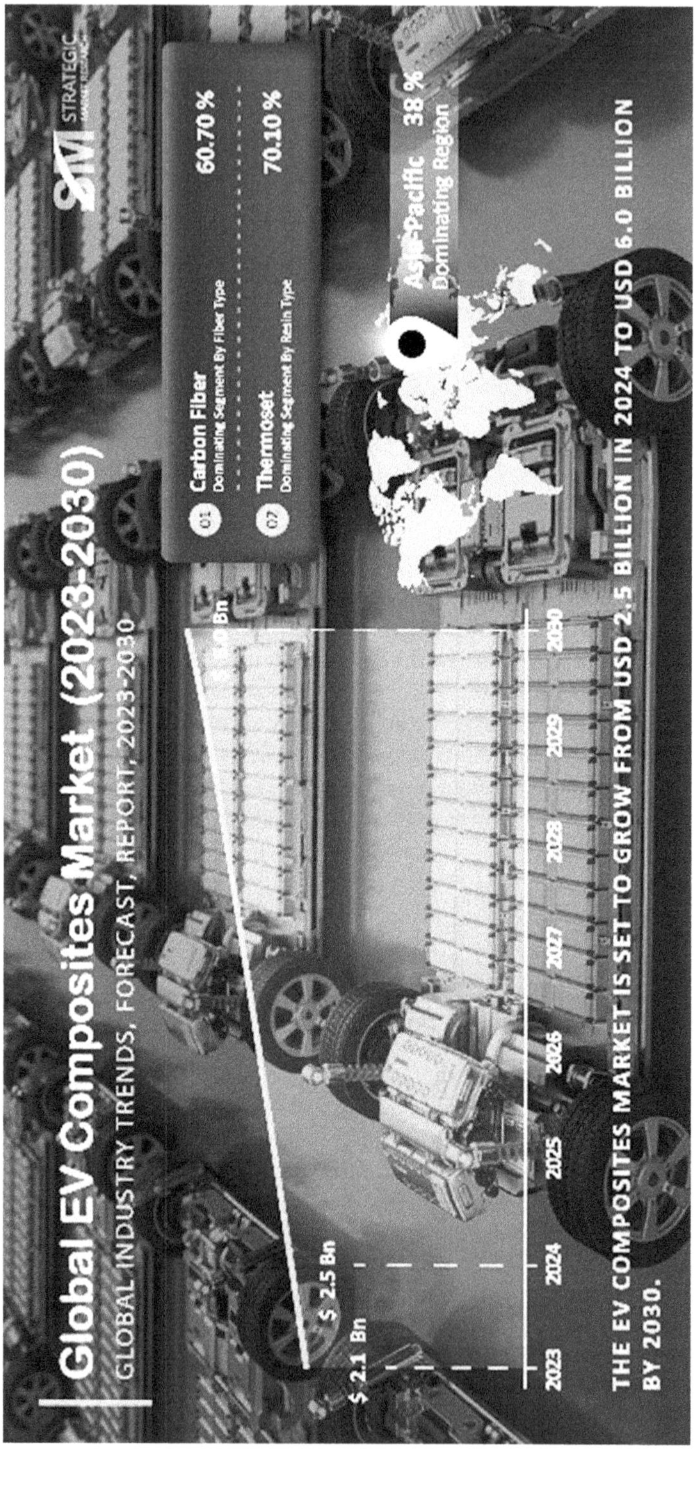

FIGURE 18.1 Market size and growth of the EV composite market. (According to strategic market research of EV composited.)

suits the trends toward increasing environmental responsibility in the automotive industry. Due to the advancements that manufacturers will make in the future, these materials should also offer new opportunities for EVs and HEVs concerning safety, energy efficiency, and concern for the environment [8,11]. Given their flexibility, efficiency, and eco-friendliness, polymers and composites represent a critically important element of the drive to create the next generation of electric and hybrid vehicles [11].

Polymers in EVs and HEVs: EVs and HEVs utilize polymers for several purposes because of their good characteristics such as lightweight, processability, and applicability. They play vital roles in increasing vehicle performance, decreasing vehicle weight, increasing safety, and decreasing manufacturing expenses. The polymers can be broadly distinguished into three categories: thermoplastics, thermosets, and elastomers, in the present case they are applied in EV and HEV forms as follows:

Thermoplastics: Thermoplastics are forms of polymers that may be melted and re-molded time and again without substantially altering their chemistry. This makes them easy to process and suitable also for mass production, particularly in the automotive industry [12].

Polypropylene (PP): PP is a low-density, corrosion-resistant, and chemically inert material that is also cost-effective. It is especially applied for car interiors including dashboards, seat backings, as well as trim pieces, battery casings, and bumpers. PP is also applied to non-structural parts of the body of both conventional EVs and HEVs due to its relatively high strength-to-weight ratio and satisfactory flexibility [13]–[15].

Polyethylene (PE): PE is another general-purpose thermoplastic material that is also used in automotive applications. It is durable and has good chemical resistance and is also not that heavy. In EVs and HEVs, PE is used in parts including fuel tanks (for HEVs), underbody shields, and electrical insulation where chemical and application resistance and hardness are essential [3].

Polycarbonate (PC): PC is a tough, clear thermoplastic commonly utilized for elements that should withstand high-impact forces. It is used in all car headlights, windows, and other exterior parts where light weight as well as impact strength are desired [16]. PC is also used for the transparent parts of interior components such as instrument panels and displays [17,18].

Acrylonitrile Butadiene Styrene (ABS): ABS is a general-purpose thermoplastic that finds wide application in automotive interiors such as door paneling, instrument paneling, consoles, etc. It is characterized by good mechanical parameters such as impact strength, stiffness, and ease of fabrication. ABS is also used on non-structural parts of the body due to zero density and capability to form almost any shape [19,20].

Polyamide (PA)—Nylon: PA, more familiarly referred to as Nylon, possesses excellent mechanical performance such as higher tensile strength and wear resistance, besides high toughness. This type of material is applied in several under-the-hood parts including engine and transmission parts and fuel lines. It is also utilized for interiors, some of the parts of the car include seat backs, this material is lightweight as well as tough [21,22].

Polyvinyl Chloride (PVC): PVC is one type of thermoplastic material that is part of the automotive industry. It is widely used in electrical devices insulation, doors, and window profiles, indoor decorative works, and floor coverings. Its distinct chemical inertness makes it useful for structural and non-structural components where they can be easily processed in a range varying between rigid or flexible materials [23].

Polyethylene Terephthalate (PET): PET also finds its application in automotive industries due to its features such as high gloss, high strength, and chemical and impact resistance. It is commonly used in automotive textiles, carpets, and upholstery. It is also used in the manufacture of small and thin parts for the interior of car and outer body parts [24–27]. The other advantage found with PET is its compatibility with the recycling aspect, another key element of sustainability objectives.

Thermosets: Thermosetting plastics are those that undergo chemical change when heated and become rigid and inflexible, having a network of covalent bonds. These are generally stronger, more heat stable, and chemically resistant than thermoplastics.

Epoxies: Epoxies are used prominently in automotive applications because of their good adhesion strength, high durability, and resistance to aggressive environments. In EVs and HEVs, epoxy resins are used in fabricating light structural parts including battery casings, and as a material in bonding automotive components. Epoxies are also used in coatings where they afford high wear and corrosion resistance [28–30].

Polyurethanes (PU): PU is another important member of the family of thermosetting polymers that are widely employed in automobile industries. It is very flexible because it is available in diverse levels of hardness, flexibility, and abrasion resistance. In the case of EVs and HEVs, polyurethanes find application in seats, interior trim, insulating materials, and bumpers [31–33]. One more example is polyurethane foams that are used for soundproofing and energy management in crash components and inner panels.

Phenolic Resins: Phenolic resins are highly resistant to heat, and therefore are employed in applications that require higher ratings of thermal stability. These include use in the production of friction materials, electrical parts, and the battery housing of EVs and HEVs due to their high-temperature capabilities which safeguard vulnerable parts [34,35].

Elastomers: Elastomers are a kind of polymer that can stretch to a large extent and recover their elasticity. They are mostly employed in sealing, cushioning, and as vibration isolators because of some of their mechanical characteristics such as flexibility and capacity to undergo large deformation but with negligible deformation damage.

Silicone: Silicone elastomers have found many uses in car manufacturing and automotive applications due to their capability to operate over a wide temperature range, flexibility, and ability to withstand the effects of the external environment including exposure to UV radiation and ozone. In EVs and HEVs, silicone is applied in seals and gaskets and electrical insulation where temperature control and sealing are important, for instance in the battery pack [36–38]. Silicone rubber is also used in hose systems for fluid management, and vibration damping as well.

Rubber: Conventional rubber products for use in EVs and HEVs include natural rubber and other synthetic grades like ethylene propylene diene monomer (EPDM) rubber. The parts include: tires, weather seals, gaskets, and hoses. Rubber is flexible, absorbs shock, and is an excellent material where sealing properties of the material are important. In the case of EVs, the rubber components are used to shield the vehicle against moisture, dust, and noise, besides guaranteeing airtight profiles. Moreover, rubber is employed in the suspension of automobiles whereas it also acts as an isolation material in drive trains [39].

Fluoroelastomers (FKM): FKM are special rubbers that provide excellent sealing characteristics that can withstand high temperatures and harsh chemical environments. They are extensively applied in high-temperature applications including engine sealing, o-rings, and gaskets. FKM are also used in EVs and HEVs to minimize all-round seals across the battery pack and electric motor to ensure heat and chemical resistance [40,41].

Chloroprene Rubber (CR): CR, commonly known as Neoprene, is suitable in the automotive industry for gaskets, seals, and vibration management. In 2010, epoxy's ranking improved due to its enhanced resistance to oil, weathering, and ozone. For car applications, Neoprene is incorporated in o-rings of EVs and HEVs, where mechanical flexibility and high-stress resistivity are essential for the automobile's body [42,43].

18.2 APPLICATION OF POLYMERS IN EVS AND HEVS

Polymers are widely used in enhancing the design, efficiency, and operations of EVs and HEVs. Their versatility, lightweight nature, and excellent properties make them indispensable in various vehicle components [29,44–46]:

Battery Components: PP, PA, and PE are used for the protection, insulation, and mechanical reinforcement of battery packs. In thermal regulation, thermally conductive materials such as epoxy resins control the temperature of the battery for its effectiveness.

Automobile Structural Components: Automotive applications of lightweight body panels, carbon fiber-reinforced polymers (CFRP), and fiber-reinforced polymer (FRP) decrease vehicle weight and hence enhance energy usage. Chassis and frames are made by thermosets like epoxy resins which improve vehicle performance due to their low density and high strength.

Interior Components: PP and PU are two materials used in products due to their wear resistance and light weight with additional comfort. Dashboards and consoles are assembled with ABS and PC, which are used for their impact-resistance and appearance.

Electrical Systems: After circuiting or hard wiring of any component, polyethylene and PVC are applied as electrical insulation and protection. Connectors and switches are made of PC and polybutylene (PBT), which are safe and reliable for use in electrical components.

Exterior Parts: In bumpers and impact absorbers, PP, ABS, and PU have been used to provide an energy-absorbing factor for better safety. They are used in headlamp lenses and trim applications, with PC, and poly methyl methacrylate (PMMA) providing light transmittance properties and mechanical strength.

Noise, Vibration, and Harshness (NVH) Control: PU and rubber-based elastomers reduce noise and vibrations, giving more comfort when driving. The advantages of polymers in EVs and HEVs are as follows [47]:

Lightweight: Polymers are comparatively lighter than regular metals; thus, the vehicular weight is also decreased and the driving range and energy efficiency are high.

Cost-Effective: Polymers are cheaper and can be easily processed, hence reducing the cost of production of both EVs and HEVs.

Durability: Polymers in vehicles are resistant to corrosion, ultraviolet radiation, and other factors that may cause wear of vehicle parts.

Design Flexibility: Polymer materials can be made into various shapes, thus they allow for freedom in the design and production of components that are quite aesthetic and unique in structure.

Impact Resistance: Both ABS and PC are used due to their high impact strength, as simpler structures run the risk of breaking during impacts or collisions.

Thermal Insulation: Polymers allow the thermostatic regulation of heat in battery and motor configurations, enhancing safety and performance.

Electrical Insulation: Most polymers possess high electrical resistivity, which is quite desirable for use in protecting the high-voltage systems featured in EVs and HEVs.

Noise and Vibration Reduction: Automotive elastomers and PU featured in NVH optimize driving comfort by lowering road noise and vibration.

Recyclability: Products with synthetic materials such as thermoplastics can be recycled and therefore can be environmentally friendly.

Corrosion Resistance: Polymers are not affected by corrosion including where used in areas such as battery trays, which increases the vehicle's working life and reduces maintenance.

Composites in EVs and HEVs: A composite material is created from two or more separate and widely distinct phases present in a single structure and working in unity to give a material certain characteristics. In the context of EVs and HEVs, composites are extensively adopted because they possess advantages such as light weight, high strength, and high durability. They are quite popular in the attempt of automotive manufacturers to reduce weight within automobiles, enhancing their performance and making them more energy-efficient [48,49].

18.3 FIBER-REINFORCED POLYMERS (FRPS)

Fiber-reinforced polymers are reinforced plastics that comprise continuously packaged fibers which include carbon fibers, glass fibers, or aramid fibers packed in polymers [44]. The fibers give the composite its strength and rigidity of structure, whereas the polymer matrix serves to cement the fibers. These composites comprise lightweight, high durability, and superior mechanical properties which make their use in the automotive industry appropriate (Figure 18.2) [50].

18.4 CARBON FIBER-REINFORCED POLYMERS (CFRPS)

CFRP composites have superior specific mechanical properties where strength is measured as a ratio of the material's weight. It provides higher tensile strength, stiffness, fatigue, and corrosion resistance compared to the conventional alloys [45,52]. Most of the CFRP applications are in automotive structural parts, body-in-white parts such as doors, hoods, fenders, battery boxes, and chassis of high-performance e-vehicles or hybrid vehicles [28]. They also help by taking advantage of the specific characteristics of CFRP to lower the vehicle weight and therefore enhance energy economy, and increase the driving range and overall performance [2,52,53].

18.5 GLASS FIBER-REINFORCED POLYMERS (GFRPS)

GFRP is cheaper than CFRP and this has made it suitable for mass production. Despite having somewhat lower specific strength and stiffness than carbon fiber, it has good mechanical properties such as impact and flexibility [54]. GFRP is also used in the nonstructural parts, such as the interior parts such as dashboards, seat backings, exterior parts including bumpers and sidelining, and underparts. GFRP is useful in decreasing the cost of production while at the same time providing lighter weight and better strength [55].

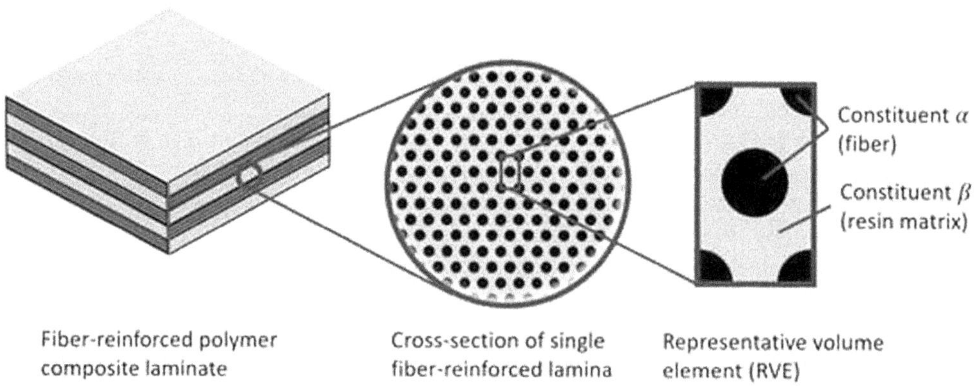

FIGURE 18.2 Fiber matrix composite [51].

18.6 METAL MATRIX COMPOSITES (MMCS)

MMCs are prepared by dispersion of a continuous phase (aluminum, magnesium, or titanium) and a discontinuous phase (carbon fiber, glass fiber, or ceramic particles) (Figure 18.3). These characteristics are overshot by the use of MMCs that retain the strength, thermal conductivity, and wear resistance of the metal while at the same delivering the light weight and stiffness of the reinforcing materials [56,57]. These composites are particularly suitable for demanding automobile applications that need strength coupled with excellent thermal stability. They are applied in parts that are always under stressing conditions, high-temperature conditions, and extreme mechanical stress conditions such as brake discs, motor housings, cylinder heads, and heat exchangers [58]. In EVs and HEVs, MMCs are most suitable for thermal management applications such as booster braking systems and electric motor housing because of the high thermal conductivity and low density of the material. This helps minimize overheating and improves the performance of the automobile [58,59].

18.7 POLYMER MATRIX COMPOSITES (PMCS)

In polymer matrix composites (PMCs), a resin that is epoxy, polyester, or thermoplastic is used as the matrix material incorporating different types of fibers including carbon, glass, and natural fibers [48]. PMCs are lighter, more resistant to corrosion, and have flexibility in design; all factors that are beneficial in automotive applications where weight reduction is paramount yet strength and fatigue-resistant materials are desired. A prime application of PMCs is in interior panels, acoustic elements, weaponry and armor trims, door linings, decks, airbag structures, and other marginally structural or non-structural components [61]. These composites play a role in lowering the amount of mass contained by the vehicle, therefore enhancing energy efficiency and operational performance in electric and hybrid vehicles. They also do not corrode easily and can be shaped into complex geometries, so being able to integrate more flexible and interesting designs into the interior of the vehicle is also more easily accomplished [62].

18.8 HYBRID COMPOSITES

Hybrid composites are obtained by using two or more different types of fibers (carbon/glass, etc.), or use fibers along with non-fiber materials. This combination makes it possible to optimize both the

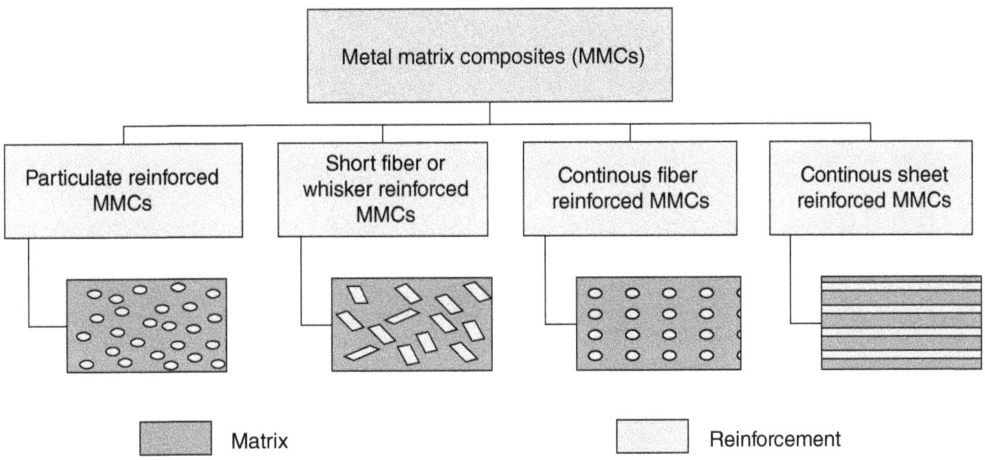

FIGURE 18.3 Classification of metal matrix composites (MMCs) [60].

strength and cost of the composite material and its weight at the same time [63]. This kind of composite is developed to fulfill some requirements in certain applications where various characteristics, including high strength and cost efficiency, are wanted. Hybrid composites find application in the areas of structural framework, automotive panels, and body parts [1,49]. They are perfect for those applications where a certain level of performance and a lower weight matter most, and allow manufacturers to achieve the required combination of weight reduction and strength for components used in vehicle construction. These are usually in mid-welfare vehicles or some parts of the car where a blend of various features is likely to be obtained [49,64,65].

18.9 NATURAL FIBER COMPOSITES

Natural fiber composites incorporate natural and renewable materials like hemp, jute, flax, or sisal as the reinforcing phase embedded in a polymer. These composite materials are recyclable, cheap, and have low density compared to other types of materials. With automakers shifting toward environment-friendly products, natural fiber composites find application owing to their environmentally friendly feature. Interior applications include fiber composites such as panels, door trims, dashboards, and sound insulations [50]. They are more commonly used in applications where high loads are not expected as the material may not be as strong as structure members. They give a lower environmental footprint during manufacture and provide a replacement for synthetic fiber composites. They are also biodegradable, which makes them fit for use in solving the problem of waste management at the end of the life cycle of a particular vehicle [44,66,67].

18.10 CERAMIC MATRIX COMPOSITES (CMCS)

Ceramic matrix composites are produced through the incorporation of ceramic fibers in a ceramic base. These composites are more well known for their high-temperature exposure tolerance; therefore they are most applicable in high temperatures (Figure 18.4). CMCs are produced to have different mechanical strength and properties to be thermally stable; therefore, their application in automotive

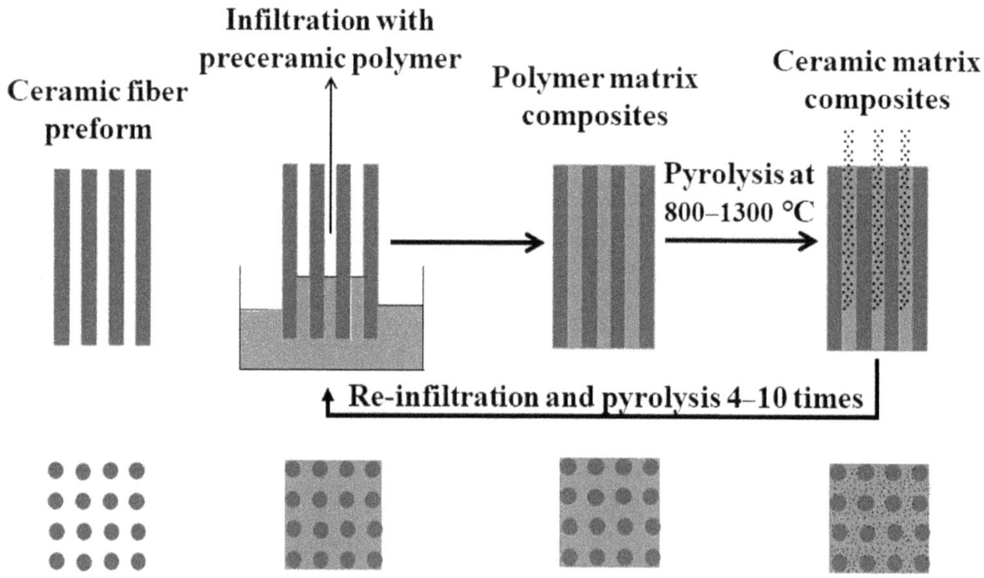

FIGURE 18.4 Ceramic matrix composites [72].

components that undergo high thermal stress is inevitable. In EV and HEV braking systems, heat shields and thermal protection systems are some of the applications of CMCs [68]. They are especially useful for controlling high temperatures, which can appear in the braking system, electric motor's touch box, and heat-exchanging equipment. Cleaning and widening of the mentioned areas through the use of CMCs assist in the dissipation of heat while enhancing vehicle performance, thereby reducing instances of overheating that may lead to degradation, wear, and tearing of essential vehicle components [69–71].

The major advantages of composites in EVs and HEVs are as follows [44,66,67]:

Weight Reduction: FRP, one of the main categories of composites, and MMCs offer some impressive advantages over conventional materials such as steel and aluminum—they are much lighter, for instance. Reduction in weight of compounds used in cars reduces the energy used by the car, increases the range of usage of batteries in electric cars, and improves the handling and performance of cars.

Improved Fuel Efficiency and Range: Composites contribute considerably toward weight reduction of the vehicle which directly has positive implications on the fuel economy and driving range of EVs/HEVs. This is even more relevant to EVs given that one of their biggest drawbacks is the limited distance a vehicle's battery can travel on a single charge.

Enhanced Durability and Corrosion Resistance: The mentioned composites, such as GFRP or natural fiber composites, do not corrode as metals do. This prolongs the service life of items like body panels, chassis, as well as underbody parts, leading to high cost savings since fewer replacements are likely to be made, and providing general vehicle durability.

Improved Safety and Impact Resistance: High-performance composites like carbon/epoxy and aramid/epoxy exhibit high impact tolerance. Such materials can have the property to dampen the energy at the time of a collision, thus enhancing the crashworthiness of a vehicle and occupant safety.

High Design Flexibility: Composites can be shaped according to varying customer demands and width/height ratios, which limits the design possibilities with metals. This is especially beneficial for manufacturers since it allows for the construction of complex shapes and thin-walled parts without affecting the mechanical characteristics of a material.

Superior Thermal Management: MMCs and CMCs have high thermal conductivity and high thermal resistance. These properties are essential in controlling the heat that is produced in several areas of any system, for instance in motors, batteries, or the braking system for electric or hybrid cars.

Cost-Effectiveness: However, advanced composites such as the CFRP are relatively costly, while other affordable materials include glass fiber and natural fiber composites. These cost-efficient materials serve to lower the assembly cost of EVs/HEVs and still contribute to decreased weight and increased functionality.

Sustainability and Eco-Friendliness: Flax, hemp, jute, and other natural fiber composites have been proven to be renewable, biodegradable, and have a lower degree of impact on the environment as compared to other synthetic materials. In use with products such as interior panels in non-structural parts, they enhance the sustainability and eco-friendliness of EVs/HEVs.

18.11 CHALLENGES

Despite the beneficial characteristics of polymer and composite applications for EVs and HEVs, there are certain difficulties. This is especially an issue with costly generation 2 materials such as carbon fiber-reinforced polymer that are relatively more expensive than some metals, although often metals can be more costly than composites. This cost spreads to production lines which call for specific tools, for instance an autoclave which is used in curing and complex molding methods.

These processes prolong the manufacturing time and add extra costs to the process, making it cumbersome to scale the application of composites for high-volume vehicles. Additionally, servicing and maintaining polymers and composites can be more difficult and costly compared to conventional materials. Composite parts are also more expensive to service since there are usually specialized equipment and skills needed for servicing. The other issue that can be added to the list of disadvantages is the low level of recyclability of many composite materials, which is an issue from the point of view of their future impact on the environment. The recycling of products has an added problem in that most composites, especially those utilized in structural applications, are difficult to recycle, hence raising concerns about waste management. However, the combination of polymers and composites with conventional metals can lead to interfacial compatibility issues that impair manufacturing efficiency and cohesion of the structures of EVs/HEVs. For the use of these materials to become commonplace in the automobile industry, all these challenges must be overcome.

Future Scope: The role of polymers and composites in EVs and HEVs has remained stable, but in the future, it may grow rapidly as there have been numerous developments in the field of material science and manufacturing. Another area of emphasis is that it aims at decreasing the production cost of expensive materials like CFRPs. Advanced technology processes such as AFP, RTM, and AM have been extended for creating these materials for application in low-cost manufacturing, which is contributing to their use in automotive applications. In the development of these technologies, CFRP and GFRPs will likely find increased use in EVs/HEVs. Furthermore, sustainable materials are now becoming more popular, with natural fiber composites such as hemp and flax being more readily marketable because of their reduced environmental impact and cost. Further, the advancement of environmentally friendly polymers and recyclable composites is expected to enhance the lifetime of EVs and HEVs along with the recycling standards in response to environmental awareness throughout the world. Further, it was found that the use of smart composites with embedded sensors may be mooted to be used in developing vehicles with enhanced monitoring capabilities of performance, safety, and durability. The convergence of polymers and composites will be combined with lightweight metals such as aluminum and magnesium, resulting in new and stronger, lighter, and cheaper materials. These will mark the progress of higher efficiency and sustainability of vehicles that are cost effective in the future.

18.12 CONCLUSION

In conclusion, polymers and composites will continue to provide significant input to the EVs and HEVs of the future. They are characterized by lightweight properties, high strength-to-weight characteristics, and multifunctional usage—all these features define their indispensability for the enhancement of vehicle efficiency, safety, and performance. Thus, polymers and composite materials used in battery enclosures, body panels, and structural parts minimize the overall weight and have a major impact on the driving range of fully EVs and even the fuel economy of hybrids. These materials are also characterized by enhanced corrosion, impact, and thermal performance, which fits well with the requirements of automotive usage. Nevertheless, the major disadvantage is high production costs, which include the complexity of the manufacturing processes; most importantly, the challenges in repair and recycling of composite materials are other drawbacks of using composite materials. These challenges will, therefore, need to be addressed if widespread application of these advanced materials is to take root. However, with these innovations appearing continuously in the realm of green materials, cheap manufacturing processes, and recycling the polymers and composites are certain to find a new home in EVs and HEVs. Most polymers and composites used in EVs reduce fuel consumption and enhance efficiency by the lightweighting of the vehicle. Thus, polymers and composites will advance as the automotive industry shifts toward environment-friendly cars, and it will expand its option of higher performance and efficient solutions for car manufacturing to support the creation of the green automobile future.

REFERENCES

[1] E. Cebe and A.B. Irez, Novel hybrid polymer composites for optimized thermal performance in electric vehicle battery cases, *Poly Compos.* 46(7), 1–12 (2024). DOI:10.1002/pc.29381.

[2] P. Gupta, B. Toksha, B. Patel, Y. Rushiya, P. Das, and M. Rahaman, Developments and research avenues for polymers in electric vehicles, *Chem Record* 22, e202200186 (2022).

[3] N. Naik, P. Suresh, S. Yadav, M.P. Nisha, J.L. Arias-Gonzáles, J.C. Cotrina-Aliaga, R. Bhat, M.D. Jalageri, Y. Kaushik, and A.B. Kunjibettu, A review on composite materials for energy harvesting in electric vehicles, *Energies* 16, 3348 (2023).

[4] A. Wazeer, A. Das, C. Abeykoon, A. Sinha, and A. Karmakar, Composites for electric vehicles and automotive sector: A review, *Green Energy Intell. Transport* 2, 100043 (2023).

[5] S. Kumar and R.S. Bharj, Emerging composite material use in current electric vehicle: A review, *Mater. Today Proc.* 5, 27946 (2018).

[6] R. Hsissou, R. Seghiri, Z. Benzekri, M. Hilali, M. Rafik, and A. Elharfi, Polymer composite materials: A comprehensive review, *Compos. Struct.* 262, 113640 (2021).

[7] M.-Y. Lyu and T.G. Choi, Research trends in polymer materials for use in lightweight vehicles, *Int. J. Precis. Eng. Manuf.* 16, 213 (2015).

[8] F. Kalhammer, Polymer electrolytes and the electric vehicle, *Solid State Ionics* 135, 315 (2000).

[9] E. Amasawa, M. Hasegawa, N. Yokokawa, H. Sugiyama, and M. Hirao, Environmental performance of an electric vehicle composed of 47% polymers and polymer composites, *Sustain. Mater. Technol.* 25, e00189 (2020).

[10] A. Nakajima, A. Shoji, K. Yonemori, and N. Seo, Novel polymer composite having diamond particles and boron nitride platelets for thermal management of electric vehicle motors, *Jpn. J. Appl. Phys.* 55, 027101 (2016).

[11] A.K. Mohanty, S. Vivekanandhan, N. Tripathi, P. Roy, M.R. Snowdon, L.T. Drzal, and M. Misra, Sustainable composites for lightweight and flame retardant parts for electric vehicles to boost climate benefits: A perspective, *Compos. C Open Access* 12, 100380 (2023).

[12] M.S. Bulut, M. Ordu, O. Der, and G. Basar, Sustainable thermoplastic material selection for hybrid vehicle battery packs in the automotive industry: A comparative multi-criteria decision-making approach, *Polymers* 16, 2768 (2024).

[13] A. Ghanbari, S. Seyedin, M. Nofar, and A. Ameli, Mechanical properties and foaming behavior of polypropylene/elastomer/recycled carbon fiber composites, *Polym. Compos.* 42, 3482 (2021).

[14] D.M. Panaitescu, R.C. Fierascu, A.R. Gabor, and C.A. Nicolae, Effect of hemp fiber length on the mechanical and thermal properties of polypropylene/SEBS/hemp fiber composites, *J. Mater. Res. Technol.* 9, 10768 (2020).

[15] R. Gogoi and G. Manik, Development of thermally conductive and high-specific strength polypropylene composites for thermal management applications in automotive, *Polym. Compos.* 42, 1945 (2021).

[16] M. Gohil and G. Joshi, in Inamuddin, Abdullah M. Asiri and Arun M. Isloor Perspective of polycarbonate composites and blends properties, applications, and future development: A review, In (Eds.) *Green Sustainable Process for Chemical and Environmental Engineering and Science* (Elsevier, 2022), pp. 393–424.

[17] S.A. Hafad, A.F. Hamood, H.A. AlSalihi, S.I. Ibrahim, A.A. Abdullah, A.A. Radhi, M.K. Al-Ghezi, and B.R. Alogaidi, Mechanical properties study of polycarbonate and other thermoplastic polymers *J. Phys.: Conf. Ser.* 1973, 012001 (2021).

[18] K. Sushmita, G. Madras, and S. Bose, The journey of polycarbonate-based composites towards suppressing electromagnetic radiation. *Funct. Composite Mater.* 2, 13 (2021).

[19] J. Jyoti, A.K. Arya, S. Chockalingam, S.K. Yadav, K.M. Subhedar, S.R. Dhakate, and B.P. Singh, Mechanical, electrical and thermal properties of graphene oxide-carbon nanotube/ABS hybrid polymer nanocomposites, *J. Polym. Res.* 27, 282 (2020).

[20] R. Zhang, L. Li, S. Long, P. Wang, F. Wen, J. Yang, and G. Wang, High-temperature dielectric polymer composite films of all-organic PVDF/ABS with excellent energy storage performance and stability *J. Mater. Chem. C* 10, 3480 (2022).

[21] A.M. Abdullah, T.N.A.T. Rahim, W.N.F.W. Hamad, D. Mohamad, H.M. Akil, and Z.A. Rajion, Mechanical and physical properties of highly ZrO2/β-TCP filled polyamide 12 prepared via fused

deposition modelling (FDM) 3D printer for potential craniofacial reconstruction application,*Dental Mater.* 34, e309 (2018).
[22] M.Y. Kondo, L.S. Montagna, G.F.D.M. Morgado, A.L.G.D. Castilho, L.A.P.D.S. Batista, E.C. Botelho, M.L. Costa, F.R. Passador, M.C. Rezende, and M.V. Ribeiro, Recent advances in the use of polyamide-based materials for the automotive industry, *Polímeros* 32, e2022023 (2022).
[23] N. Ahmad, A. Kausar, and B. Muhammad, Perspectives on polyvinyl chloride and carbon nanofiller composite: A review, *Polym. Plast. Technol. Eng.* 55, 1076 (2016).
[24] V. Galstyan, P. D'Angelo, G. Tarabella, D. Vurro, and T. Djenizian, High versatility of polyethylene terephthalate (PET) waste for the development of batteries, biosensing and gas sensing devices, *Chemosphere* 359, 142314 (2024).
[25] S. Kuciel, K.E. Mazur, M. Robakowska, and D. Paukszta, Mechanical, thermal and performance evaluation of hybrid basalt/carbon fibers reinforced bio-based polyethylene terephthalate (BioPet) composites. *Int. J. Precis. Eng. Manuf. Green Tech.* 11, 1557 (2024).
[26] A.K. Singh, R. Bedi, and B.S. Kaith, Composite materials based on recycled polyethylene terephthalate and their properties – A comprehensive review, *Compos. B Eng.* 219, 108928 (2021).
[27] H.-J. Um, Y.-T. Hwang, I.-J. Bae, and H.-S. Kim, Design and manufacture of thermoplastic carbon fiber/polyethylene terephthalate composites underbody shield to protect the lithium-ion batteries for electric mobility from ground impact, *Compos. B Eng* 238, 109892 (2022).
[28] S. Zhu, J. Lu, M. Fan, Y. Chernysh, Y. Pan, R. Dang, J. Guan, L. Yu, B. Lin, and X. Shen, Effectively improve the mechanical properties of carbon fabric/epoxy composites by oxidized carbon nitride, *Polym. Compos.* 45, 4338 (2024).
[29] A. Kausar, I. Rafique, and B. Muhammad, Review of applications of polymer/carbon nanotubes and epoxy/CNT composites, *Polym. Plast. Technol. Eng.* 55, 1167 (2016).
[30] N. Saba and M. Jawaid, Epoxy resin based hybrid polymer composites, In Vijay Kumar Thakur, Manju Kumari Thakur, Asokan Pappu (Eds.) *Hybrid Polymer Composite Materials* (Elsevier, 2017), pp. 57–82.
[31] S.K. Bhaskaran, K. Boga, R. Arukula, and S.K. Gaddam, Natural fibre reinforced vegetable-oil based polyurethane composites: A review, *J. Polym. Res.* 30, 325 (2023).
[32] A. Kausar, Polyurethane composite foams in high-performance applications: A review, *Polym. Plast. Technol. Eng.* 57, 346 (2018).
[33] M.S.M. Misenan, A.S.A. Khiar, and T. Eren, Polyurethane-based polymer electrolyte for lithium ion batteries: A review, *Polym. Int.* 71, 751 (2022).
[34] M. Asim, N. Saba, M. Jawaid, M. Nasir, M. Pervaiz, and O.Y. Alothman, A review on phenolic resin and its composites, *CAC* 14, 185 (2018).
[35] S. S, M. S, J. S, and V. J, Improving fire resistance in E-vehicles: A studyon MPP-enhanced S-glass/phenolic hybrid composites, *E3S Web of Conf.* 529, 02001 (2024).
[36] A.M. Joshi and A.A. Athawale, Electrically conductive silicone/organic polymer composites, *Silicon* 6, 199 (2014).
[37] A.E. Vlastos and S.M. Gubanski, Surface structural changes of naturally aged silicone and Epdm composite insulators, *IEEE Trans. Power Delivery* 6, 888 (1991).
[38] E.L. Warrick, O.R. Pierce, K.E. Polmanteer, and J.C. Saam, Silicone elastomer developments 1967–1977, *Rubber Chem. Technol.* 52, 437 (1979).
[39] B. Yoon, J.Y. Kim, U. Hong, M.K. Oh, M. Kim, S.B. Han, J.-D. Nam, and J. Suhr, Dynamic viscoelasticity of silica-filled styrene-butadiene rubber/polybutadiene rubber (SBR/BR) elastomer composites. *Compos. B Eng.* 187, 107865 (2020).
[40] B. Ameduri, Fluorinated Elastomers, in *Encyclopedia of Polymer Science and Technology*, edited by H.F. Mark, 3rd ed. (Wiley, 2023), pp. 1–22.
[41] D. Kumar, S. Dolui, S. Banerjee, and B. Ameduri, Fluorinated thermoplastic elastomers: Current status and future applications, in *Advances in Thermoplastic Elastomers* (Elsevier, 2024), pp. 335–361.
[42] M.G. Maya, S.C. George, T. Jose, L. Kailas, and S. Thomas, Development of a flexible and conductive elastomeric composite based on chloroprene rubber, *Polym. Testing* 65, 256 (2018).
[43] A. Saritha Chandran and S.K. Narayanankutty, An elastomeric conducting composite based on polyaniline coated nylon fiber and chloroprene rubber, *Eur. Polym. J.* 44, 2418 (2008).

[44] Z. Sun, Progress in the research and applications of natural fiber-reinforced polymer matrix composites, *Sc. Eng. Compos. Mater.* 25, 835 (2018).

[45] K.B.M. Ismail, M.A. Kumar, S. Mahalingam, B. Raj, and J. Kim, Carbon fiber-reinforced polymers for energy storage applications, *J. Energy Storage* 84, 110931 (2024).

[46] M.M. El Sayed, Production of polymer hydrogel composites and their applications, *J. Polym. Environ.* 31, 2855 (2023).

[47] J.L. Gómez-Urquiza, A. Velando-Soriano, M.J. Membrive-Jiménez, L. Ramírez-Baena, R. Aguayo-Estremera, E. Ortega-Campos, G.A. Cañadas-De la Fuente. Prevalence and levels of burnout in nursing students: A systematic review with meta-analysis. *Nurse Educ Pract.* 72, 103753 (2023 Oct). doi: 10.1016/j.nepr.2023.103753. Epub 2023 Aug 24. PMID: 37651959.

[48] S. Kangishwar, N. Radhika, A.A. Sheik, A. Chavali, and S. Hariharan, *Polym. Bull.* 80, 47 (2023).

[49] C. Dong, Review of natural fibre-reinforced hybrid composites, *J. Reinforced Plast. Compos.* 37, 331 (2018).

[50] Z. Sun, Y. Duan, H. An, X. Wang, S. Liang, and N. Li, Research progress and application of natural fiber composites, *J. Natural Fibers* 20, 2206591 (2023).

[51] E. Anderson, B. Gunawan, J. Nicholas, M. Ingraham, and B.A. Hernandez-Sanchez, A multicontinuum-theory-based approach to the analysis of fiber-reinforced polymer composites with degraded stiffness and strength properties due to moisture absorption, *JMSE* 11, 421 (2023).

[52] N. Forintos and T. Czigany, Reinforcing carbon fibers as sensors: The effect of temperature and humidity, *Compos. B: Eng.* 162, 331 (2019).

[53] J.A. Butenegro, M. Bahrami, J. Abenojar, and M.Á. Martínez, Recent progress in carbon fiber reinforced polymers recycling: A review of recycling methods and reuse of carbon fibers, *Materials* 14, 6401 (2021).

[54] D.K. Rajak, P.H. Wagh, and E. Linul, Manufacturing technologies of carbon/glass fiber-reinforced polymer composites and their properties: A review, *Polymers* 13, 3721 (2021).

[55] C. Zhu, Z. Qiao, H. Wang, and C. Huang, The study of functional glass fiber veils for composites protection: Flame resistance and mechanical performance, *J. Compos. Sci.* 8, 268 (2024).

[56] X. Wu and W. Zhang, A review on aluminum matrix composites' characteristics and applications for automotive sector, *Heliyon* 10, e38576 (2024).

[57] A. Kumar Sharma, R. Bhandari, A. Aherwar, R. Rimašauskienė, and C. Pinca-Bretotean, A study of advancement in application opportunities of aluminum metal matrix composites, *Mater. Today Proc.* 26, 2419 (2020).

[58] A. Macke, B.F. Schultz, P.K. Rohatgi, and N. Gupta, Metal matrix composites for automotive applications, in *Advanced Composite Materials for Automotive Applications*, edited by A. Elmarakbi, 1st ed. (Wiley, 2013), pp. 311–344.

[59] H. Singh, G. Singh Brar, H. Kumar, and V. Aggarwal, A review on metal matrix composite for automobile applications, *Mater. Today Proc.* 43, 320 (2021).

[60] E.M. Gutema and H.G. Lemu, Conventional Machining of Metal Matrix Composites towards Sustainable Manufacturing—Present Scenario and Future Prospects, *J. Compos. Sci.* 8, 356 (2024).

[61] J. Lee, J. Son, J. Lim, I. Kim, S. Kim, N. Cho, W. Choi, and D. Shin, Transformer-based mechanical property prediction for polymer matrix composites, *Korean J. Chem. Eng.* 41, 3005 (2024).

[62] R.O. Ogunleye and S. Rusnakova, A review of prestressed fibre-reinforced polymer matrix composites, *Polymers* 14, 60 (2021).

[63] R. Chandel, N. Sharma, and S.A. Bansal, A review on recent developments of aluminum-based hybrid composites for automotive applications, *Emergent Mater.* 4, 1243 (2021).

[64] S. Arya, S. Chauhan, R. Kumar, and B. Kelkar, Development of natural fiber reinforced thermoplastic bonded hybrid wood veneer composite, *Maderas. Cienc. Tecnol.* 25, (2023).

[65] P. Zuo, D.V. Srinivasan, and A.P. Vassilopoulos, Review of hybrid composites fatigue, *Compos. Struct.* 274, 114358 (2021).

[66] S. Arumugam, J. Kandasamy, S. Venkatesan, R. Murugan, V. Lakshmi Narayanan, M.T.H. Sultan, F.S. Shahar, A.U.M. Shah, T. Khan, and T.A. Sebaey, A review on the effect of fabric reinforcement on strength enhancement of natural fiber composites, *Materials* 15, 3025 (2022).

[67] M.R. Zakaria, M.Z. Ahmad Thirmizir, M.S. Zainol Abidin, H. Md Akil, M.F. Omar, A. Anjang Ab Rahman, N. Nosbi, and N. Ab Ghafar, A comparison process between wet lay-up, single vacuum

bagging and bouble vacuum bagging toward natural fibre (Palm, coconut and Kenaf) reinforced epoxy composite laminates, *Fibers Polym.* 24, 2107 (2023).
[68] G. Fantozzi and P. Reynaud, Mechanical hysteresis in ceramic matrix composites, *Mater. Sci. Eng. A* 521–522, 18 (2009).
[69] K. Konopka, Alumina composites with metal particles in ceramic matrix, *Powder. Metall. Met. Ceram.* 54, 374 (2015).
[70] J. Sun, D. Ye, J. Zou, X. Chen, Y. Wang, J. Yuan, H. Liang, H. Qu, J. Binner, and J. Bai, A review on additive manufacturing of ceramic matrix composites, *J. Mater. Sci. Technol.* 138, 1 (2023).
[71] P.O. Guglielmi, Processing of all-oxide ceramic matrix composites with RBAO matrices *J. Ceram. Sci. Tech.* 7(1), 87–96 (2016).
[72] S. Shrivastava, D.K. Rajak, T. Joshi, D.K. Singh, and D.P. Mondal, Ceramic matrix composites: Classifications, manufacturing, properties, and applications, *Ceramics* 7, 652 (2024).

Index

Note: Page numbers in *italics* and **bold** denote figures and tables, respectively.

F

G

N

O

W

X

Y

Z

For Product Safety Concerns and Information please contact our EU representative GPSR@taylorandfrancis.com
Taylor & Francis Verlag GmbH, Kaufingerstraße 24, 80331 München, Germany

www.ingramcontent.com/pod-product-compliance
Ingram Content Group UK Ltd.
Pitfield, Milton Keynes, MK11 3LW, UK
UKHW052219190926
475702UK00006B/1103
9781032975054